Firefly

WORLD FACTBOOK

A Firefly Book

Published by Firefly Books Ltd, 2003

Cartography by Philip's
Text: Keith Lye

First printing

National Library of Canada Cataloguing in Publication Data

Lye, Keith
Firefly world factbook / Keith Lye. – 1st North American ed.
Previous eds. published under title: Philip's world factbook.
Includes index.
ISBN 1-55297-839-7
1. Geography–Handbooks, manuals, etc. 2. History–Handbooks, manuals, etc. 3. Economics–Handbooks, manuals, etc. 4. Political science–Handbooks, manuals, etc. I. Lye, Keith Philip's world factbook.
II. Title.
III. Title: World factbook.

G123.L94 2003 909 C2003-901959-4

Publisher Cataloging-in-Publication Data
(Library of Congress standards)

Lye, Keith.
Firefly world factbook / Keith Lye. _1st American ed.
[352] p. : col. ill. , col. maps ; cm.
Originally published as: Philip's world factbook ; London : George Philip, 2000.
Includes index.
Summary: Illustrated reference guide to every country, territory, principality and dependency in the world, summarizing basic statistics, climate, vegetation, history, politics and economy.
ISBN 1-55297-839-7 (pbk.)
1. Geography – Handbooks, manuals, etc. 2. History – Handbooks, manuals, etc.
3. Economics – Handbooks, manuals, etc. 4. Political science – Handbooks, manuals, etc. I. Title.
909 21 G123.L94 2003

Published in Great Britain by Philip's,
a division of Octopus Publishing Group Ltd,
2–4 Heron Quays, London E14 4JP

Published in Canada in 2003 by
Firefly Books Ltd
3680 Victoria Park Avenue
Toronto, Ontario, M2H 3K1

Published in the United States in 2003 by
Firefly Books (U.S.) Inc.
P.O. Box 1338, Ellicott Station
Buffalo, New York 14205

Printed in Hong Kong

Firefly

WORLD FACTBOOK

KEITH LYE

Firefly Books

CONTENTS

INTRODUCTION

The *Firefly World Factbook* contains articles and maps for all the countries of the world, arranged in alphabetical order. A simple locator map shows the position of the country within its region, and the national flag is illustrated with an explanation of its origin. The smallest countries or islands, such as the islands of the Pacific Ocean, are grouped together within their region, with a shared map. Other small countries are found alongside their larger neighbors; for example, Bhutan and Nepal are included in the entry for India.

The country maps are physically colored and are usually a column width in size, but a country's shape, area, and importance has sometimes merited a larger map. Where neighboring countries are in dispute, the maps show the *de facto* boundary between nations; that is, the boundary that exists in the real world rather than the boundary that a particular country may wish to be shown on a map.

Each map contains a number of symbols indicating the major towns and cities, principal roads and railroads, with the main border crossings. Major international airports and the highest point in most countries are also shown. A scale bar indicates the distances between places, and the lines of latitude and longitude show the country's global position.

The spellings of the placenames on the maps match the forms that are used in everyday English; for example, Rome (not Roma), Munich (not München), Moscow (not Moskva). Only the more common accented characters of the Western European languages have been used.

At the end of the book is a short index. Each entry has a page number and the quadrant of the map where that name begins; in a clockwise direction, "a" indicates the top left quarter, "b" is the top right, "c" is the bottom right, and "d" is the bottom left quarter.

KEY

City or town
Capital city
Major airport
Highest point in country
International boundary
Railroad
Road
River

THE WORLD: PHYSICAL

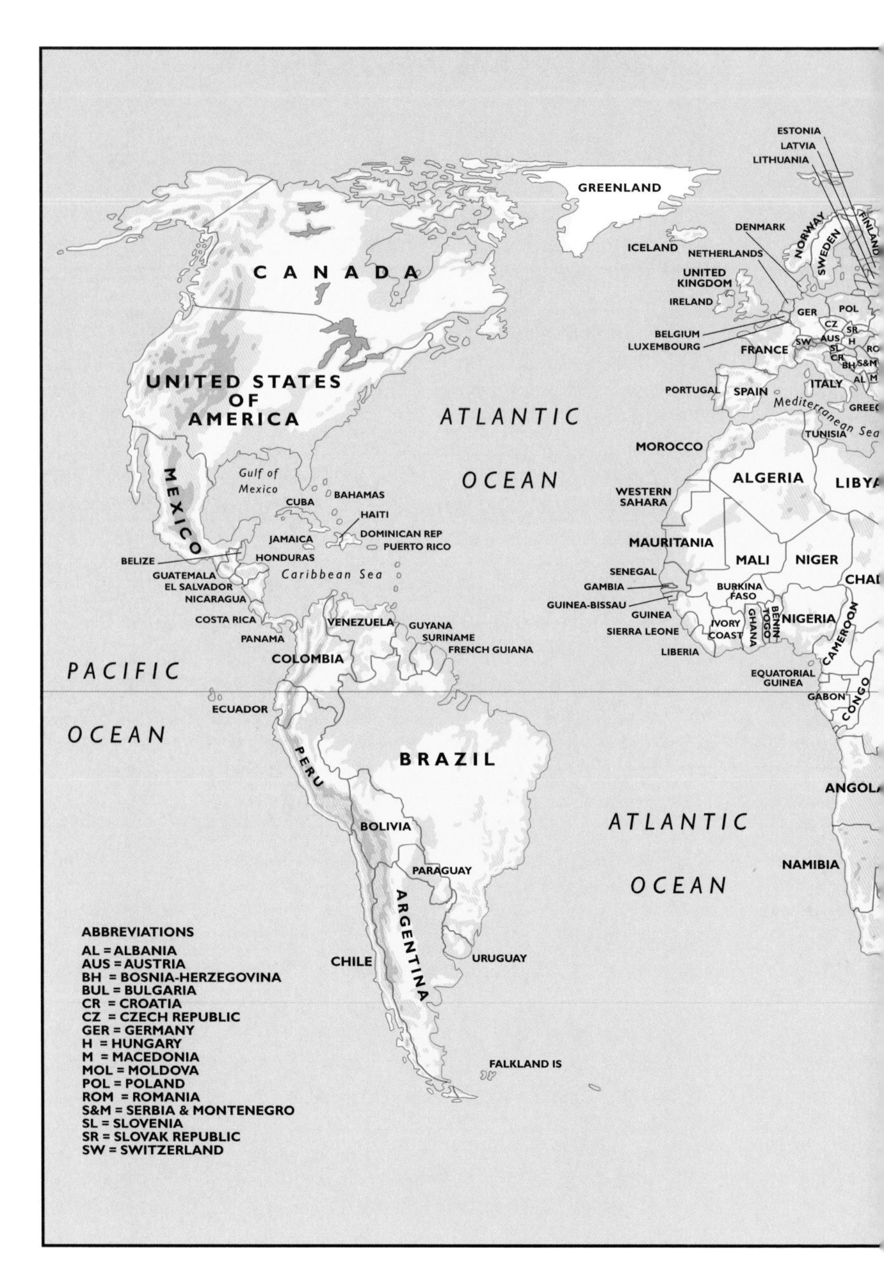

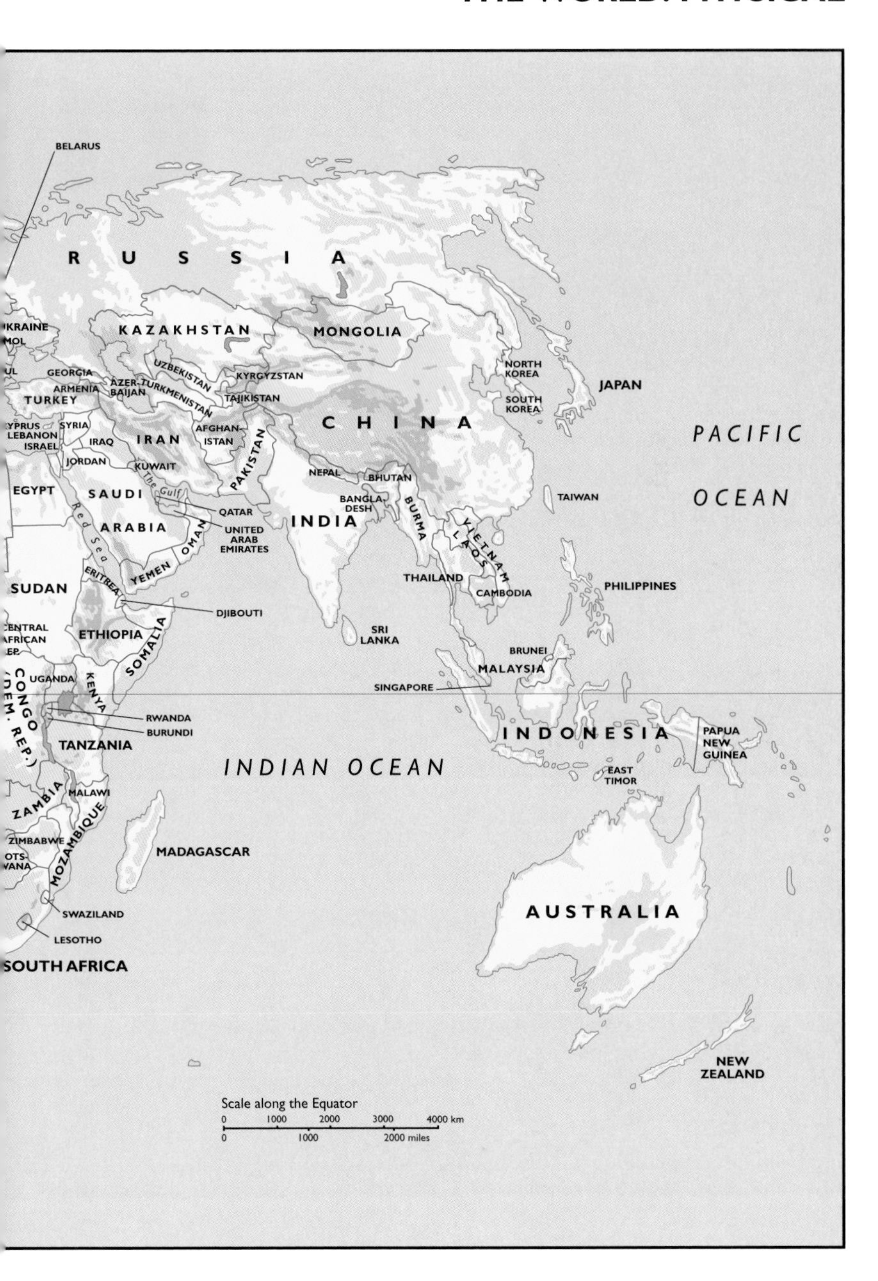
BELARUS
R U S S I A
KRAINE
MOL
KAZAKHSTAN
MONGOLIA
GEORGIA
ARMENIA
AZER-BAIJAN
UZBEKISTAN
TURKMENISTAN
KYRGYZSTAN
TAJIKISTAN
TURKEY
NORTH KOREA
SOUTH KOREA
JAPAN
SYRIA
LEBANON
ISRAEL
IRAQ
IRAN
AFGHAN-ISTAN
C H I N A
PACIFIC OCEAN
JORDAN
KUWAIT
PAKISTAN
NEPAL
BHUTAN
EGYPT
SAUDI ARABIA
The Gulf
QATAR
UNITED ARAB EMIRATES
BANGLA-DESH
INDIA
BURMA
TAIWAN
Red Sea
OMAN
VIETNAM
LAOS
YEMEN
ERITREA
THAILAND
CAMBODIA
PHILIPPINES
SUDAN
DJIBOUTI
ETHIOPIA
SOMALIA
SRI LANKA
BRUNEI
MALAYSIA
UGANDA
KENYA
CONGO (DEM. REP.)
SINGAPORE
RWANDA
BURUNDI
INDONESIA
PAPUA NEW GUINEA
TANZANIA
INDIAN OCEAN
EAST TIMOR
ZAMBIA
MALAWI
MOZAMBIQUE
ZIMBABWE
MADAGASCAR
AUSTRALIA
SWAZILAND
LESOTHO
SOUTH AFRICA
NEW ZEALAND
Scale along the Equator
0 1000 2000 3000 4000 km
0 1000 2000 miles

AFGHANISTAN

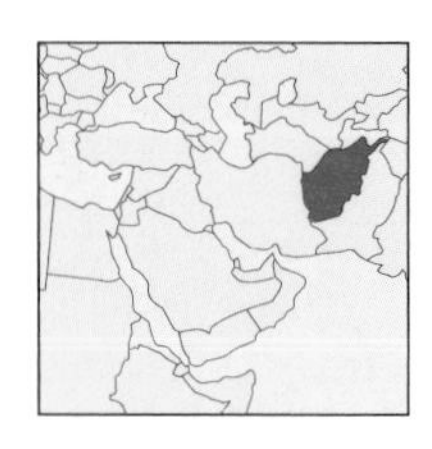

Introduced in January 2002, this flag replaces that of the *Mujaheddin* ("holy warriors"), who defeated Afghanistan's socialist government but lost power at the end of 2001. The flag is the 19th different design used by Afghanistan since 1901.

The Republic of Afghanistan is a landlocked country in southern Asia. It is bordered by Turkmenistan, Uzbekistan, Tajikistan, China, Pakistan, and Iran. The central highlands, reach a height of more than 22,966 ft [7,000 m] in the east and make up nearly three-quarters of Afghanistan. The main range is the Hindu Kush, which is cut by deep, fertile valleys.

North of the central highlands are broad plateaux and hilly areas. To the south are lowlands, consisting largely of desert or semidesert.

CLIMATE

The height of the land and the country's remote position have a great effect on the climate. In winter, northerly winds bring cold, snowy weather to the mountains. But summers are hot and dry. The rainfall decreases to the south, but temperatures are higher throughout the year.

VEGETATION

Grasslands cover much of the north, while the vegetation in the dry south is sparse. Trees are rare in both regions. But forests of such coniferous trees as pine and fir grow on the higher mountain slopes, with cedars lower down. Alder, ash, juniper, oak, and walnut grow in the mountain valleys.

AREA 251,772 sq miles [652,090 sq km]

POPULATION 26,813,000

CAPITAL (POPULATION) Kabul (1,565,000)

GOVERNMENT Transitional regime

ETHNIC GROUPS Pashtun ("Pathan") 38%, Tajik 25%, Hazara 19%, Uzbek 6%, others 12%

LANGUAGES Pashtu, Dari/Persian (both official), Uzbek

RELIGIONS Islam (Sunni Muslim 84%, Shiite Muslim 15%)

CURRENCY Afghani = 100 puls

HISTORY

In ancient times, Afghanistan was invaded by Aryans, Persians, Greeks, and Macedonians, and warrior armies from central Asia. Arab armies introduced Islam in the late 7th century.

The modern history of Afghanistan began in 1747, when the various tribes in the area united for the first time. In the 19th century, Russia and Britain struggled for control of Afghanistan. Russia was seeking an outlet to the Indian Ocean, while Britain wanted to protect its Indian territories against Russia.

British troops invaded Afghanistan in 1839–42 and again in 1878. In 1919, however, Afghanistan became fully independent.

POLITICS

After World War II (1939–45), the country remained neutral during the Cold War which developed between the United States and the Soviet Union and their allies. In 1973, army leaders overthrew the government and made the country a republic.

In 1978, the government was taken over by a socialist group. It then turned to the Soviet Union for aid. Many people rebelled against the pro-Communist government and, in late 1979, Soviet troops invaded the country. During the 1980s, the Soviet troops attempted to put down the rebel Muslim forces, who were called the *Mujaheddin* ("holy warriors"). The Soviet troops withdrew in 1989, but the civil war went on. In 1992, the Mujaheddin entered Kabul and set up an Islamic government. By the end of 1996, a fundamentalist group called Taliban (meaning "students") had become the leading Islamic faction.

By 2001, the Taliban regime controlled 90% of the country. But, in October 2001, after the country's refusal to surrender Osama bin Laden, the suspected leader of the bombings in New York City and Washington, DC, in September 2001, international action led by the US overthrew the Taliban regime. A coalition group led by Hamid Karzai formed the government until elections could be held in 2004.

DID YOU KNOW

- that the narrow Khyber Pass leading from Pakistan to Afghanistan once had great strategic importance
- that northern Afghanistan was a province of the ancient Persian empire; it was then called Bactria
- that Afghanistan is the world's second largest producer of the drug opium, after Burma (Myanmar)
- that Alexander the Great conquered Afghanistan in about 330 BC

ECONOMY

Afghanistan is one of the world's poorest countries. About 60% of the people make their living by farming. Many people are seminomadic herders of sheep and other animals. Wheat is the chief crop. Natural gas is produced, but most mineral deposits await development. There are few factories. Exports include karakul skins (which are used to make hats and jackets), cotton, dried fruit and nuts, fresh fruit, and natural gas.

ALBANIA

Albania's official name, *Shqiperia,* means "Land of the Eagle," and the black double eagle was the emblem of the 15th-century hero Scanderbeg. A star placed above the eagle in 1946 was removed in 1992 when a non-Communist government was formed.

The Republic of Albania is a country in the Balkan peninsula. It faces the Adriatic Sea in the west and is bordered by Serbia and Montenegro (former Yugoslavia), Macedonia, and Greece. About 70% of the country is mountainous, with the highest point, Korab, reaching 9,068 ft [2,764 m] on the Macedonian border. Most people live on the coastal lowlands in the west. This is the main farming region. Albania lies in an earthquake zone and severe earthquakes occur occasionally.

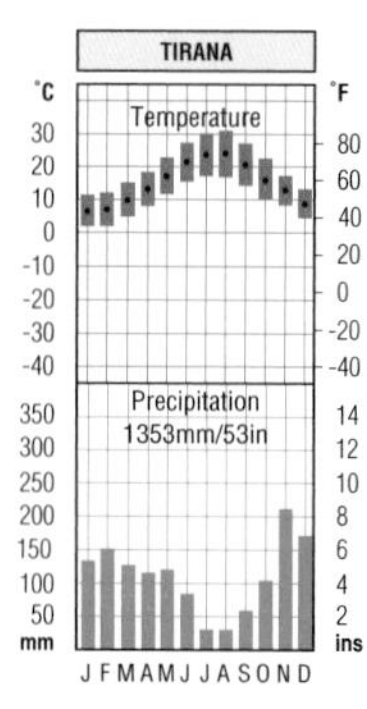

CLIMATE

The coastal areas of Albania have a typical Mediterranean climate, with fairly dry, sunny summers and cool, moist winters. Inland, Tirana sometimes has freezing temperatures in the months of winter. The mountains have a severe climate, with heavy winter snowfalls.

VEGETATION

Maquis, the typical scrubland of Mediterranean areas, covers much of the lowlands which are not farmed. Oaks also grow here, with beech and chestnut higher up, and conifers on the steeper mountain slopes.

AREA 11,100 sq miles [28,750 sq km]

POPULATION 3,510,000

CAPITAL (POPULATION) Tirana (251,000)

GOVERNMENT Multiparty republic

ETHNIC GROUPS Albanian 95%, Greek 3%, Macedonian, Vlachs, Gypsy

LANGUAGES Albanian (official)

RELIGIONS Many people say they are non-believers; of the believers, 65% follow Islam and 33% follow Christianity (Orthodox 20%, Roman Catholic 13%)

CURRENCY Lek = 100 qindars

HISTORY AND POLITICS

In ancient times, Albania was part of a region called Illyria. In 167 BC, it became part of the Roman empire. In the 15th century, a leader named Scanderbeg successfully led the Albanians against the invading Ottoman Turks. But after he died in 1468, the Turks took over. Albania became part of the Ottoman empire until 1912.

Italy invaded Albania in 1939, but German forces occupied Albania in 1943. In 1944 Albanian Communists, led by a resistance leader called Enver Hoxha, took power. The Communists enjoyed good relations with Yugoslavia until 1948, when the Soviet Union split with Yugoslavia.

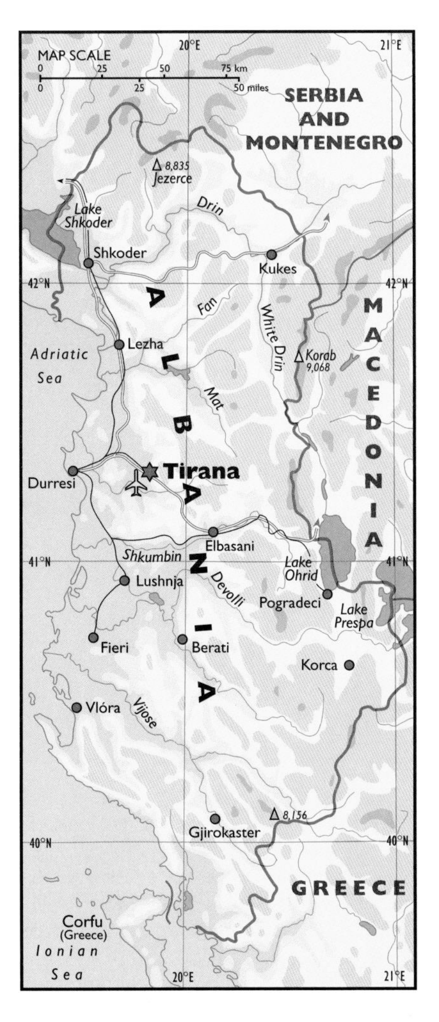

In the early 1960s, Albania broke with the Soviet Union when the Russians criticized China's policies. Albania was allied to China until the late 1970s, when it accused the Chinese of abandoning Communist principles. In the early 1990s, the Albanian government introduced reforms, permitting opposition parties. In 1991, the Communists won a majority in national elections. In 1992, the non-Communist Democratic Party won power, but in 1997, a financial crisis led to the formation of a socialist government. The socialists were re-elected in 2001.

ECONOMY

Albania is Europe's poorest country. In the mid-1990s, 59% of Albanians worked on large state and collective farms, but the government has encouraged private ownership since 1991. Major crops include fruits, maize, olives, potatoes, sugar beet, vegetables, and wheat. Livestock farming is also important.

Albania has some mineral reserves, such as chromite, copper, and nickel, which are exported. The country also has some oil, brown coal, and hydroelectricity, and a few heavy industries.

Albania's official name, Shqiperia, appears on this stamp, one of a set depicting birds, issued in 1968, which shows a waxwing.

ALGERIA

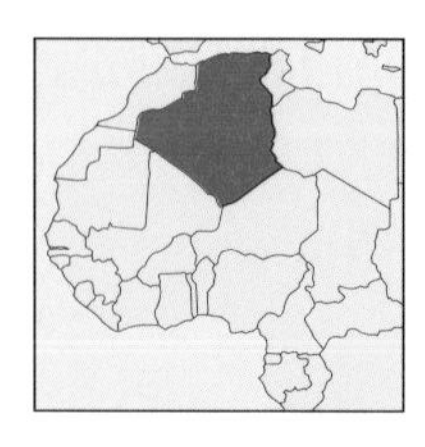

The star and crescent and the color green on Algeria's flag are traditional symbols of the Islamic religion. The liberation movement which fought for independence from French rule from 1954 used this flag. It became the national flag when Algeria became independent in 1962.

The People's Democratic Republic of Algeria is Africa's second largest country after Sudan. Most Algerians live in the north, on the fertile coastal plains and hill country bordering the Mediterranean Sea. South of this region lie high plateaux and ranges of the Atlas Mountains. Four-fifths of Algeria is in the Sahara, the world's largest desert. Most people in the Sahara live at oases, where springs and wells supply water.

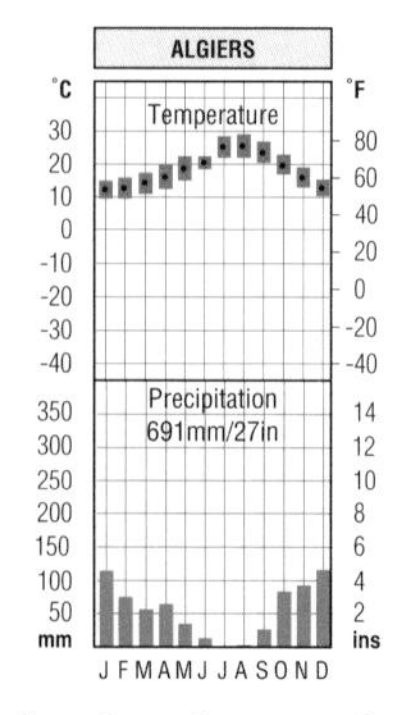

CLIMATE

Algiers, the capital, has a Mediterranean climate; summers here are warm and dry, and winters are mild and moist. But the northern highlands have warmer summers and colder winters. The Sahara is hot by day, but cool at night. The yearly rainfall is less than 8 inches [200 mm].

VEGETATION

The northern region has areas of scrub and farmland, with forests of Aleppo pine and cork oak on mountain slopes. The Sahara Desert contains regions of sand dunes, gravel-strewn plains and bare rock. Few plants grow outside of oases and wetter highland areas.

AREA 919,590 sq miles [2,381,740 sq km]

POPULATION 31,736,000

CAPITAL (POPULATION) Algiers (1,722,000)

GOVERNMENT Socialist republic

ETHNIC GROUPS Arab-Berber 99%

LANGUAGES Arabic and Berber (both official), French

RELIGIONS Sunni Muslim 99%

CURRENCY Algerian dinar = 100 centimes

HISTORY AND POLITICS

In early times, the region came under such rulers as the Phoenicians, Carthaginians, Romans, and Vandals. Arabs invaded the area in the AD 600s, converting the local Berber to Islam. Arabic became the chief language, though Berber dialects are still spoken in the highlands.

France ruled Algeria from 1830 until 1962, when the socialist FLN (National Liberation Front) formed a one-party government. Opposition parties were permitted in 1989. In 1991, the Islamic Salvation Front (FIS) won a general election. The FLN canceled the election results and declared a state of emergency. Conflict led to the deaths of about 100,000 in 1991–99. Abdelaziz Bouteflika, supported by the army, was elected

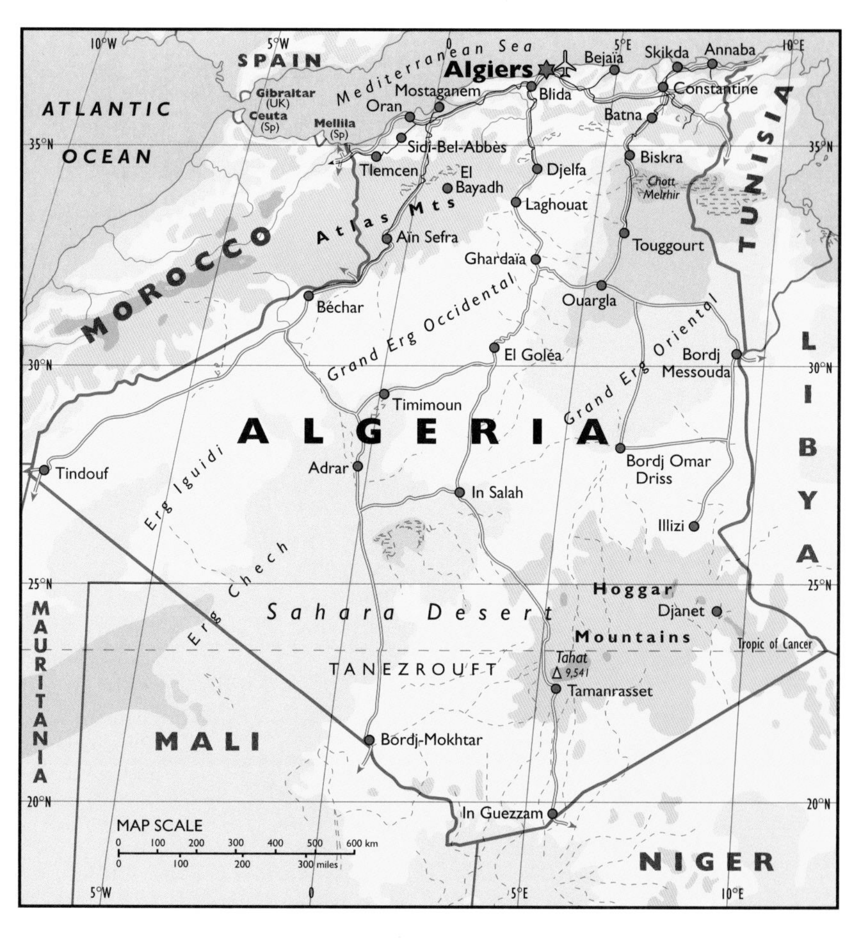

president in 1999. The scale of violence was reduced in the early 2000s. But clashes between Islamists and the army continued.

ECONOMY

Algeria is a developing country. Its chief resources are oil and gas, discovered under the Sahara Desert in 1956. Its natural gas reserves are among the largest in the world. Gas and oil account for more than 90% of Algeria's exports. Other manufactured products include cement, iron and steel, textiles, and vehicles. Farming employs 12% of the population. Major crops include barley, citrus fruits, dates, olives, potatoes, and wheat. In the northeast, Berber nomads raise cattle, goats, and sheep. Because of high unemployment, many people work abroad, especially in France.

ANGOLA

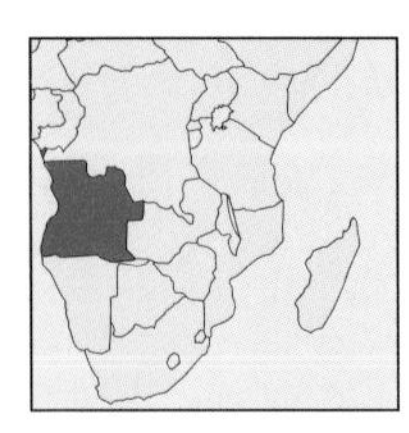

The flag is based on the flag of the MPLA (the Popular Movement for the Liberation of Angola) during the independence struggle. The emblem includes a star symbolizing socialism, one half of a gearwheel to represent industry, and a machete symbolizing agriculture.

The Republic of Angola is a large country, more than twice the size of France, on the southwestern coast of Africa. Most of the country is part of the huge plateau which makes up the interior of southern Africa. In the west is a narrow coastal plain.

Angola has many rivers. In the northeast, several rivers flow northward to the River Congo. In the south, some rivers, including the Cubango (Okavango) and the Cuanda, flow southeastward into the interior of Africa.

AREA 481,351 sq miles [1,246,700 sq km]
POPULATION 10,366,000
CAPITAL (POPULATION) Luanda (2,250,000)
GOVERNMENT Multiparty republic
ETHNIC GROUPS Ovimbundu 37%, Kimbundu 25%, Bakongo 13%, others 25%
LANGUAGES Portuguese (official), many others
RELIGIONS Traditional beliefs 47%, Christianity (Roman Catholic 38%, Protestant 15%)
CURRENCY Kwanza = 100 lwei

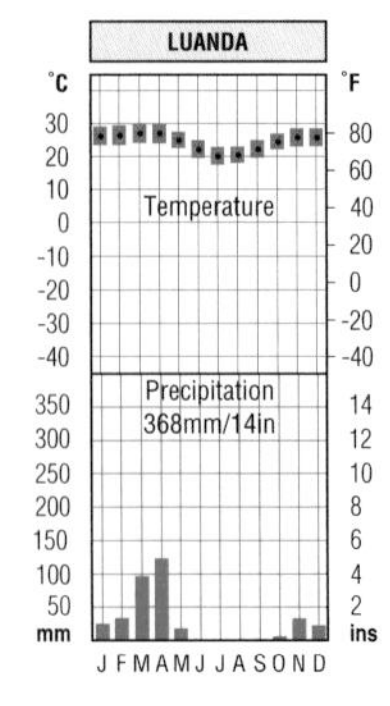

CLIMATE

Angola has a tropical climate, with temperatures of over 68°F [20°C] throughout the year, though the highest areas are cooler. The coastal regions are dry, increasingly so to the south of Luanda, but the rainfall increases to the north and east. The rainy season is between November and April.

VEGETATION

Grasslands cover much of the country. The coastal plain has little vegetation and the southern coast is a desert region that merges into the bleak Namib Desert. Some rain forest grows in the north.

HISTORY

Bantu-speaking peoples from the north settled in Angola around 2,000 years ago. In the late 15th century, Portuguese navigators, seeking a route to Asia around Africa, explored the coast and, in the early 16th century, the Portuguese set up bases.

Angola became important as a source of slaves for Brazil, Portugal's huge colony in South America. After the decline of the slave trade, Portuguese settlers began to develop the land. The Portuguese population increased greatly in the 20th century.

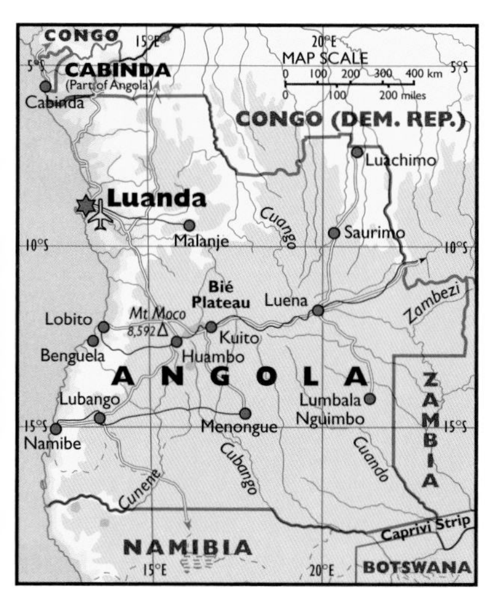

In the 1950s, local nationalists began to demand independence. In 1956, the MPLA (Popular Movement for the Liberation of Angola) was founded with support from the Mbundu and mestizos (people of African and European descent). The MPLA led a revolt in Luanda in 1961, but it was put down by Portuguese troops.

Other opposition groups developed. In the north, the Kongo set up the FNLA (Front for the Liberation of Angola), while, in 1966, southern peoples, including many Ovimbundu, formed UNITA (National Union for the Total Independence of Angola).

POLITICS

After Angola became independent in 1975, the rival groups struggled for power. The MPLA formed the government, but UNITA guerrillas, with support from South Africa, fought against government troops. A peace treaty in 1991 was followed by elections, which were won by the MPLA. UNITA's leaders refused to accept the result. Efforts to achieve peace failed until the UNITA leader, Jonas Savimbi, was killed in action in 2002. UNITA's army was disbanded and a peace agreement officially ended the 27-year-long civil war.

ECONOMY

Angola is a developing country where more than 70% of the people make a meager living by farming. The main food crops are cassava and maize, while coffee is exported.

But Angola has much economic potential. It has oil reserves, near Luanda and in the Cabinda enclave, which is separated from Angola by a strip of land belonging to Congo (Dem. Rep.). Oil is the leading export.

Angola produces diamonds and has reserves of copper, manganese, and phosphates. The country has a growing manufacturing sector.

These stamps, from an "Animals" set issued in 1953, show two Angolan mammals – a sable antelope and a leopard.

ARGENTINA

The "celeste" (sky blue) and white stripes were the symbols of independence around the city of Buenos Aires, where an independent government was set up in 1810. It became the national flag in 1816. The gold May Sun was added two years later.

The Argentine Republic is South America's second largest and the world's eighth largest country. The high Andes range in the west contains Mount Aconcagua, the highest peak in the Americas. In southern Argentina, the Andes Mountains overlook Patagonia, a plateau region. In east-central Argentina lies a fertile plain called the *pampas*. The northeast also contains lowland plains. The Gran Chaco lies west of the River Paraná, while Mesopotamia is the fertile plain between two rivers, the Paraná and the Uruguay.

AREA 1,068,296 sq miles [2,766,890 sq km]
POPULATION 37,385,000
CAPITAL (POPULATION) Buenos Aires (10,990,000)
GOVERNMENT Federal republic
ETHNIC GROUPS European 97%, Mestizo, Amerindian
LANGUAGES Spanish (official)
RELIGIONS Roman Catholic 92%, Protestant 2%, Jewish 2%
CURRENCY Peso = 10,000 australs

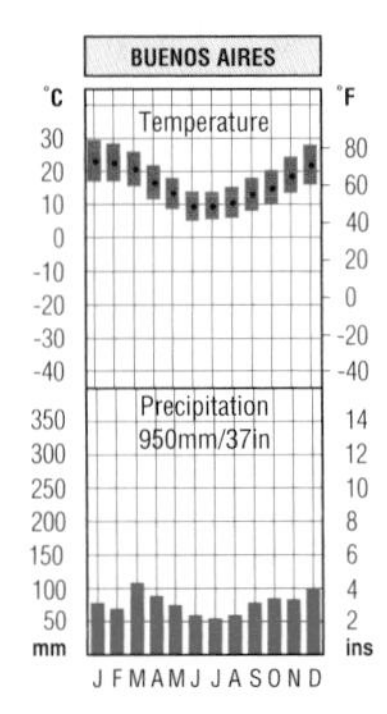

CLIMATE

Argentina's climates range from subtropical in the north to temperate in the south, with extreme conditions in the high Andes. The rainfall is abundant in the northeast, but is lower to the west and south. Patagonia is a dry region, crossed by rivers that rise in the Andes Mountains.

VEGETATION

The Gran Chaco is a forested region, known for its quebracho trees, which yield tannin used in the leather industry. Mesopotamia and the *pampas* are grassy regions, though large areas are farmed. Patagonia is too dry for crops and the main activity on the grassy tablelands is sheep raising.

HISTORY AND POLITICS

Spanish explorers first reached the coast in 1516. Spanish settlers followed, first in search of gold. But soon the settlers began to establish huge farms and ranches. Spanish rule continued until Buenos Aires declared itself independent in 1810, followed by the provinces in 1816. In 1853, the country adopted a federal constitution, giving the provinces the power to control their own affairs.

Argentina's economy developed steadily in the late 19th and early 20th centuries. But since the 1930s, political instability has marred the country's progress, with frequent coups, high unemployment and inflation. From 1976, during the so-called "dirty war," torture, political murders and wrongful imprisonment became common. In 1982, partly to divert attention from the poor state of the economy, Argentina invaded the Falkland Islands, which it called the Islas Malvinas and which it had long claimed. Britain retook the islands later in 1982. In 1983, the military rulers were forced to hold elections. Constitutional government was restored. However, following a long recession, a major financial crisis in 2001 threatened Argentina's stability.

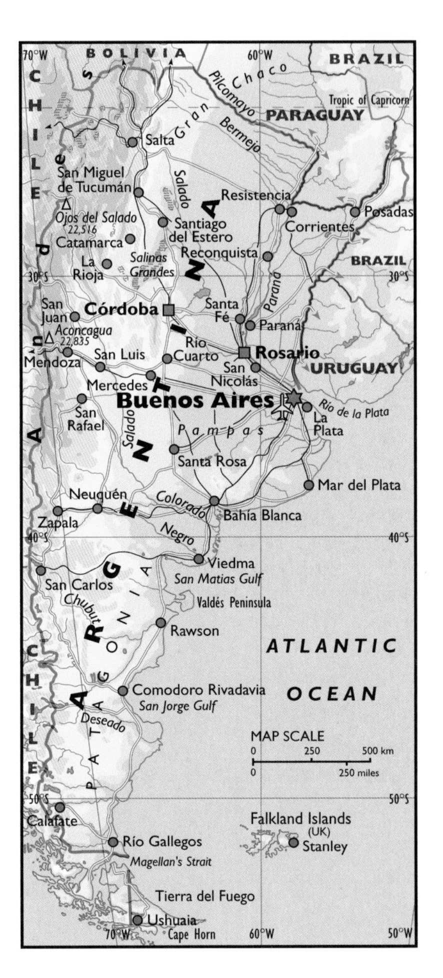

ECONOMY

The World Bank describes Argentina as an "upper-middle-income" developing country. There are large areas of fertile farmland. The main products are beef, maize, and wheat. Other products include citrus fruits, cotton, grapes, sorghum, soya beans, sugarcane, tea, and wood.

About 90% of the people live in cities and towns, where many factories process farm products. But the country has also set up many other industries, including the manufacture of cars, electrical equipment, and textiles. Oil is the leading mineral resource. The leading exports are meat, wheat, maize, vegetable oils, hides and skins, and wool. In 1991, Argentina, Brazil, Paraguay, and Uruguay set up Mercosur, an alliance aimed at creating a common market.

This stamp, issued in 1988, shows a Roman Catholic church. It reflects the country's Christian and Spanish culture.

ARMENIA

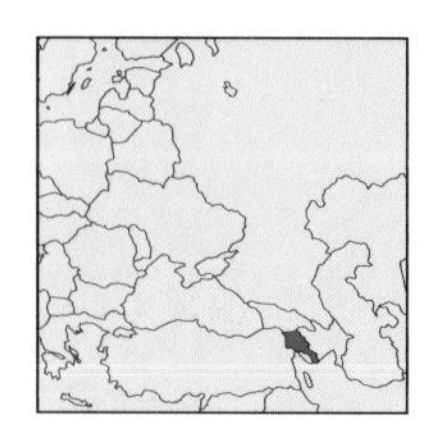

Armenia's flag was first used between 1918 and 1922, when the country was an independent republic. It was readopted on 24 August 1990. The red represents the blood shed in the past, the blue the land of Armenia, and the orange the courage of the people.

The Republic of Armenia is a landlocked country in southwestern Asia. Most of Armenia consists of a rugged plateau, crisscrossed by long faults (cracks). Movements along the faults cause earthquakes. The highest point is Mount Aragats, at 13,419 ft [4,090 m] above sea level. The lowest land is in the northwest, where the capital Yerevan is situated, and the northeast. The largest lake is Ozero (Lake) Sevan. Armenia has many fast-flowing rivers which have cut deep gorges in the plateau.

AREA 11,506 sq miles [29,800 sq km]
POPULATION 3,336,000
CAPITAL (POPULATION) Yerevan (1,256,000)
GOVERNMENT Multiparty republic
ETHNIC GROUPS Armenian 93%, Azerbaijani 3%, Russian, Kurd
LANGUAGES Armenian (official)
RELIGIONS Armenian Orthodox
CURRENCY Dram = 100 couma

CLIMATE

The height of the land, which averages 4,920 ft [1,500 m] above sea level, helps to give Armenia severe winters and cool summers. The highest peaks are snow-capped, but the total yearly rainfall is generally low, between about 8 inches and 31 inches [200 mm to 800 mm].

VEGETATION

The vegetation in Armenia ranges from semidesert to grassy steppe, forest, mountain pastures, and treeless tundra at the highest levels. Oak forests are found in the southeast, with beech being the most common tree in the forests of the northeast.

HISTORY

Armenia, which once covered a much larger area than it does today, became part of the Roman empire in 55 BC. Christianity became the state religion in the 4th century AD. Later, the area came under first Arab and then Turkish control.

By the early 16th century, Armenia was part of the Turkish Ottoman empire. In the 19th century, Armenians suffered great hardships under Turkish rule and, in the 1890s, the Turks slaughtered hundreds of thousands of Armenians. During World War I (1914–18), the Turks deported many Armenians, fearing that the Armenians would support Russia.

In 1918, an independent Armenian

republic was set up in the area held by Russia, though the western part of historic Armenia remained in Turkey, and another area was held by Iran. In 1920, Armenia became a Communist republic and, in 1922, it became, with Azerbaijan and Georgia, part of the Transcaucasian Republic within the Soviet Union. However, in 1936, the three territories became separate Soviet Socialist Republics.

POLITICS

Armenia has long disputed the status of an area, called Nagorno-Karabakh, inside Azerbaijan, because the majority of the people there were Armenians. After the breakup of the Soviet Union in 1991, fighting broke out between Armenia and Azerbaijan. In 1992, Armenia occupied the land between its eastern border and Nagorno-Karabakh. The fighting led to large movements of Armenians and Azerbaijanis. A ceasefire in 1994 left Armenia in control of about 20% of Azerbaijan's land area. Attempts to resolve the dispute failed in 2001.

ECONOMY

The World Bank classifies Armenia, whose economy was badly hit by the conflict with Azerbaijan in the early 1990s, as a "lower-middle-income" economy. Under Communist rule, the government controlled most economic activities. But since 1991, the government has encouraged free enterprise, selling farmland and state-owned industries.

The leading economic activities are manufacturing and mining, and the chief exports are machinery and transport equipment, metals, and chemicals. Copper is the chief metal mined in Armenia, but gold, lead, and zinc are also mined. About 10% of the land is used to grow crops, including barley, fruits, grapes for wine-making, potatoes, and wheat.

The airport at the capital city of Yerevan is the subject of this recent airmail stamp, issued by Armenia in 1992.

ATLANTIC OCEAN

ASCENSION

Ascension is a volcanic island in the South Atlantic Ocean with an area of 34 sq miles [88 sq km]. It lies about 700 miles [1,130 km] north-west of St Helena, from which it has been administered since 1922. It has no native population and the 700 or so people who live there are involved with such matters as satellite research, telecommunications, and servicing a mid-ocean air strip.

AZORES

The Azores are a group of nine large and several other small islands rising from the Mid-Atlantic Ridge in the North Atlantic Ocean. They are mostly mountainous, being of fairly recent volcanic origin. The islands cover an area of 868 sq miles [2,247 sq km], with a population of about 243,000. The capital is Ponta Delgada (population 21,000). The Azores, which lie about 745 miles [1,200 km] west of Lisbon, have been Portuguese since the middle of the 15th century. Today, they form an autonomous region of Portugal.

BERMUDA

Bermuda is a group of about 150 small islands consisting of coral, which forms caps on the tops of ancient volcanoes rising from the sea bed. Situated about 570 miles [920 km] east of Cape Hatteras in the United States, they cover an area of 20 sq miles [53 sq km]. The population is 64,000 and the capital is Hamilton (population 6,000). Bermuda is Britain's oldest colony. Tourism is a major industry. More than 60% of the people are of African descent. People of European origin make up about 37% of the population.

CANARY ISLANDS

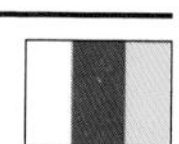

The Canary Islands comprise seven large and numerous small, mostly mountainous, islands about 62 miles [100 km] off the coast of northwest Africa. The islands have a total area of 2,807 sq miles [7,273 sq km]. The highest point is Pico de Teide, which rises to 12,202 ft [3,718 m]. The population is 1,577,000.

Claimed by Portugal in 1341, the Canary Islands became Spanish in 1479. Since 1927, they have formed two provinces of Spain. The capital of the province of Santa Cruz de Tenerife, also called Santa Cruz de Tenerife, has a population of 223,000. The province of Las Palmas has its capital at Las Palmas de Gran Canaria. With a population of 342,000, it is one of Spain's largest cities. With fertile soils and a mild subtropical climate, farming and fruit growing are major activities on the Canary Islands.

CAPE VERDE

The Republic of Cape Verde is an independent island nation situated about 350 miles [560 km] west of Dakar in West Africa. It consists of ten large and five small islands. Cape Verde has a total area of 1,556 sq miles

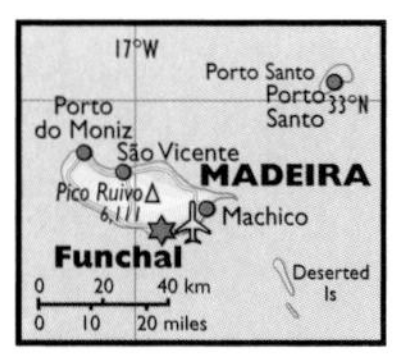
17°W
Porto Santo
Porto Santo
33°N
Porto do Moniz
São Vicente
MADEIRA
Pico Ruivo
6,111
Machico
Funchal
Deserted Is
0 20 40 km
0 10 20 miles

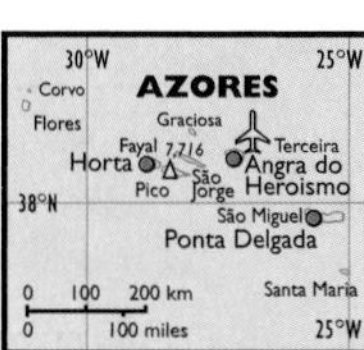
30°W
25°W
Corvo
AZORES
Flores
Graciosa
Fayal
7,716
Terceira
Horta
Angra do Heroismo
Pico
São Jorge
38°N
São Miguel
Ponta Delgada
0 100 200 km
Santa Maria
0 100 miles
25°W

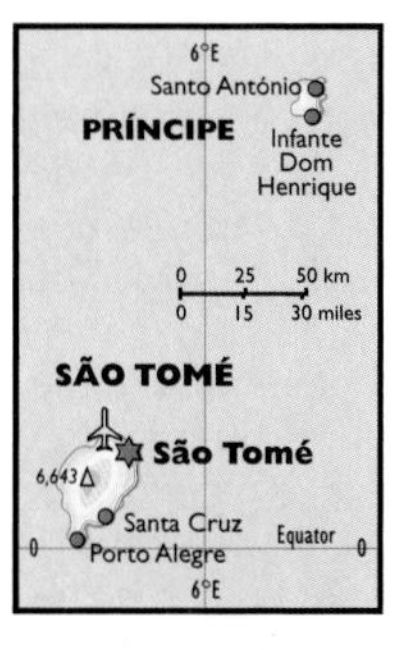
6°E
Santo António
PRÍNCIPE
Infante Dom Henrique
0 25 50 km
0 15 30 miles
SÃO TOMÉ
São Tomé
6,643
Santa Cruz
Porto Alegre
Equator
0
0
6°E

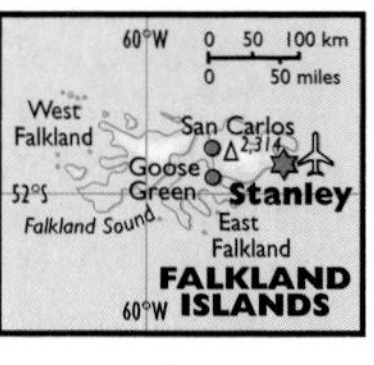
60°W
0 50 100 km
0 50 miles
West Falkland
San Carlos
2,314
Goose Green
Stanley
52°S
Falkland Sound
East Falkland
FALKLAND ISLANDS
60°W

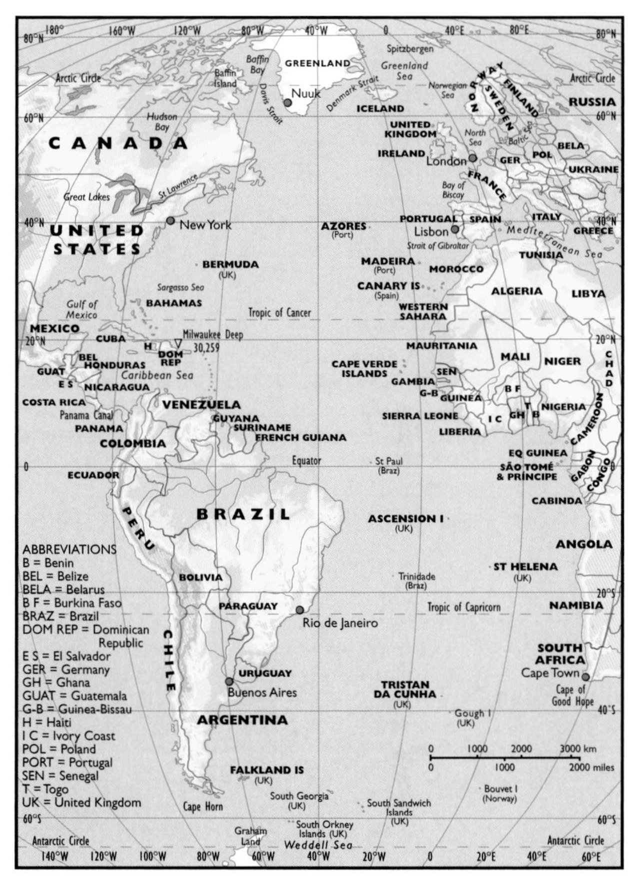
80°N
180°
160°W
120°W
80°W
40°W
0
40°E
80°E
80°N
Spitzbergen
Baffin Bay
GREENLAND
Greenland Sea
Arctic Circle
Baffin Island
Norwegian Sea
NORWAY
FINLAND
SWEDEN
Arctic Circle
Nuuk
Denmark Strait
Davis Strait
RUSSIA
ICELAND
60°N
Hudson Bay
UNITED KINGDOM
North Sea
Baltic
60°N
CANADA
BELA
IRELAND
London
GER
POL
UKRAINE
FRANCE
Bay of Biscay
St Lawrence
Great Lakes
PORTUGAL
SPAIN
ITALY
40°N
40°N
UNITED STATES
New York
AZORES
(Port)
Lisbon
Mediterranean Sea
GREECE
Strait of Gibraltar
TUNISIA
BERMUDA
(UK)
MADEIRA
(Port)
MOROCCO
CANARY IS
(Spain)
ALGERIA
LIBYA
Sargasso Sea
Gulf of Mexico
BAHAMAS
Tropic of Cancer
WESTERN SAHARA
MEXICO
Milwaukee Deep
CUBA
30,259
20°N
H
DOM REP
MAURITANIA
20°N
BEL
HONDURAS
Caribbean Sea
CAPE VERDE ISLANDS
MALI
NIGER
GUAT
SEN
GAMBIA
E S
NICARAGUA
G-B
GUINEA
B F
CHAD
COSTA RICA
VENEZUELA
NIGERIA
Panama Canal
SIERRA LEONE
T
PANAMA
GUYANA
SURINAME
I C
GH
B
LIBERIA
CAMEROON
COLOMBIA
FRENCH GUIANA
EQ GUINEA
0
Equator
St Paul
(Braz)
SÃO TOMÉ & PRÍNCIPE
GABON
CONGO
ECUADOR
CABINDA
PERU
BRAZIL
ASCENSION I
(UK)
ABBREVIATIONS
B = Benin
BEL = Belize
BELA = Belarus
B F = Burkina Faso
BRAZ = Brazil
DOM REP = Dominican Republic
E S = El Salvador
GER = Germany
GH = Ghana
GUAT = Guatemala
G-B = Guinea-Bissau
H = Haiti
I C = Ivory Coast
POL = Poland
PORT = Portugal
SEN = Senegal
T = Togo
UK = United Kingdom
ANGOLA
BOLIVIA
ST HELENA
(UK)
Trinidade
(Braz)
20°S
PARAGUAY
Tropic of Capricorn
NAMIBIA
Rio de Janeiro
CHILE
SOUTH AFRICA
Cape Town
URUGUAY
TRISTAN DA CUNHA
(UK)
Buenos Aires
Cape of Good Hope
ARGENTINA
Gough I
(UK)
40°S
0 1000 2000 3000 km
0 1000 2000 miles
FALKLAND IS
(UK)
Bouvet I
(Norway)
South Georgia
(UK)
Cape Horn
South Sandwich Islands
(UK)
60°S
60°S
South Orkney Islands (UK)
Graham Land
Weddell Sea
Antarctic Circle
Antarctic Circle
140°W
120°W
100°W
80°W
60°W
40°W
20°W
0
20°E
40°E
60°E

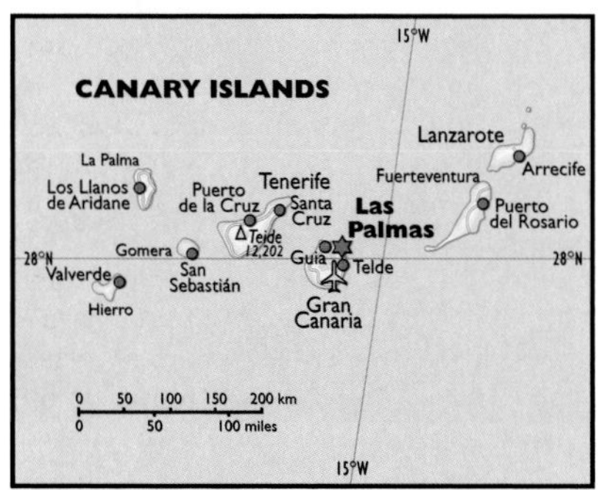
15°W
CANARY ISLANDS
Lanzarote
La Palma
Arrecife
Fuerteventura
Los Llanos de Aridane
Puerto de la Cruz
Tenerife
Santa Cruz
Las Palmas
Puerto del Rosario
Teide
12,202
28°N
Gomera
Guia
28°N
Valverde
San Sebastián
Telde
Hierro
Gran Canaria
0 50 100 150 200 km
0 50 100 miles
15°W

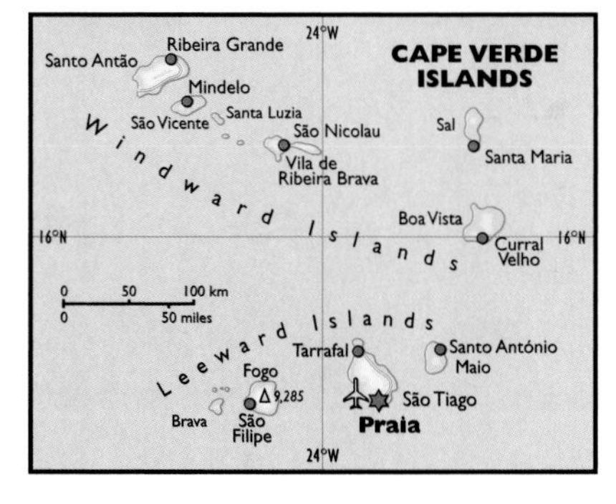
24°W
Ribeira Grande
Santo Antão
CAPE VERDE ISLANDS
Mindelo
Santa Luzia
São Vicente
São Nicolau
Sal
Windward Islands
Vila de Ribeira Brava
Santa Maria
Boa Vista
16°N
16°N
Curral Velho
0 50 100 km
0 50 miles
Leeward Islands
Tarrafal
Santo António
Maio
Fogo
9,285
São Tiago
Brava
São Filipe
Praia
24°W

[4,030 sq km] and a population of 405,000. Its capital, situated on the island of São Tiago, is Praia (population 69,000).

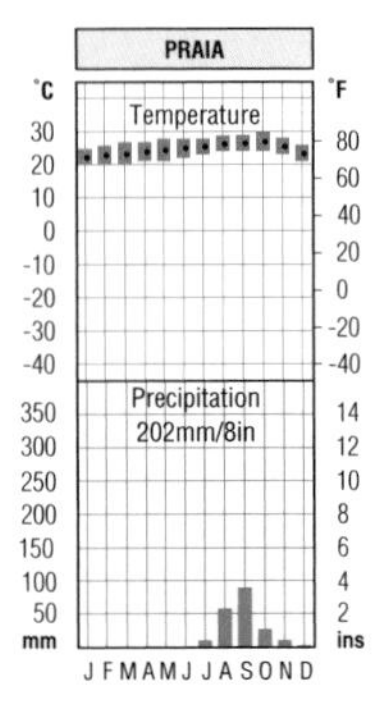

Cape Verde has a tropical climate, with high temperatures all year round. Praia in the south is one of the wettest places, but the country is arid throughout. Cape Verde was ruled by Portugal from the 15th century, but it became independent in 1975. The official language is Portuguese, but most people speak a Creole dialect. About 70% of the people are of mixed African and Portuguese descent. Most of the rest are Africans. Poor soils and lack of water have hindered development and many people emigrate to find work. The World Bank rates Cape Verde as a "low-income" developing country. The main activities are farming and fishing.

FALKLAND ISLANDS

The Falkland Islands consist of two large and more than 200 small islands. They lie about 310 miles [500 km] east of the Strait of Magellan in South America. The islands cover 4,699 sq miles [12,170 sq km]. The population is 2,895 and the capital is Stanley. The Falkland Islands are a British dependency, but they are claimed by Argentina, where they are known as the Islas Malvinas. In April 1982, Argentine forces invaded the islands, but Britain reconquered them in June 1982. In normal times, sheep rearing is the only activity and wool is the main export. South Georgia and South Sandwich Islands ceased to be part of the Falkland Islands Dependencies in 1985.

GREENLAND

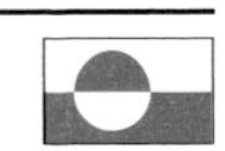

Greenland is the world's largest island, covering an area of 838,999 sq miles [2,175,600 sq km]. It is geographically part of North America (*see Canada map, page 67*), although, politically, it is a self-governing division of Denmark. An ice sheet, the world's second largest after Antarctica's, covers more than four-fifths of the island and settlements are confined to the rocky coasts. The population is 56,000. The capital is Nuuk (or Godthaab).

Vikings founded a colony in Greenland in the AD 980s. The colony disappeared in the 15th century, but it was recolonized when a Norwegian mission was set up in 1721. When the union between Norway and Denmark ended in 1814, Greenland remained with Denmark. Greenland became a Danish province in 1953. But, after Denmark joined the EEC in 1973, many Greenlanders wanted self-government, which they achieved in 1981, and in 1997 Danish place-names were superseded by Inuit forms. Fishing is the main industry, but Greenland's economy depends on assistance from Denmark. Greenland has a 27-member Home Rule Parliament.

MADEIRA

Madeira is the name for a group of volcanic islands situated 350 miles [550 km] west of the coast of Morocco. Madeira is also the name of the largest island in this group, which forms an autonomous region of Portugal. The islands, including Porto Santo, the only other inhabited island in the group, cover an area of 314 sq miles [813 sq km]. The population is 259,000 and the capital is Funchal (population 45,000). The Portuguese first reached Madeira in 1419, and Funchal was founded in 1421. With a warm temperate climate and fertile soils, farming and wine production are important activities. Madeira is known for its wicker furniture and baskets.

ST HELENA

St Helena is a volcanic island in the South Atlantic Ocean, nearly 1,240 miles [2,000 km] west of Africa. It covers an area of 47 sq miles [122 sq km], with a population of 7,266. The capital is Jamestown (population 1,500). St Helena became famous when the French emperor Napoleon I was held there from 1815 to 1821. This British dependency is now the administrative center for Ascension and Tristan da Cunha.

SÃO TOMÉ & PRÍNCIPE

The Democratic Republic of São Tomé and Príncipe, an island country 186 miles [300 km] west of Gabon, consists of two main islands, which are volcanic and mountainous, and several small ones. The country has a total land area of 372 sq miles [964 sq km]. The population is 165,000, and the capital is São Tomé (population 36,000). Lying just to the north of the Equator, the temperatures are high throughout the year. During the dry season (from June to September), practically no rain falls. The total rainfall is moderate. Forests cover about three-quarters of the land. Portuguese navigators reached the islands, which were uninhabited, in 1470. In 1522, the islands became a Portuguese colony whose economy depended on slaves brought from mainland Africa. The islands became independent in 1975 and the country became a one-party socialist state. Marxist policies were abandoned in 1990 and multiparty elections were held in 1991, 1996 and 2001. The main product is cocoa.

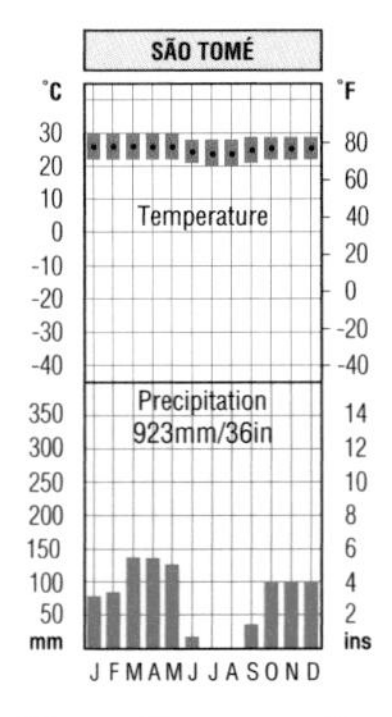

TRISTAN DA CUNHA

Tristan da Cunha is the largest of a group of four scattered islands in the South Atlantic Ocean. The others are Gough, Inaccessible, and Nightingale islands. Covering an area of 40 sq miles [104 sq km], the territory has a population of about 300. Tristan da Cunha is administered as a dependency of St Helena.

AUSTRALIA

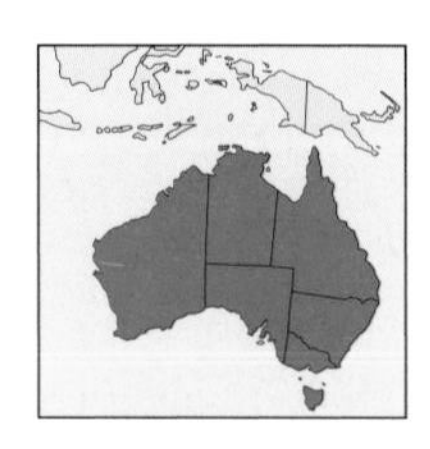

The national flag, above right, was adopted in 1901. It includes the British Union Flag, revealing Australia's historic links with Britain. In 1995, the Australian government put the flag used by the country's Aboriginal people, below right, on the same footing as the national flag.

The Commonwealth of Australia, the world's sixth largest country, is also a continent. Australia is the flattest continent. The main highland zone is the Great Dividing Range in the east. This range extends from the Cape York Peninsula to Victoria. The mountains of Tasmania are an extension of the range. The highest peak is Mount Kosciuszko in New South Wales. The Great Dividing Range separates the eastern coastal plains from the Central Lowlands. The southeastern lowlands are drained by the Murray and Darling, Australia's longest rivers. Lake Eyre, to the west, is Australia's largest lake. But Lake Eyre is dry for much of the time. The Western Plateau makes up two-thirds of Australia. It is generally flat, but a few low mountain ranges break the monotony of the landscape.

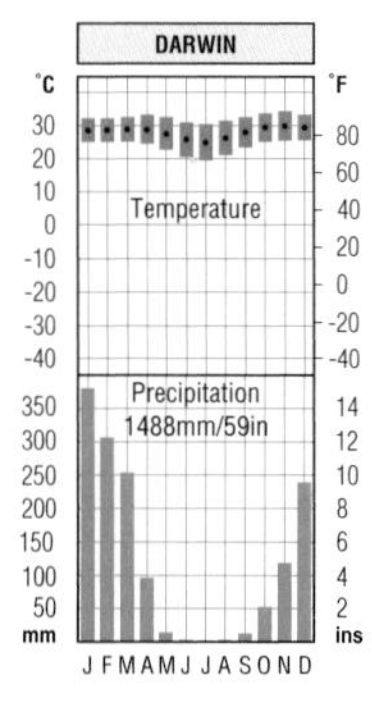

CLIMATE

Only 10% of Australia has an average yearly rainfall of more than 39 inches [1,000 mm]. These areas include the tropical north, where Darwin is situated, the northeast coast and the southeast, where Sydney is located. Parts of southern Australia, including Perth, have dry summers and moist winters. The interior is dry, where water is quickly evaporated in the heat.

AREA 2,967,893 sq miles [7,686,850 sq km]

POPULATION 19,358,000

CAPITAL (POPULATION) Canberra (325,000)

GOVERNMENT Federal constitutional monarchy

ETHNIC GROUPS Caucasian 92%, Asian 7.5%, Aboriginal 1.5%

LANGUAGES English (official)

RELIGIONS Roman Catholic 26%, Anglican 24%, other Christian 24%, non-Christian 24%

CURRENCY Australian dollar = 100 cents

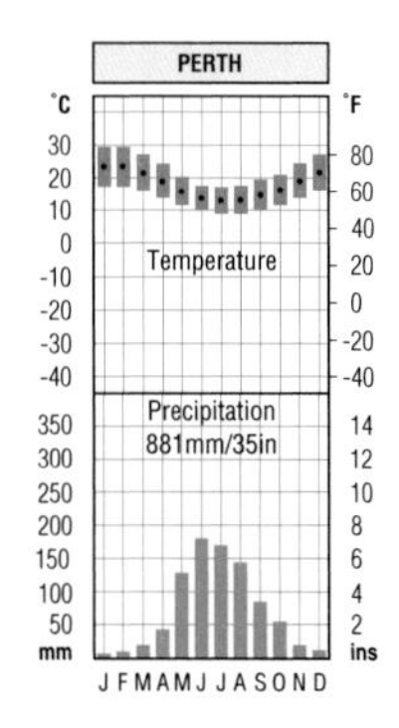

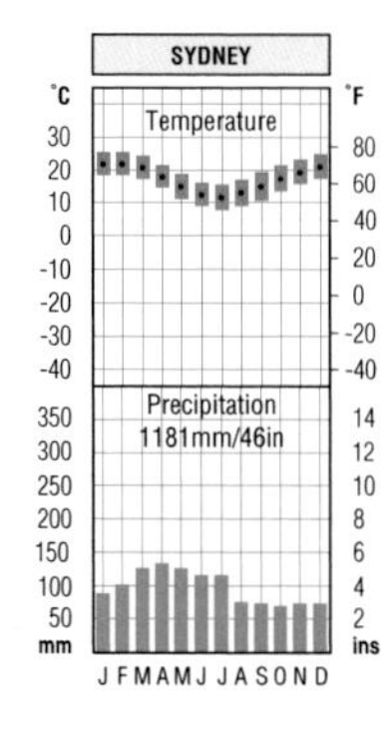

DID YOU KNOW

- that the name Australia comes from the Latin *australis*, meaning "southern"
- that Australia has more sheep than any other country
- that the aboriginal name for Ayers Rock in central Australia is *Uluru*, which means "great pebble"
- that 59% of Australians live in the five largest cities (Sydney, Melbourne, Brisbane, Perth, and Adelaide)
- that the Great Barrier Reef, off the northeast coast of Australia, is the world's largest coral reef
- that Australia leads the world in producing bauxite, diamonds, lead ore, top-quality opals, and wool

VEGETATION

Much of the Western Plateau is desert. Around the desert are areas of grass and low shrubs. The Central Lowlands are grasslands used to raise farm animals. Water comes from artesian wells that tap underground rocks which contain water.

The north has areas of savanna (tropical grassland with scattered trees) and rain forest. In dry areas, shrubs called acacias are common. In wetter areas, eucalyptus (or gum) trees are found. Australia has many flowering plants. When heavy rains occur in dry areas, flowering plants grow quickly, transforming the dry land into a carpet of colors.

HISTORY

The Aboriginal people of Australia entered the continent from Southeast Asia more than 50,000 years ago. The first European explorers were Dutch in the 17th century, but they did not settle. In 1770, the British Captain Cook explored the east coast and, in 1788, the first British settlement was established for convicts on the site of what is now Sydney. The first free settlers arrived in 1791.

In the 19th century, the continent was divided into several colonies, which later became states. In 1901, the states united to create the Commonwealth of Australia. The federal capital was established at Canberra, in the Australian Capital Territory, in 1927.

POLITICS

Australia has strong ties with the British Isles. But in the last 50 years, people from other parts of Europe and, most recently, from Asia have settled in Australia. Ties with Britain were also weakened by Britain's membership of the European Union. Many Australians now believe that they should become more involved economically and politically with their

Matthew Flinders, a British explorer, sailed around Australia in 1801–2, accurately mapping much of its coastline. This stamp depicting Flinders was issued to commemorate Australia Day in 1980.

neighbors in eastern Asia and the Americas rather than with Europe. In 1999, Australia held a referendum on whether the country should become a republic or remain a constitutional monarchy. By a majority of about 55 to 45, the country retained its status as a monarchy.

Another important political issue concerns the position of the Aboriginal people. During the 1990s, the government took steps to grant the Aboriginal people land rights over their traditional areas.

ECONOMY

Australia is a prosperous country. Its economy was based on agriculture. Crops can be grown on only 6% of the land, though dry pasture covers another 58%.

The country remains a major producer and exporter of farm products, particularly cattle, wheat, and wool, followed by dairy products, many kinds of fruits, and sugarcane. Grapes grown for wine-making are also important. Australian wines are now sold around the world.

The country is also rich in natural resources. It is a major producer of minerals, including bauxite, coal, copper, diamonds, gold, iron ore, manganese, nickel, silver, tin, tungsten, and zinc. Australia also produces some oil and natural gas. Metals, minerals and farm products account for the bulk of Australia's exports.

Australia's imports are mostly manufactured products, although the country makes many factory products, especially consumer goods, such as foods and household articles. Major imports include machinery and other goods used by factories.

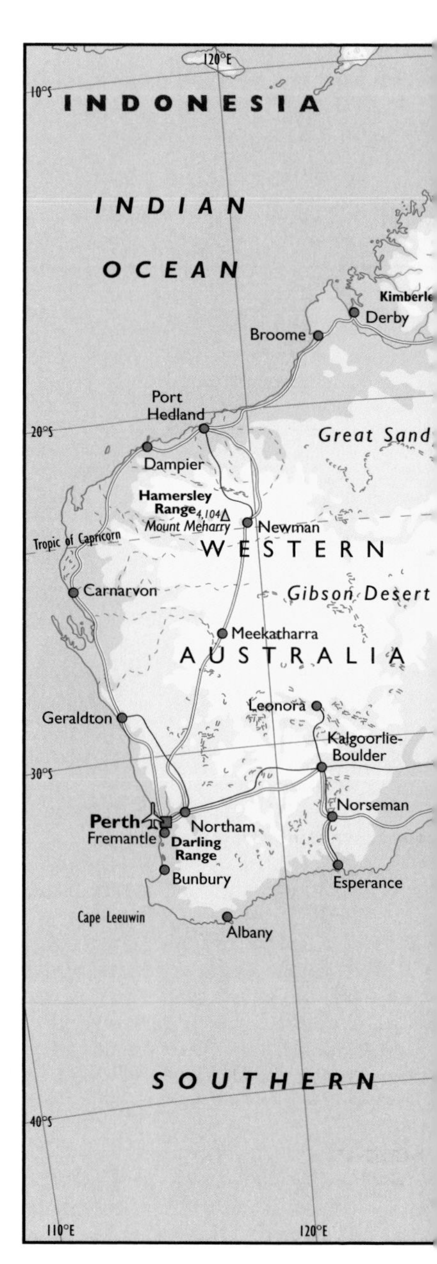

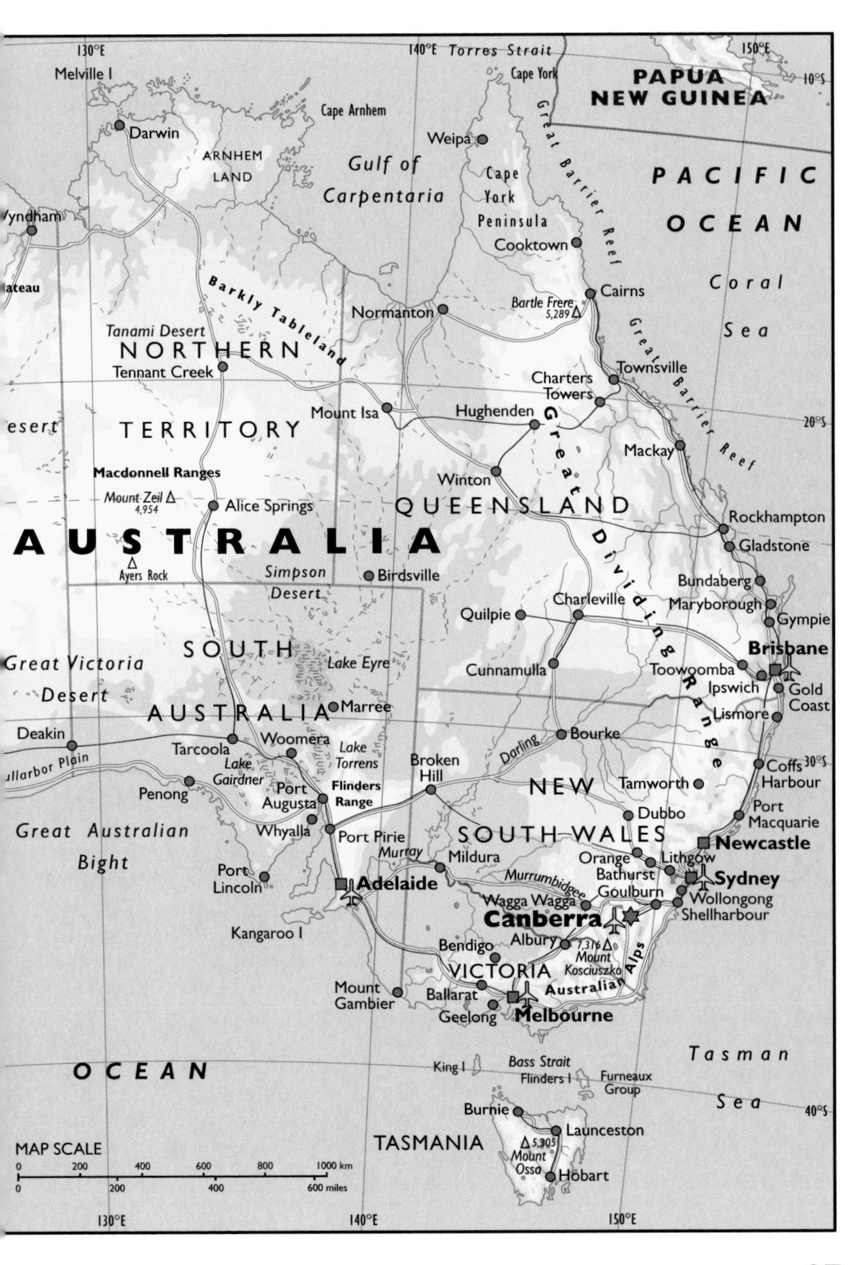

130°E
140°E
150°E
Torres Strait
Cape York
PAPUA NEW GUINEA
10°S
Melville I
Darwin
Cape Arnhem
ARNHEM LAND
Weipa
Gulf of Carpentaria
Cape York Peninsula
Great Barrier Reef
PACIFIC OCEAN
Wyndham
Cooktown
Coral Sea
Barkly Tableland
Cairns
Normanton
Bartle Frere 5,289
Tanami Desert
NORTHERN TERRITORY
Tennant Creek
Townsville
Charters Towers
Mount Isa
Hughenden
20°S
Mackay
Great Dividing Range
Macdonnell Ranges
Winton
Mount Zeil 4,954
Alice Springs
QUEENSLAND
Rockhampton
Gladstone
AUSTRALIA
Ayers Rock
Simpson Desert
Birdsville
Bundaberg
Maryborough
Charleville
Quilpie
Gympie
Brisbane
Great Victoria Desert
SOUTH AUSTRALIA
Lake Eyre
Cunnamulla
Toowoomba
Ipswich
Gold Coast
Marree
Lismore
Deakin
Tarcoola
Woomera
Bourke
Nullarbor Plain
Lake Torrens
Broken Hill
Darling
30°S
Coffs Harbour
Lake Gairdner
Penong
Port Augusta
Flinders Range
NEW SOUTH WALES
Tamworth
Port Macquarie
Dubbo
Great Australian Bight
Whyalla
Port Pirie
Newcastle
Murray
Mildura
Orange
Lithgow
Bathurst
Port Lincoln
Adelaide
Murrumbidgee
Sydney
Goulburn
Wagga Wagga
Wollongong
Shellharbour
Canberra
Kangaroo I
Albury
Bendigo
7,316 Mount Kosciuszko
Australian Alps
VICTORIA
Mount Gambier
Ballarat
Geelong
Melbourne
Tasman Sea
OCEAN
King I
Bass Strait
Flinders I
Furneaux Group
40°S
Burnie
Launceston
TASMANIA
5,305 Mount Ossa
Hobart
MAP SCALE
0 200 400 600 800 1000 km
0 200 400 600 miles

AUSTRIA

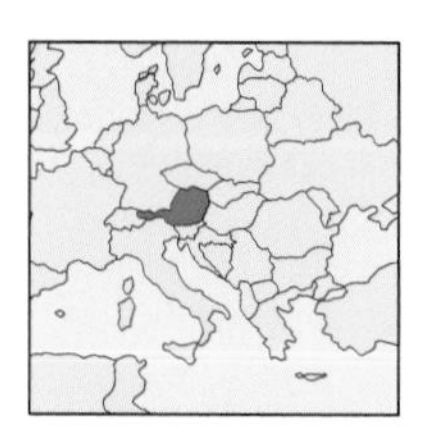

According to legend, the colors on Austria's flag date back to a battle in 1191, during the Third Crusade, when an Austrian duke's tunic was stained with blood, except under his swordbelt, where it remained white. The flag was officially adopted in 1918.

The Republic of Austria is a landlocked country in the heart of Europe. About three-quarters of the land is mountainous, and tourism and winter sports are major activities in this scenic country. Northern Austria contains the valley of the River Danube, which rises in Germany and flows to the Black Sea, and the Vienna basin. This is Austria's main farming region.

Southern Austria contains ranges of the Alps, which rise to their highest point at Grossglockner, at 12,457 ft [3,797 m] above sea level.

AREA 32,374 sq miles [83,850 sq km]

POPULATION 8,151,000

CAPITAL (POPULATION) Vienna (or Wien, 1,560,000)

GOVERNMENT Federal republic

ETHNIC GROUPS Austrian 93%, Croatian, Slovene

LANGUAGES German (official)

RELIGIONS Roman Catholic 78%, Protestant 5%, Islam

CURRENCY Euro = 100 cents

CLIMATE

The climate of Austria is influenced both by westerly and easterly winds. The moist westerly winds bring rain and snow. They also moderate the temperatures. But dry easterly winds bring cold weather in winter and hot weather in summer. This gives eastern Austria a more continental climate than the western part of the country.

VEGETATION

Woods and meadows cover much of Austria. The forests include such trees as beech, larch, oak, and spruce. Austria has a higher proportion of forest than any other European country.

HISTORY AND POLITICS

Austria was once part of the Holy Roman Empire and, under rulers from the Habsburg family, it became the most important state in the empire. When the empire ended in 1806, the Habsburg ruler became emperor of Austria. In 1867, Austria and Hungary set up a powerful dual monarchy, which finally collapsed in 1918, at the end of World War I. In 1938, Germany annexed Austria and Austria supported Germany in World War II. Austria was partitioned in 1945 but became a neutral federal republic in 1955. Austria joined the European Union on January 1, 1995,

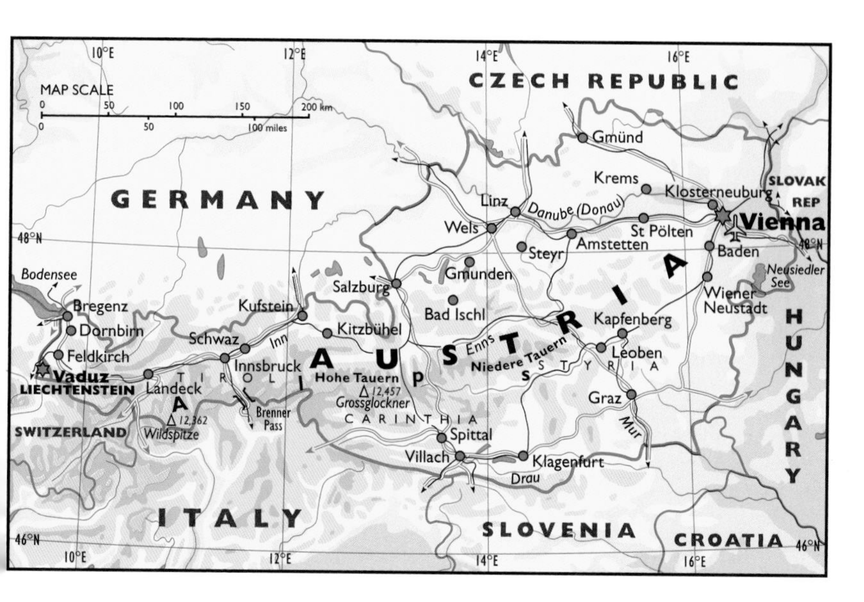

but it became a focus of controversy when, in 2000, a coalition government was formed by the right-wing People's Party and the extreme right-wing Freedom Party. However, the vote for the Freedom Party was greatly reduced in elections in 2002.

ECONOMY

Austria is a prosperous country. It has plenty of hydroelectric power, some oil and gas, and reserves of lignite (brown coal), but these do not meet its needs and fossil fuels are imported.

The country's leading economic activity is manufacturing metals and metal products, including iron and steel, vehicles, machines, machine tools, and ships. Vienna is the main industrial center, though factories are found throughout the country. Craft industries produce fine glassware, jewelry, and porcelain.

Crops are grown on 18% of the land and another 24% is pasture. Dairy and livestock farming are the leading activities. Major crops include barley, potatoes, rye, sugar beet, and wheat. In 2001, about 27 million tourists visited the country.

DID YOU KNOW

- that Austria's longest river, the Danube, is Europe's second longest at 1,770 miles [2,850 km], after the River Volga (2,300 miles [3,700 km])
- that Vienna's Schönbrunn Zoo, which opened in 1765, is the world's oldest zoo still in existence
- that Joseph Haydn, Wolfgang Amadeus Mozart, Franz Schubert, and Johann Strauss, jun., were all born in Austria
- that every summer people come from all over Europe to Salzburg, to a music festival celebrating Mozart

AZERBAIJAN

Azerbaijan's flag was adopted in 1991. The blue stands for the sky, the red for freedom, and the green for the land and for Islam. The crescent and the star also symbolize Islam. The points of the star represent the eight groups of people in Azerbaijan.

The Azerbaijani Republic is a country, facing the Caspian Sea to the east, in the southwest of Asia. It includes an area called the Naxçivan Autonomous Republic, which is completely cut off from the rest of Azerbaijan by Armenian territory. The Caucasus Mountains border Russia in the north. Another highland region, including the Little Caucasus Mountains and part of the rugged Armenian plateau, is in the southwest. Between these regions lies a broad plain drained by the River Kura which flows into the Caspian Sea.

AREA 33,436 sq miles [86,600 sq km]
POPULATION 7,771,000
CAPITAL (POPULATION) Baki (or Baku, 1,713,000)
GOVERNMENT Federal multiparty republic
ETHNIC GROUPS Azeri 90%, Dagestani 3%, Russian, Armenian
LANGUAGES Azerbaijani (official)
RELIGIONS Islam 93%, Russian Orthodox 2%, Armenian Orthodox 2%
CURRENCY Manat = 100 gopik

CLIMATE

Azerbaijan has hot summers and cool winters. The rainfall is low on the plains, ranging from about 5 to 15 inches [130 mm to 380 mm] per year. The annual rainfall is much higher in the highlands and on the subtropical southeastern coast.

VEGETATION

Forests of beech, oak, and pine trees grow on the mountain slopes, with pastures at higher levels. The dry lowlands contain grassy steppe and semidesert.

HISTORY

In ancient times, the area which is now Azerbaijan was invaded many times. In AD 642, Arab armies introduced Islam. But most modern Azerbaijanis are descendants of Persians and Turkic peoples who migrated to the area from the east by the 9th century. The area later came under Mongol rule between the 13th and 15th centuries, and then the Persian Safavid dynasty.

In the 18th century, Russia began to expand. By the early 19th century, Azerbaijan was under Russian rule. After the Russian Revolution of 1917, attempts were made to form a Transcaucasian Federation made up of Armenia, Azerbaijan, and Georgia. When this failed, Azerbaijanis set up an independent state. But Russian forces occupied the area in 1920. In 1922, the Communists set up a Transcaucasian Republic consisting of Armenia, Azerbaijan, and Georgia under Russian control. In 1936, the three areas became separate Soviet Socialist Republics within the Soviet Union.

POLITICS

In the late 1980s, the government of the Soviet Union began to introduce reforms, giving the people more freedom. In 1991, the Soviet Union was voted out of existence and Azerbaijan became an independent nation.

Since independence, the country's economic progress has been slow, partly because of civil unrest in Nagorno-Karabakh, a region within Azerbaijan, where the majority of the people are Christian Armenians.

In 1992, Armenia occupied the area between its eastern border and Nagorno-Karabakh, while ethnic Armenians took over Nagorno-Karabakh itself. This conflict led to large migrations of Armenians and Azerbaijanis. A ceasefire was agreed in 1994, with about 20% of Azerbaijan territory under Armenian control. An attempt to resolve the dispute failed in 2001.

> **DID YOU KNOW**
> - that at the start of the 20th century, Azerbaijan was the world's leading oil producer; in 1901, it produced half of the world's oil
> - that the Caspian Sea coast is 66 ft [20 m] below sea level
> - that the Azerbaijani language closely resembles Turkish
> - that the world's first oil pipeline (which was made from wood) and the first oil tanker were made in Azerbaijan

ECONOMY

With its economy in disarray since the breakup of the Soviet Union, Azerbaijan now ranks among the world's "lower-middle-income" countries. Its chief resource is oil and the chief oilfields are in the Baki region, on the shore of the Caspian Sea and in the sea itself. In 1994, Azerbaijan invited Western companies to develop the offshore oil deposits.

Manufacturing, including oil refining and the production of chemicals, machinery, and textiles, is the most valuable activity. Large areas are irrigated and crops include cotton, fruit, grains, tea, tobacco, and vegetables. Fishing is also important, though the Caspian Sea has become increasingly polluted. Under Communism, the government controlled economic activity, but, since 1991, governments have encouraged private enterprise.

BANGLADESH

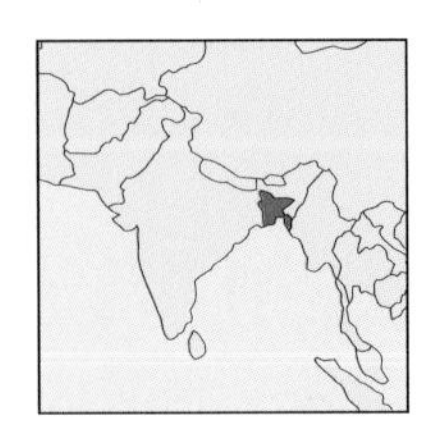

Bangladesh adopted this flag in 1971, following the country's break from Pakistan. The green is said to represent the fertility of the land. The red disk is the sun of independence. It commemorates the blood shed during the struggle for freedom.

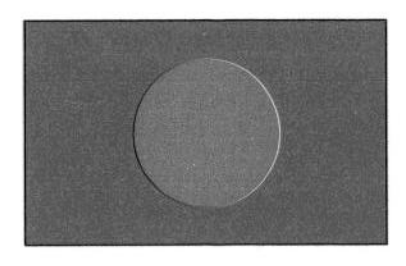

The People's Republic of Bangladesh is one of the world's most densely populated countries. Apart from hilly regions in the far northeast and southeast, most of the land is flat and covered by fertile alluvium spread over the land by the Ganges, Brahmaputra, and Meghna rivers. These rivers overflow when they are swollen by the annual monsoon rains. Floods also occur along the coast, 357 miles [575 km] long, when cyclones (hurricanes) drive seawater inland. These periodic storms cause great human suffering.

AREA 55,598 sq miles [144,000 sq km]
POPULATION 131,270,000
CAPITAL (POPULATION) Dhaka (or Dacca, 7,832,000)
GOVERNMENT Multiparty republic
ETHNIC GROUPS Bengali 98%, tribal groups
LANGUAGES Bengali and English (both official)
RELIGIONS Islam 83%, Hinduism 16%
CURRENCY Taka = 100 paisas

DHAKA
Temperature
Precipitation 2673mm/105in
J F M A M J J A S O N D

CLIMATE

Bangladesh has a tropical monsoon climate. In the winter, dry winds blow from the north. But in the spring, the land heats up and moist winds blow from the south. They bring heavy rain. The coldest month is January and the hottest is April, though it remains hot throughout the monsoon season.

VEGETATION

Most of Bangladesh is cultivated, but forests cover about 16% of the land. They include bamboo forests in the northeast and mangrove forests in the swampy Sundarbans region in the southwest, which is a sanctuary for the Royal Bengal tiger.

HISTORY AND POLITICS

Islam, the chief religion today, was introduced into Bengal in the 13th century. In 1576, Bengal became part of the Mughal empire, which also included most of Afghanistan, Pakistan, and India. European influence increased in the 16th century. In 1858, Bengal became part of British India.

In 1947, British India was partitioned between the mainly Hindu India and the Muslim Pakistan. Pakistan consisted of two parts, West and East Pakistan, which were separated by about 1,000 miles [1,600 km] of Indian territory. Differences devel-

DID YOU KNOW

- that the name Bangladesh means "the land of the Bengalis"
- that Bangladesh ranks second to India in the production of jute
- that the world's most devastating recorded cyclone (hurricane) killed an estimated 1 million people in Bangladesh in 1970
- that Bangladesh, which is smaller than the US state of Florida, ranks eighth in population among the countries of the world
- that Bangladesh has the largest river delta in the world, where the waters of the River Ganges and the River Brahmaputra flow into the Bay of Bengal. The flat delta is composed of alluvium deposited by the rivers

oped between West and East Pakistan. In 1971, the East Pakistanis rebelled. After a nine-month civil war, they declared East Pakistan to be a separate nation named Bangladesh.

Bangladesh became a one-party state in 1975, but military leaders seized control and dissolved parliament. In the years that followed, several military coups occurred. But in 1991, Bangladesh held its first free elections since independence.

ECONOMY

Bangladesh is one of the world's poorest countries. Its economy depends mainly on agriculture, which employs more than half of the people. Rice is the chief crop and Bangladesh is the world's fourth largest producer.

Other important crops include jute, sugarcane, tobacco, and wheat. Jute processing is the leading manufacturing industry, and jute is the leading export. Other manufactures include leather, paper, and textiles. Some 60% of the internal trade is carried by boat.

Many Muslim women in Bangladesh have little social contact outside the home and they usually cover their heads with veils in the presence of strangers. This stamp, showing a Bengali woman, was issued in 1987 to commemorate the Bengali New Year.

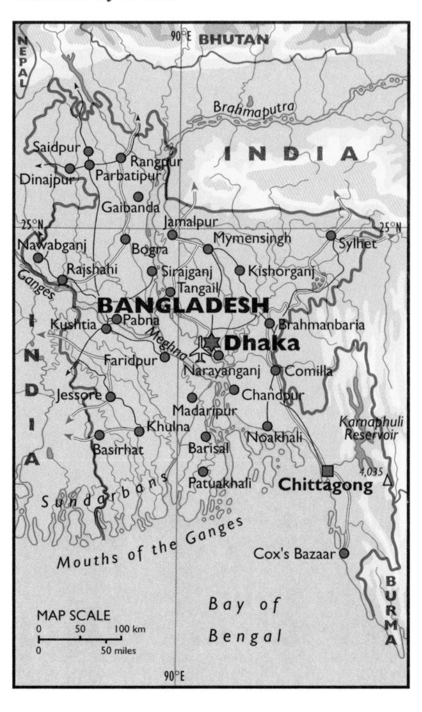

BELARUS

In September 1991, Belarus adopted a red and white flag to replace the flag used in the Soviet era. But, in June 1995, following a referendum in which Belarussians voted to improve relations with Russia, this was replaced with a design similar to the flag of 1958, but without the hammer and sickle.

The Republic of Belarus, or Belorussia as it is also known, is a landlocked country in Eastern Europe. It was formerly part of the Soviet Union. The land is low-lying and mostly flat. In the south, much of the land is marshy. This area contains Europe's largest marsh and peat bog, the Pripet Marshes. A hilly region extends from northeast to southwest through the center of the country. It includes the highest point in Belarus, which is situated near the capital Minsk. This hill reaches a height of only 1,135 ft [346 m] above sea level.

AREA 80,154 sq miles [207,600 sq km]
POPULATION 10,350,000
CAPITAL (POPULATION) Minsk (1,717,000)
GOVERNMENT Multiparty republic
ETHNIC GROUPS Belarussian 81%, Russian 11%, Polish, Ukrainian
LANGUAGES Belarussian and Russian (both official)
RELIGIONS Eastern Orthodox 80%, other (including Roman Catholic, Protestant, Jewish, Islam) 20%
CURRENCY Belarussian rouble = 100 kopecks

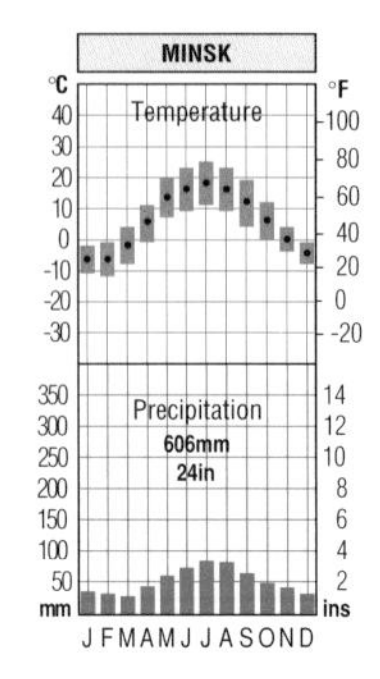

CLIMATE

The climate of Belarus is affected by both the moderating influence of the Baltic Sea and continental conditions to the east. The winters are cold and the summers warm. The average annual rainfall is between about 22 and 28 inches [550 mm to 700 mm].

VEGETATION

Forests cover about a third of Belarus. The colder north has such trees as alder, birch, and pine. Ash, oak, and hornbeam grow in the warmer south. Farmland and pasture have replaced most of the original forest.

HISTORY AND POLITICS

Slavic people settled in what is now Belarus about 1,500 years ago. In the 9th century, the area became part of the first East Slavic state, Kievan Rus. In the 13th century, Mongol armies overran the area and, later, Belarus became part of Lithuania. In 1569, Lithuania, including Belarus, became part of Poland. In the 18th century, Russia took over most of eastern Poland, including Belarus. Yet the people of Belarus still maintained their individuality.

In 1918, Belarus became an independent republic, but Russia invaded the country and, in 1919, a Communist state was set up. In 1922, Belarus became a founder republic of the Soviet Union. In 1991, Belarus again became an independent republic, although Belarus continued to support reunification with Russia. A union treaty aiming at a merger of the two countries was signed in 1999, though it was largely symbolic. Russia insisted that a referendum would have to take place before any merger.

DID YOU KNOW

- that Belarus is also called Belorussia or Byelorussia
- that about one in four of the people of Belarus were killed during World War II (1939–45)
- that the name Belarus comes from the Russian words *Belaya Rus*, meaning "White Russia"
- that Belarus and Poland jointly administer a World Heritage Site, Belorezha Forest (*Bialowieza* in Polish); it contains European bison (*wisent*)

ECONOMY

The World Bank classifies Belarus as an "upper-middle-class" economy. However, it faces many problems in changing a government-run economy into a free market one. Manufacturing became important under Communist rule and the country produces chemicals, vehicles, machine tools, and textiles. Farming is important and major products include barley, eggs, flax, meat, potatoes, and other vegetables. Exports include machinery and transport equipment, chemicals, and food products.

The former national flag, adopted in 1991 but replaced in 1995, flying above an outline map of Belarus, was the subject of this stamp issued in 1992. The site of the capital city, Minsk, whose name is written in the Cyrillic alphabet, is marked on the center of the map.

BELGIUM

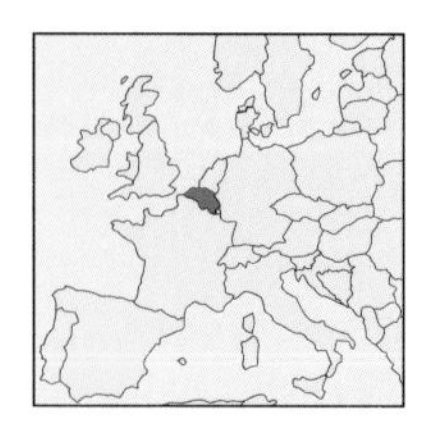

Belgium's national flag was adopted in 1830, when the country won its independence from the Netherlands. The colors come from the arms of the province of Brabant, in central Belgium, which rebelled against Austrian rule in 1787.

The Kingdom of Belgium is a densely populated country in western Europe. Behind the coastline on the North Sea, which is 39 miles [63 km] long, lie coastal plains. Some low-lying areas, called polders, are protected from the sea by dykes (or sea walls). Central Belgium consists of low plateaux and the only highland region is the Ardennes in the southeast. The chief rivers are the Schelde in the west, and the Sambre and Meuse which flow between the central plateau and the Ardennes. The river valleys are fertile.

AREA 11,780 sq miles [30,510 sq km]

POPULATION 10,259,000

CAPITAL (POPULATION) Brussels (Brussel or Bruxelles, 948,000)

GOVERNMENT Federal constitutional monarchy

ETHNIC GROUPS Fleming 58%, Walloon 31%, Italian, French, Dutch, Turkish, Moroccan

LANGUAGES Dutch, French and German (all official)

RELIGIONS Roman Catholic 75%, Protestant or other 25%

CURRENCY Euro = 100 cents

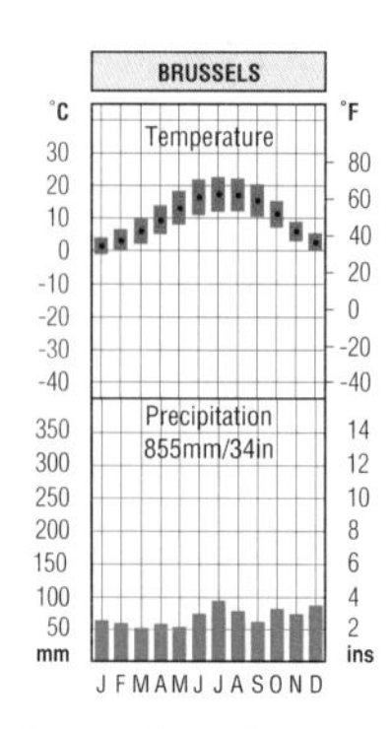

CLIMATE

Belgium has a cool, temperate climate. Moist winds from the Atlantic Ocean bring fairly heavy rain, especially in the Ardennes. In January and February much snow falls on the Ardennes. Brussels has mild winters and warm summers. The highlands are much cooler.

VEGETATION

Farmland and pasture cover about half of Belgium. The forests, especially in the Ardennes, contain such trees as beech, birch, elm, and oak. In the north, the birch forests and heathland have been largely replaced by plantations of evergreen trees.

HISTORY

In the Middle Ages, Belgium was split into small states, but the country was united by the dukes of Burgundy in the 14th and 15th centuries. Belgium later came, at various times, under Austrian, Spanish, and French rule.

In 1815, Belgium and the Netherlands united as the "low countries," but Belgium became independent in 1830. In 1885, Belgium became a major power when the Congo Free State (now Congo [Dem. Rep.]) became a Belgian colony.

Belgium's economy was weakened by the two World Wars, but, from 1945, the country recovered quickly, first through collaboration with the Netherlands and Luxembourg, which formed a customs union called Benelux, and later through its membership of the European Economic Community.

POLITICS

A central political problem in Belgium has been the tensions between the Dutch-speaking Flemings and the French-speaking Walloons. In the 1970s, the government divided the country into three economic regions: Dutch-speaking Flanders, French-speaking Wallonia, and bilingual Brussels. In 1993, Belgium adopted a federal system of government. Each of the regions now has its own parliament, which is responsible for local matters. Elections under this new system were held in 1995 and 1999.

DID YOU KNOW

- that Belgium has two official names: *Royaume de Belgique* in French, and *Koninkrijk België* in Dutch
- that, between 1885 and 1960, Belgium ruled the Congo Free State (now Congo [Dem. Rep.]), a country 80 times as large as Belgium
- that in Brussels every notice and street sign is printed in both Flemish (Dutch) and French
- that Belgium was the site of several great battles, including Waterloo (1815), Passchendaele (Passendale), and Ypres (Ieper) in World War I (1914–18), and the Battle of the Bulge (1944)
- that several organizations, including the headquarters of the European Union and the North Atlantic Treaty Organization (NATO), are situated in Brussels

ECONOMY

Belgium is a major trading nation, with a highly developed economy. It has coal reserves, though most of its mines have been closed because they are uneconomic. In 2002, its government announced that it would close its seven nuclear reactors by 2005. Belgium imports fuels and many raw materials. However, the leading activity is manufacturing. It produces chemicals, processed food, and steel. The textile industry has been important since medieval times.

Agriculture employs only 6% of the people, but Belgian farmers produce most of the food needed by the people. Barley and wheat are the chief crops, followed by flax, hops, potatoes, and sugar beet, but the most valuable activities are dairy farming and livestock rearing.

BELIZE

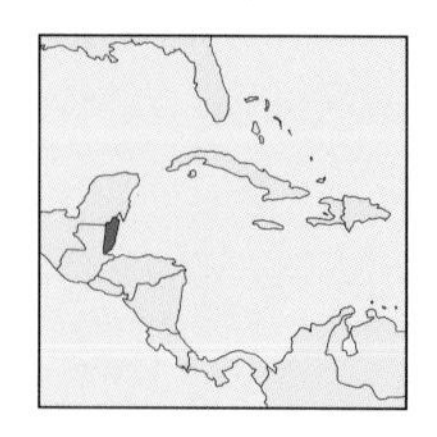

The badge in the center of the flag shows loggers, carrying axes and tools. Forestry is a major industry and Belize has valuable hardwoods, including mahogany, the national tree. The motto reads *Sub Umbra Floreo,* meaning "Flourish in the Shade."

Belize, a small country in Central America, is a monarchy, whose head of state is Britain's monarch. A governor-general represents the monarch, while an elected government, headed by a prime minister, actually rules the country.

Behind the swampy coastal plain in the south, the land rises to the low Maya Mountains, which reach a height of 3,675 ft [1,120 m] at Victoria Peak. Northern Belize is mostly low-lying and swampy. The main river, called the River Belize, flows across the center of the country.

AREA 8,865 sq miles [22,960 sq km]
POPULATION 256,000
CAPITAL (POPULATION) Belmopan (4,000)
GOVERNMENT Constitutional monarchy
ETHNIC GROUPS Mestizo (Spanish-Indian) 44%, Creole (mainly African American) 30%, Mayan Indian 11%, Garifuna (Black-Carib Indian) 7%, other 8%
LANGUAGES English (official), Creole, Spanish
RELIGIONS Roman Catholic 62%, Protestant 30%
CURRENCY Belize dollar = 100 cents

BELMOPAN
Temperature
Precipitation 1890mm/74in
J F M A M J J A S O N D

CLIMATE

Belize has a tropical, humid climate. Temperatures are high throughout the year and the average yearly rainfall ranges from 51 inches [1,300 mm] in the north to over 150 inches [3,800 mm] in the south. February to May are the driest months. Hurricanes sometimes hit the coast.

VEGETATION

Swamp vegetation and rain forest cover large areas. In the north, ironwood, mahogany, and sapote (date plum) are common trees, while cedar, oak, and pine are the main trees in the south. The coastal plains are covered by savanna (tropical grassland with scattered trees), while mangrove swamps line the coast.

HISTORY

Between about 300 BC and AD 1000, Belize was part of the Maya empire, which extended into Mexico, Guatemala, Honduras, and El Salvador. But this civilization had declined long before Spanish explorers reached the coast in the early 16th century.

Spain claimed the area, although no Spaniards settled there. Instead,

Spain ruled the area as part of Guatemala. Shipwrecked British sailors founded the first European settlement in the area in 1638. Over the next 150 years, Britain gradually took control of Belize, using slaves to open up the territory. Spain's last attempt to reassert its rule by force was defeated in 1798. In 1862, the area, then known as British Honduras, became a British colony. The country was renamed Belize in 1973 and it became fully independent in 1981.

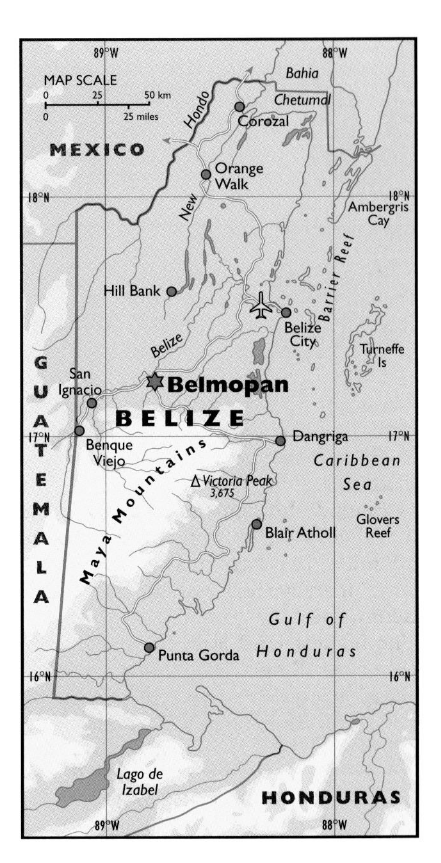

POLITICS

Guatemala, which had claimed Belize since the early 19th century, objected to the country becoming independent, and British troops remained in the country to prevent a possible invasion. In 1983, Guatemala reduced its claim to the southern fifth of Belize. In the early 1990s, improvements in relations with Guatemala, which finally agreed to recognize Belize's independence in 1992, led Britain to announce its decision to withdraw its troops by the end of 1993.

DID YOU KNOW

- that Belmopan became capital of Belize in 1970, because the old capital, Belize City, had been severely damaged by a hurricane in 1961
- that the world's second longest coral reef after Australia's Great Barrier Reef lies off the coast of Belize
- that Belize contains sites built by the Maya people more than 1,000 years ago; the largest site is Altún Ha, north of Belize City
- that the world's first jaguar reserve was set up in 1986 in Belize's Maya Mountains

ECONOMY

The World Bank classifies Belize as a "lower-middle-income" developing country. Its economy is based on agriculture and sugarcane is the chief commercial crop and export. Other crops include bananas, beans, citrus fruits, maize, and rice. Forestry, fishing, and tourism are other important activities.

BENIN

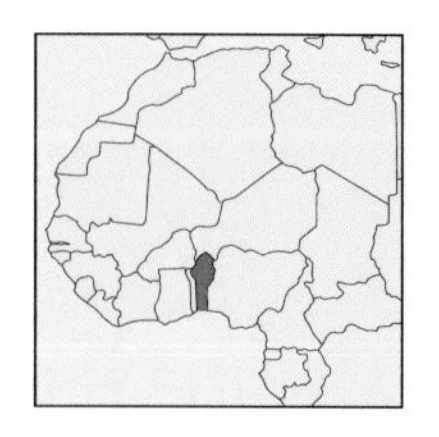

The colors on this flag, used by Africa's oldest independent nation, Ethiopia, symbolize African unity. Benin adopted this flag after independence in 1960. A flag with a red (Communist) star replaced it between 1975 and 1990, after which Benin dropped its Communist policies.

AREA 43,483 sq miles [112,620 sq km]
POPULATION 6,591,000
CAPITAL (POPULATION) Porto-Novo (179,000)
GOVERNMENT Multiparty republic
ETHNIC GROUPS Fon, Adja, Bariba, Yoruba, Fulani
LANGUAGES French (official), Fon, Adja, Yoruba
RELIGIONS Traditional beliefs 50%, Christianity 30%, Islam 20%
CURRENCY CFA franc = 100 centimes

The Republic of Benin is one of Africa's smallest countries. Its coastline is only 62 miles [100 km] long. It extends north–south for about 390 miles [620 km]. Lagoons line the coast, and the country has no natural harbors. The harbor at Cotonou, Benin's main port and chief commercial center, is artificial.

Behind the lagoons is a flat plain partly covered by rain forests. About 50 miles [80 km] inland, there is a large marshy depression. In central Benin, the land rises to low plateaux. The highest land is in the northwest. Northern Benin is covered by savanna, the habitat of such animals as buffaloes, elephants, and lions. The north has two national parks, the Penjari and the "W," which Benin shares with Burkina Faso and Niger.

CLIMATE

Benin has a hot, wet climate. The average annual temperature on the coast is about 77°F [25°C], and the average rainfall is about 52 inches [1,330 mm]. The inland plains are wetter than the coast. But the rainfall then decreases to the north, which is hot throughout the year, with a rainy summer season and a very dry winter.

HISTORY

The ancient kingdom of Dahomey had its capital at Abomey. In the 17th century, the kings of Dahomey became involved in supplying slaves to European slave traders, including the Portuguese who shipped many Dahomeans to the Americas, particularly to Portugal's huge territory of Brazil.

After slavery was ended in the 19th century, the French began to gain influence in the area. Around 1851, France signed a treaty with the kingdom of Dahomey and, in the 1890s, they made Dahomey a colony. From 1904, they ruled it as part of a region called French West Africa, a huge federation that also included what are now Burkina Faso, Guinea, Ivory Coast, Mauritania, Mali, Niger, and Senegal.

POLITICS

Benin became self-governing in 1958

and fully independent in 1960. After much instability and many changes of government, a military group took over in 1972. The country, renamed Benin in 1975, became a one-party socialist state. Socialism was abandoned in 1989. Multiparty elections were held in 1991, 1996, and 2001.

ECONOMY

Benin is a developing country. About 55% of the people earn their living by farming, though many farmers grow little more than they need to feed

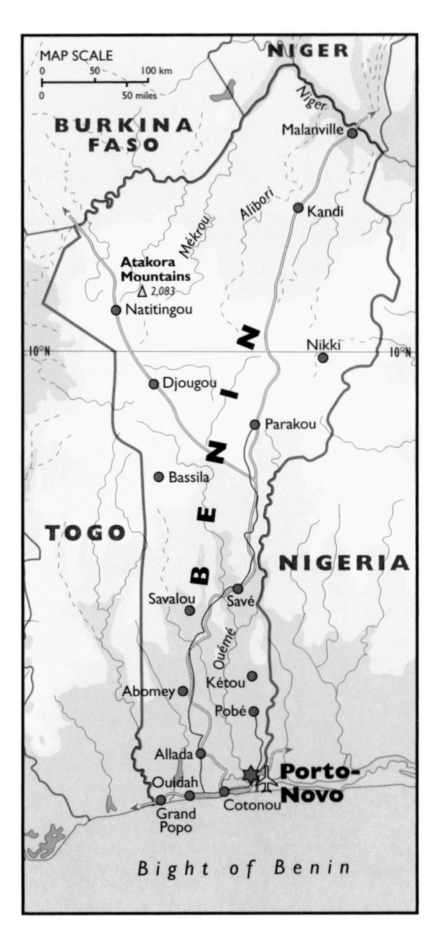

their own families. Major food crops include beans, cassava, maize, millet, rice, sorghum, and yams. The chief crops grown for export are cotton, palm oil, and palm kernels. Forestry is also important.

Benin produces some oil and limestone, but there is little manufacturing. Tourism is growing, though Benin lacks the facilities required by visitors outside the coastal towns.

DID YOU KNOW

- that Benin was called Dahomey until 1975
- that, between the 17th and 19th centuries, slaves were the main exports of Dahomey
- that many Dahomean customs, including voodoo, were preserved by slaves taken to the Americas and they are still practised today by black groups in Brazil
- that Benin was once the western part of the Kingdom of Benin, an ancient civilization based in Nigeria. The capital, Benin City, is now capital of Bendel, a state in Nigeria
- that beautiful brass, bronze, and ivory sculptures and plaques made in Benin between the 15th and 17th centuries now fetch high prices at art auctions around the world
- that when the Dutch expelled the Portuguese from the Gold Coast (now Ghana) in 1642, the Portuguese made Ouidah on the Benin coast their main headquarters; Ouidah remained a tiny Portuguese enclave until 1961

BOLIVIA

This flag, which has been Bolivia's national and merchant flag since 1888, dates back to 1825 when Bolivia became independent. The red stands for Bolivia's animals and the courage of the army, the yellow for its mineral resources, and the green for its agricultural wealth.

The Republic of Bolivia is a landlocked country which straddles the Andes Mountains in central South America. The Andes rise to a height of 21,463 ft [6,542 m] at Nevado Sajama in the west. The Andes range contains part of Lake Titicaca in the north, with Lake Poopó to the south.

About 40% of Bolivians live on a high plateau called the Altiplano in the Andean region. Eastern Bolivia is a vast lowland plain, drained by headwaters of the Madeira River, a tributary of the mighty Amazon.

LA PAZ
°C Temperature °F
30 20 10 0 -10 -20 -30 -40
80 60 40 20 0 -20 -40
Precipitation 575mm/23in
350 300 250 200 150 100 50 mm
14 12 10 8 6 4 2 ins
J F M A M J J A S O N D

CLIMATE

The climate of La Paz in the Andean region is greatly affected by the altitude. La Paz has frosts in winter. The highest Andean peaks are always covered by snow. The eastern plains have a hot and humid climate. The main rainy season in Bolivia is between December and February.

VEGETATION

The windswept Altiplano is a treeless grassland, which merges higher up into the *puna*, a region of scrub. Tropical rain forests cover much of the east, together with areas of open grassland and swamps.

AREA 424,162 sq miles [1,098,580 sq km]
POPULATION 8,300,000
CAPITAL (POPULATION) La Paz (1,126,000)
GOVERNMENT Multiparty republic
ETHNIC GROUPS Mestizo 30%, Quechua 30%, Aymara 25%, White 15%
LANGUAGES Spanish, Aymara and Quechua (all official)
RELIGIONS Roman Catholic 95%
CURRENCY Boliviano = 100 centavos

HISTORY AND POLITICS

American Indians have lived in Bolivia for at least 10,000 years. The main groups today are the Aymara and Quechua people.

When Spanish soldiers arrived in the early 16th century, Bolivia was part of the Inca empire. Following the defeat of the Incas, Spain ruled from 1532 to 1825, when Antonio José de Sucre, one of revolutionary leader Simón Bolívar's generals, defeated the Spaniards. Since independence, Bolivia has lost much territory to its neighbors. As recently as 1932, Bolivia fought with Paraguay

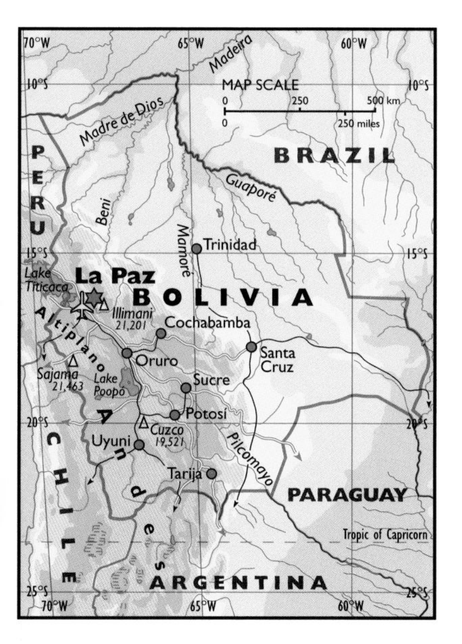

for control of the Gran Chaco region. Bolivia lost and most of this area passed to Paraguay in 1938.

In the last 50 years, Bolivia has been ruled by a succession of civilian and military governments, which violated human rights. Constitutional government was restored in 1982. In the 1980s and 1990s, Bolivia pursued free enterprise policies.

DID YOU KNOW

- that Bolivia was named after Simón Bolívar, leader of the struggle for the independence of Spain's South American colonies
- that La Paz is the world's highest capital city, at about 11,893 ft [3,625 m] above sea level
- that Bolivia ranks fifth among the world's tin producers, eighth among tungsten producers, and ninth in silver production
- that Bolivia shares with Peru the world's highest navigable lake, Titicaca

ECONOMY

Bolivia is one of the poorest countries in South America. It has several natural resources, including tin, silver, and natural gas, but the chief activity is agriculture, which employs 42% of the people. Potatoes, wheat, and a grain called *quinoa* are important crops on the Altiplano, while bananas, cocoa, coffee, and maize are grown at lower, warmer levels. Manufacturing is on a small scale, and the chief exports are mineral ores, especially tin. But experts believe that the main export may be coca, which is used to make the drug cocaine. Coca is exported illegally. However, by the early 2000s, natural gas was replacing coca as the chief export.

Bolivia's coat of arms, shown on this 1972 stamp, contains a central scene including a mountain, where silver is mined, a breadfruit tree, and an alpaca, which is raised for its fine wool. At the top is the magnificent Andean condor, the world's heaviest bird of prey.

BOSNIA-HERZEGOVINA

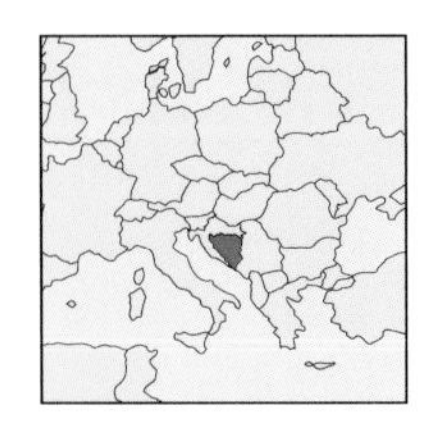

Bosnia-Herzegovina adopted a new flag in 1998, because the previous flag was thought by Croats and Serbs to be synonymous with the wartime Muslim regime. The blue background and white stars represent the country's links with the EU; the triangle stands for the three ethnic groups within the country.

The Republic of Bosnia-Herzegovina, which is often called Bosnia for short, is one of the five republics to emerge from the former Federal People's Republic of Yugoslavia. Much of the country is mountainous or hilly. An arid limestone plateau is in the southwest. The River Sava, which forms most of the northern border with Croatia, is a tributary of the River Danube. Because of the country's odd shape, the coastline is limited to a short stretch of 13 miles [20 km] on the Adriatic coast.

AREA 19,745 sq miles [51,129 sq km]
POPULATION 3,922,000
CAPITAL (POPULATION) Sarajevo (526,000)
GOVERNMENT Federal republic
ETHNIC GROUPS Bosnian 49%, Serb 31%, Croat 17%
LANGUAGES Serbo-Croatian
RELIGIONS Islam 40%, Serbian Orthodox 31%, Roman Catholic 15%, Protestant 4%
CURRENCY Convertible mark = 100 paras

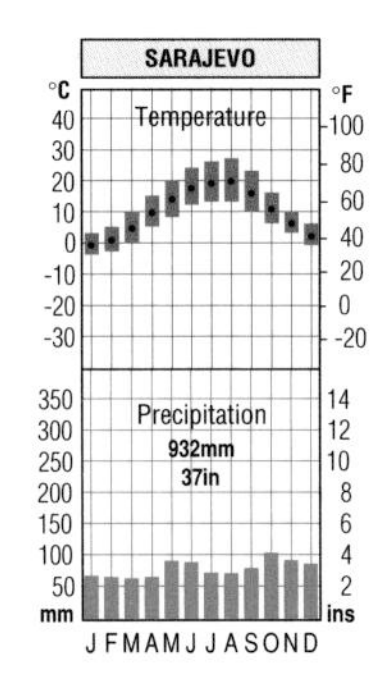

CLIMATE

A Mediterranean climate, with dry, sunny summers and moist, mild winters, prevails only near the coast. Inland, the weather becomes more severe, with hot, dry summers and bitterly cold, snowy winters. The north experiences the most severe weather.

VEGETATION

Forests of beech, oak, and pine grow in the north. The south has bare limestone landscapes, interspersed with farmland and pasture, which together cover nearly half of the country.

HISTORY

Slavs settled in the area that is now Bosnia-Herzegovina around 1,400 years ago. In the late 15th century, the area was taken by the Ottoman Turks.

In 1878, Austria-Hungary gained temporary control over Bosnia-Herzegovina. Austria-Hungary formally took over the area in 1908. In 1914, World War I (1914–18) began when Austria-Hungary's Archduke Francis Ferdinand was assassinated in Sarajevo.

In 1918, Bosnia-Herzegovina became part of the Kingdom of the Serbs, Croats, and Slovenes, which was renamed Yugoslavia in 1929. Germany occupied the area during World War II (1939–45). From 1945, Communist governments ruled Yugoslavia as a federation containing six

republics, one of which was Bosnia-Herzegovina. In the 1980s, the country faced problems as Communist policies proved unsuccessful and differences arose between the country's ethnic groups.

POLITICS

In 1990, free elections were held in Bosnia-Herzegovina and the non-Communists won a majority. A Muslim, Alija Izetbegovic, was elected president. In 1991, Croatia and Slovenia, other parts of the former Yugoslavia, declared themselves independent. In 1992, Bosnia-Herzegovina held a vote on independence. Most Bosnian Serbs boycotted the vote, while the Muslims and Bosnian Croats voted in favor.

Many Bosnian Serbs, opposed to independence, started a war against the non-Serbs. They soon occupied more than two-thirds of the land. The Bosnian Serbs were accused of "ethnic cleansing" – that is, the killing or expulsion of other ethnic groups from Serb-occupied areas. The war was later extended when Croat forces seized other parts of the country.

In 1995, the warring parties agreed to a solution to the conflict. This involved keeping the present boundaries of Bosnia-Herzegovina, but dividing it into two self-governing provinces, one Bosnian Serb and one Muslim-Croat, under a central, multi-ethnic government. Elections were held in 1996, 1998, and 2002.

ECONOMY

The economy of Bosnia-Herzegovina, the least developed of the six republics of the former Yugoslavia apart from Macedonia, was shattered by the war in the early 1990s. Before the war, manufactures were the main exports, including electrical equipment, machinery and transport equipment, and textiles. Farm products include fruits, maize, tobacco, vegetables, and wheat, but the country has to import food.

Sarajevo, capital of Bosnia-Herzegovina, was the subject of one of a set of six stamps issued by Yugoslavia in 1965. The stamps showed views of the capitals of the country's six republics.

BOTSWANA

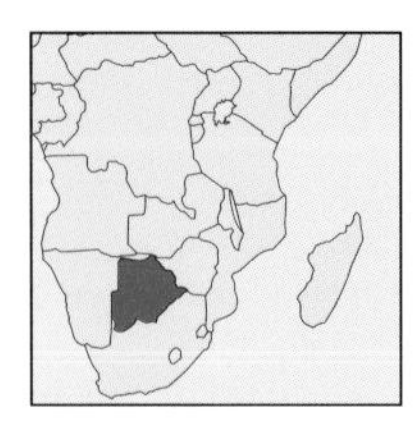

The black and white zebra stripe on the center of Botswana's flag symbolizes racial harmony. The blue represents rainwater, because water supply is the most vital need in this dry country. This flag was adopted in 1966, when Botswana became independent from Britain.

The Republic of Botswana is a landlocked country which lies far from the sea in the heart of southern Africa. Most of the land is flat or gently rolling, with an average height of about 3,280 ft [1,000 m]. More hilly country lies in the east. The Kalahari, a semidesert area mostly comprising grasses and thorn scrub, covers much of Botswana.

Most of the south has no permanent streams. But large depressions occur in the north. In one of them, the Okavango River, which flows from Angola, forms a large delta, an area of swampland. Another depression contains the Makgadikgadi Salt Pans. During floods, the Botletle River drains from the Okavango Swamps into the Makgadikgadi Salt Pans.

AREA 224,606 sq miles [581,730 sq km]
POPULATION 1,586,000
CAPITAL (POPULATION) Gaborone (133,000)
GOVERNMENT Multiparty republic
ETHNIC GROUPS Tswana 75%, Shona 12%, San (Bushmen) 3%
LANGUAGES English (official), Setswana
RELIGIONS Traditional beliefs 50%, Christianity 50%
CURRENCY Pula = 100 thebe

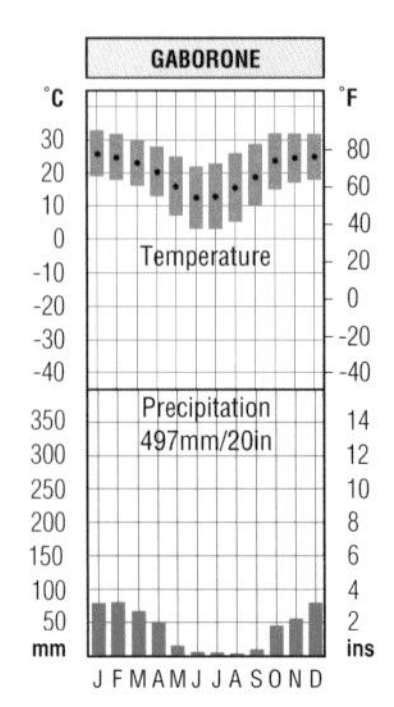

CLIMATE

Gaborone, the capital of Botswana, lies in the wetter eastern part of the country, where most people live. Temperatures are high in the summer months (October to April), but the winter months are much cooler. In winter, night-time temperatures sometimes drop below freezing point.

The average yearly rainfall varies from more than 16 inches [400 mm] in eastern Botswana to less than 8 inches [200 mm] in the southwest.

HISTORY AND POLITICS

The earliest inhabitants of the region were the San, who are also called Bushmen. They had a nomadic way of life, hunting wild animals and collecting wild plant foods.

The Tswana, who speak a Bantu language, now form the majority of the population. They are cattle owners, who settled in eastern Botswana more than 1,000 years ago. Their arrival led the San to move into the dry Kalahari region. Today, the San form a tiny minority of the population. Most of them live in

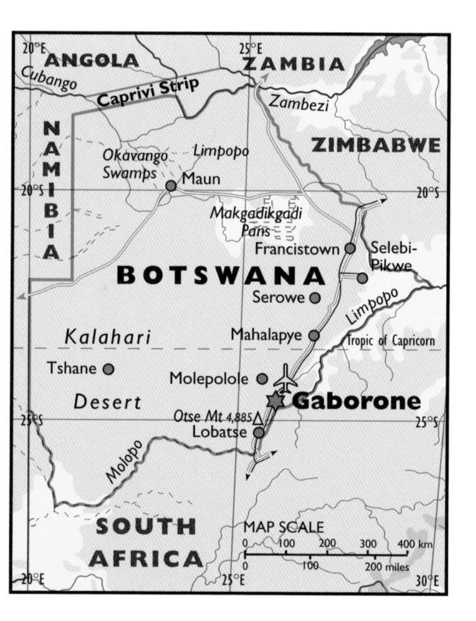

permanent settlements and work on cattle ranches.

Britain ruled the area as the Bechuanaland Protectorate between 1885 and 1966. When the country became independent, it adopted the name of Botswana. Since then, unlike many African countries, Botswana has been a stable multiparty democracy.

ECONOMY

In 1966, Botswana was one of Africa's poorest countries, depending on meat and live cattle for its exports. But the discovery of minerals, including coal, cobalt, copper, diamonds, and nickel, has boosted the economy. However, about a fifth of the people still work as farmers, raising cattle and growing such crops as millet, maize, beans, and vegetables.

Botswana also has some food-processing plants and factories that manufacture such things as soap and textiles.

This stamp is one of a set of four issued in 1979 to show some of the traditional handicrafts of Botswana. This stamp illustrates basketry. Other traditional crafts include beadwork, clay modeling, and pottery, the subjects of the other stamps in this set. Botswanan arts and crafts have suffered in recent years because of competition from cheap imported products sold in stores.

DID YOU KNOW

- that Botswana was called the Bechuanaland Protectorate until it became independent from Britain in 1966
- that Botswana has one of the world's highest HIV/AIDS infection rates. By the early 2000s, the average life expectancy had fallen from 60 years to 40 years
- that about 17% of the land in Botswana is protected in national parks and wildlife reserves. This is a higher proportion than in any other African country. Around half a million tourists visit Botswana every year

BRAZIL

The green on the flag symbolizes Brazil's rain forests and the yellow diamond its mineral wealth. The blue sphere bears the motto "Order and Progress." The 27 stars, arranged in the pattern of the night sky over Rio de Janeiro, represent the states and the federal district.

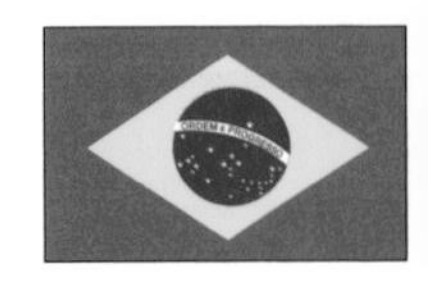

The Federative Republic of Brazil is the world's fifth largest country. Brazil contains three main regions. The Amazon basin in the north covers more than half of the country. The Amazon is the world's second longest river, though it has a far greater volume than any other river.

Brazil's second region is the northeast, which consists of a coastal plain and the *sertão*, the inland plateaux and hill country. The main river in this region is the São Francisco.

The third region is made up of the plateaux in the southeast. This region, which covers about a quarter of the country, is the most developed and densely populated part of Brazil. Its main river is the Paraná, which flows south through Argentina.

AREA 3,286,472 sq miles [8,511,970 sq km]

POPULATION 174,469,000

CAPITAL (POPULATION) Brasília (2,051,000)

GOVERNMENT Federal republic

ETHNIC GROUPS White 55%, Mulatto 38%, African American 6%, other 1%

LANGUAGES Portuguese (official), Spanish, English, French

RELIGIONS Roman Catholic 80%

CURRENCY Real = 100 centavos

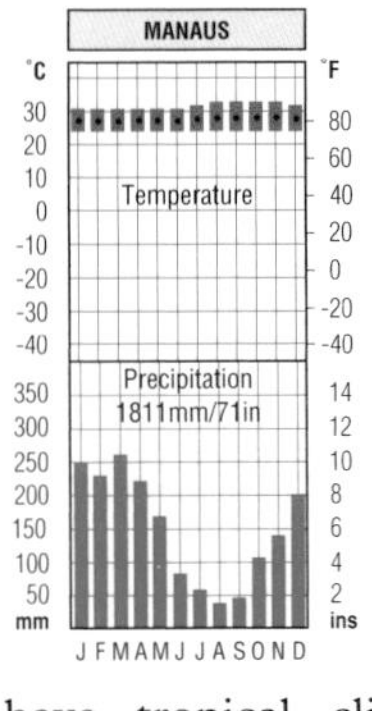

CLIMATE

Manaus has high temperatures all through the year. The rainfall is heavy, though the period from June to September is drier than the rest of the year. The capital, Brasília, and the city Rio de Janeiro also have tropical climates, with much more marked dry seasons than Manaus. The far south has a temperate climate.

The driest region is the northeastern interior, where the average yearly rainfall is only 10 inches [250 mm] in places. The rainfall is unreliable. Severe droughts are common in this region of Brazil.

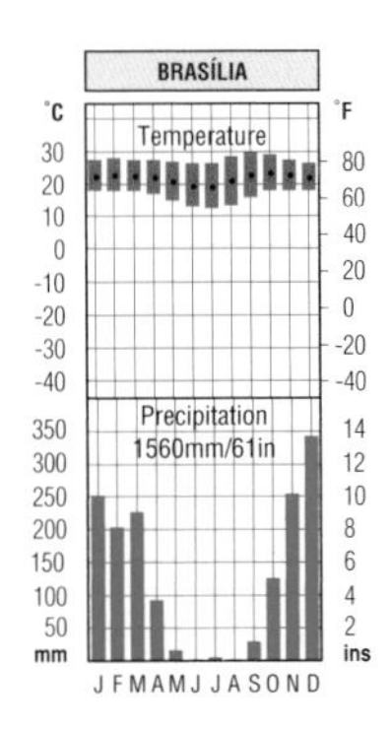

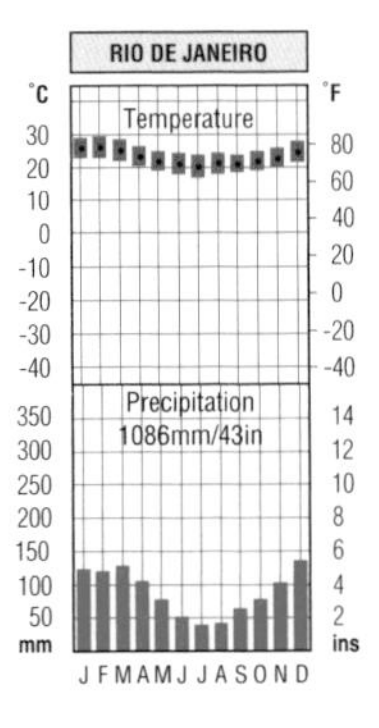

DID YOU KNOW
- that Brazil covers about 48% of South America
- that Brazil leads the world in the production of coffee, oranges, and sugarcane
- that the Amazon and its tributaries drain the world's largest river basin
- that Brazil has more Roman Catholics than any other country
- that Brazil is the only country to have appeared in every one of the soccer World Cup finals competition since it began in 1930
- that oceangoing ships can travel the entire length of the River Amazon in Brazil

VEGETATION

The Amazon basin contains the world's largest rain forests, which the Brazilians call the *selvas*. The forests contain an enormous variety of plant and animal species. But many species are threatened by loggers and others who want to exploit the forests. Forest destruction is also ruining the lives of the last surviving groups of Amazonian Indians.

Forests grow on the northeastern coasts, but the dry interior has large areas of thorny scrub. The southeast contains fertile farmland and large ranches. A large swampy area lies along Brazil's borders with Bolivia and Paraguay.

HISTORY

The Portuguese explorer Pedro Alvarez Cabral claimed Brazil for Portugal in 1500. With Spain occupied in western South America, the Portuguese began to develop their colony, which was more than 90 times as big as Portugal. To do this, they enslaved many local Amerindian people and introduced about 4 million African slaves to work on their plantations and in the mines.

Brazil declared itself an independent empire in 1822. The first emperor, Pedro I, was the son of King Joãs VI of Portugal. He was forced to resign in 1831 and the throne passed to his son, Pedro II. During his reign, slavery was gradually abolished. It finally came to an end in 1888. Brazil adopted a federal system of government in 1881. In 1889, Brazil became a republic and began a program of economic development.

POLITICS

From the 1930s, Brazil faced many political problems, including social unrest, corruption, and frequent spells of dictatorial government by military leaders.

Education has long been regarded as an essential part of any development strategy in the under-developed world, because progress is impossible without a skilled work force. Brazil has a good record in education. Around 80% of adults can read and write. The stamp, issued in 1974, commemorates a college's bicentenary.

BRAZIL

A new constitution, which came into force in 1988, took powers from the president and transferred many of them to parliament, paving the way to a return to democracy in 1990. In 1991, Brazil, Argentina, Paraguay, and Uruguay set up Mercosur, an alliance aimed at creating a common market.

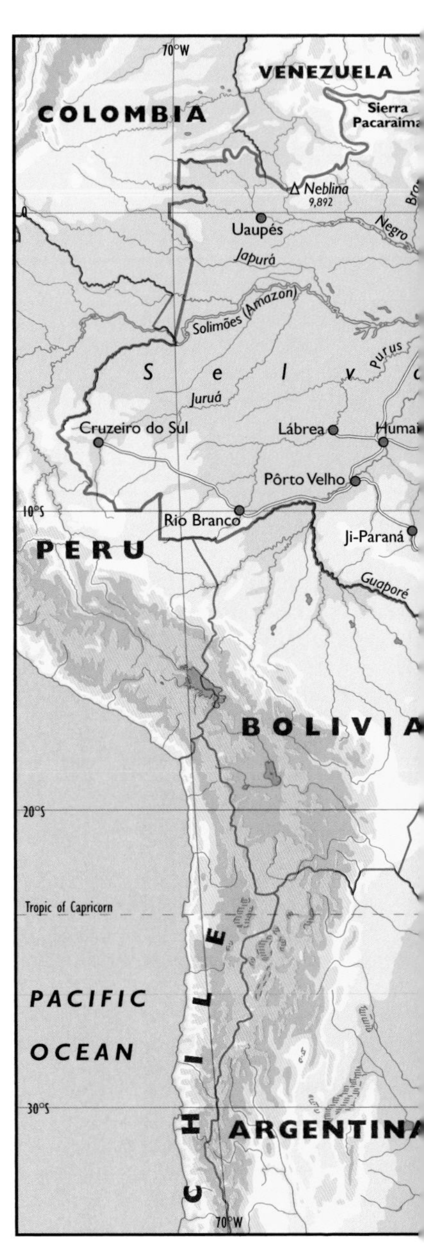

ECONOMY

The United Nations has described Brazil as a "Rapidly Industrializing Country," or RIC. Its total volume of production is one of the largest in the world. But many people, including poor farmers and residents of the *favelas* (city slums), do not share in the country's fast economic growth. Widespread poverty, together with high inflation and unemployment, cause political problems.

By the late 1990s, industry was the most valuable activity, employing 20% of the people. Brazil is among the world's top producers of bauxite, chrome, diamonds, gold, iron ore, manganese, and tin. It is also a major manufacturing country. Its products include aircraft, cars, chemicals, processed food, including raw sugar, iron and steel, paper, and textiles.

Brazil is one of the world's leading farming countries and agriculture employs 28% of the people. Coffee is a major export. Other leading products include bananas, citrus fruits, cocoa, maize, rice, soya beans, and sugarcane. Brazil is also the top producer of eggs, meat, and milk in South America.

Forestry is a major industry, though many people fear that the exploitation of the rain forests, with 1.5% to 4% of Brazil's forest being destroyed every year, is a disaster for the entire world.

GUYANA
SURINAME
FRENCH GUIANA
ATLANTIC OCEAN
Mouths of the Amazon
Equator
Macapá
Marajo I.
Amazon
Belém
São Luís
Camocim
Manaus
Santarém
Fortaleza
Parnaíba
Tucuruí Reservoir
Marabá
Teresina
Cape Sao Roque
Natal
João Pessoa
Cape Branco
Tapajós
Xingu
Floriano
Campina Grande
Recife
Tocantins
Araguaia
Juàzeiro
Maceió
BRAZIL
Sobradinho Reservoir
Mato Grosso
Aracaju
Feira de Santana
Salvador
São Francisco
Itabuna
Cuiabá
Brasília
Goiâna
Pirapora
Brazilian Highlands
Uberlândia
Governador Valadares
Belo Horizonte
Campo Grande
Uberaba
Vitória
9,482
Aquidauana
São José
Três Lagoas
Furnas Reservoir
Juiz de Fora
Ribeirão Prêto
Campos
Paraná
Ponta Porã
Petrópolis
Campinas
São Paulo
Niterói
Rio de Janeiro
PARAGUAY
Tropic of Capricorn
Sorocaba
Santos
Itaipu Reservoir
Guarapuava
Ponta Grossa
Curitiba
Iguaçu Falls
ATLANTIC OCEAN
Florianópolis
Passo Fundo
Caxias do Sul
Santa Maria
Pôrto Alegre
Patos Lagoon
MAP SCALE
0
250
500
750
1000 km
250
500 miles
Pelotas
Rio Grande
Lake Mirim
URUGUAY
50°W
40°W
30°W
10°S
20°S
30°S

BULGARIA

This flag, first adopted in 1878, uses the colors associated with the Slav people. The national emblem, incorporating a lion – a symbol of Bulgaria since the 14th century – was first added to the flag in 1947. It is now added only for official government occasions.

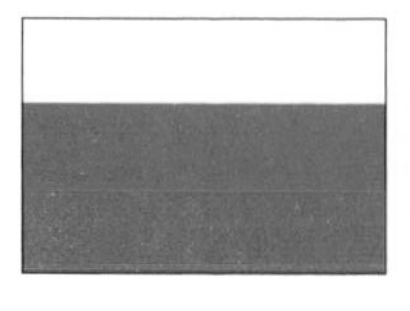

The Republic of Bulgaria is a country in the Balkan peninsula, facing the Black Sea in the east. Northern Bulgaria consists of a plateau which falls to the valley of the River Danube, which forms most of Bulgaria's frontier with Romania. The heart of Bulgaria is mountainous. The main ranges are the Balkan Mountains in the center and the Rhodope (or Rhodopi) Mountains in the south. Between these ranges is the River Maritsa valley, which forms an east–west route between the coast and the interior.

AREA 42,822 sq miles [110,910 sq km]
POPULATION 7,707,000
CAPITAL (POPULATION) Sofia (1,139,000)
GOVERNMENT Multiparty republic
ETHNIC GROUPS Bulgarian 83%, Turkish 8%, Gypsy 3% Macedonian, Armenian, other
LANGUAGES Bulgarian (official), Turkish
RELIGIONS Christianity (Eastern Orthodox 87%), Islam 13%
CURRENCY Lev = 100 stotinki

CLIMATE

There are hot summers and cold winters in Bulgaria, though they are seldom severe. The rainfall is moderate. The Black Sea coast is a resort area. The east has drier and warmer summers than the west. In winter, cold winds sometimes blow from the northeast, bringing bitterly cold spells.

VEGETATION

More than half of Bulgaria is under crops or pasture. Forests, including beech and spruce, cover about 35% of the country. Trees swathe the mountain slopes, with grassy meadows and alpine plants above the tree line. In the warmer south, the plants are similar to those in the lands around the Mediterranean Sea.

HISTORY

Most of the Bulgarian people are descendants of Slavs and nomadic Bulgar tribes who arrived from the east in the 6th and 7th centuries. A powerful Bulgar kingdom was set up in 681, but the country became part of the Byzantine empire in the 11th century.

Ottoman Turks ruled Bulgaria from 1396 and ethnic Turks still form a sizable minority in the country. In 1879, Bulgaria became a monarchy, and in 1908 it became fully independent. Bulgaria was an ally of Germany in World War I (1914–18) and again in World War II (1939–45). In 1944, Soviet troops invaded Bulgaria and, after the war, the monarchy was abolished and the country became a Communist ally of the Soviet Union.

POLITICS

In the late 1980s, reforms in the Soviet Union led Bulgaria's government to introduce a multiparty system in 1990. A non-Communist government was elected in 1991 in the first free elections in 44 years. Since 1991, Bulgaria has faced problems in trying to reform the country and its economy. In 2001, the former king Siméon became prime minister.

Bulgaria has a coastline, 175 miles [282 km] long, on the Black Sea. The 1969 stamp showing a deep-sea trawler highlights the country's fishing industry. The top stamp, issued in 1971, is one of a set entitled "Health Resorts." It is a reminder of the importance of tourism to the country.

ECONOMY

According to the World Bank, Bulgaria in the 1990s was a "lower-middle-income" developing country. Bulgaria has some deposits of minerals, including brown coal, manganese, and iron ore. But manufacturing is the leading economic activity, though problems arose in the early 1990s, because much industrial technology is outdated. The main products are chemicals, processed foods, metal products, machinery, and textiles. Manufactures are the leading exports. Bulgaria trades mainly with countries in Eastern Europe.

Wheat and maize are the chief crops. Fruit, oilseeds, tobacco, and vegetables are also important; these are grown in the south-facing valleys overlooking the Maritsa plains. A valuable product is attar of roses, an oil obtained from rose petals for use in perfumes. Livestock farming, especially the rearing of dairy and beef cattle, sheep, and pigs, is important. More than 2 million tourists visited Bulgaria in 2000.

BURKINA FASO

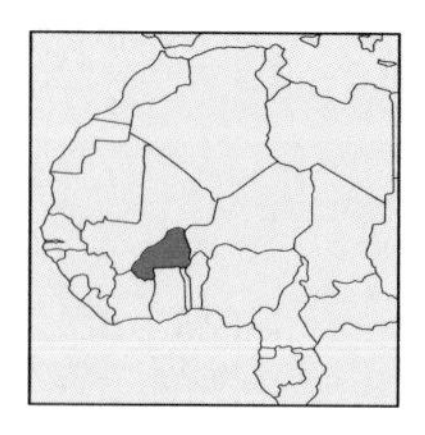

This flag was adopted in 1984, when Upper Volta was renamed Burkina Faso. The red, green, and yellow colors used on this flag symbolize the desire for African unity. This is because they are used on the flag of Ethiopia, Africa's oldest independent country.

The Democratic People's Republic of Burkina Faso is a landlocked country, a little larger than the United Kingdom, in West Africa. But Burkina Faso has only one-sixth of the population of the UK.

Burkina Faso consists of a plateau, between about 650 ft and 2,300 ft [200 m to 700 m] above sea level. The plateau is cut by several rivers. Most rivers flow south into Ghana or east into the River Niger. During droughts, some of the rivers stop flowing, becoming marshes.

AREA 105,869 sq miles [274,200 sq km]
POPULATION 12,272,000
CAPITAL (POPULATION) Ouagadougou (690,000)
GOVERNMENT Multiparty republic
ETHNIC GROUPS Mossi 48%, Gurunsi, Senufo, Lobi, Bobo, Mande, Fulani
LANGUAGES French (official)
RELIGIONS Islam 50%, traditional beliefs 40%, Christianity 10%
CURRENCY CFA franc = 100 centimes

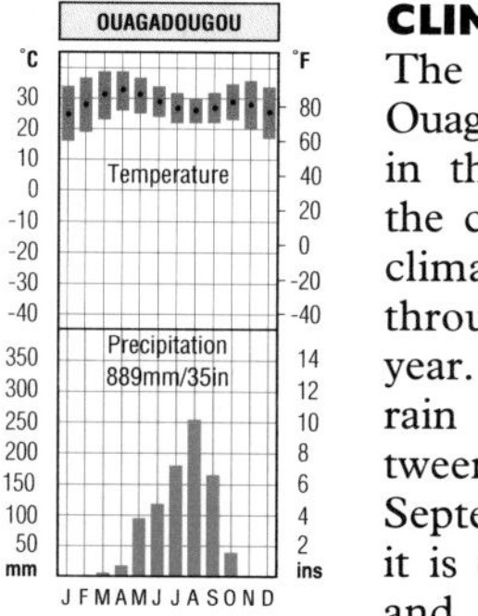

CLIMATE

The capital city, Ouagadougou, lies in the center of the country. The climate is hot throughout the year. Most of the rain occurs between May and September, when it is often humid and uncomfortable. The rainfall is erratic and droughts are common.

VEGETATION

The northern part of the country is covered by savanna, consisting of grassland with stunted trees and shrubs. It is part of a region called the Sahel, where the land merges into the Sahara Desert. Overgrazing of the land and deforestation (that is, the chopping down of trees to clear land) are common problems here.

Woodlands border the rivers and parts of the southeast are swampy. The southeast contains the "W" National Park, which Burkina Faso shares with Benin and Niger, and the Arly Park. A third wildlife area is the Po Park, south of Ouagadougou.

HISTORY

The people of Burkina Faso are divided into two main groups. The Voltaic group includes the Mossi, who form the largest single group, and the Bobo. The other main group is the Mande family. Burkina Faso also contains some Fulani herders and Hausa traders, who are related to the people of northern Nigeria.

In early times, the ethnic groups in Burkina Faso were divided into kingdoms and chiefdoms. The leading kingdom, which was ruled by an absolute monarch called the Moro Naba, was that of the Mossi. It has existed since the 13th century.

The French conquered the Mossi capital of Ouagadougou in 1897 and they made the area a protectorate. In 1919, the area became a French colony called Upper Volta. Upper Volta remained under French rule until 1960, when the country became fully independent.

POLITICS

After independence, Upper Volta became a one-party state. But it was unstable. Military groups seized control several times and a number of political killings took place.

In 1984, the country's name was changed to Burkina Faso. Elections were held in 1991 – for the first time in more than ten years – but the military kept an important role in the government.

DID YOU KNOW

- that the name Burkina Faso means "land of the upright people"
- that Burkina Faso's former name, Upper Volta, referred to three rivers, the Black Volta, Red Volta, and White Volta, which flow south through the country into Ghana where they empty into Lake Volta
- that the Mossi people of Burkina Faso have a royal dynasty which dates back more than 800 years. The Mossi people are headed by the Moro Naba (Mossi emperor), who still holds court in the capital city of Ouagadougou

ECONOMY

Burkina Faso is one of the world's 20 poorest countries and has become very dependent on foreign aid. About 90% of the people earn their living by farming or raising livestock. Grazing land covers about 37% of the land and farmland 10%.

Most of Burkina Faso is dry with thin soils. The country's main food crops are beans, maize, millet, rice, and sorghum. Cotton, groundnuts, and shea nuts, whose seeds produce a fat used to make cooking oil and soap, are grown for sale abroad. Livestock are also an important export.

The country has few resources and manufacturing is on a small scale. There are some deposits of manganese, zinc, lead, and nickel in the north of the country, but there is not yet a good enough transport route there. Many young men seek jobs abroad in Ghana and Ivory Coast. The money they send home to their families is important to the country's economy.

BURMA (MYANMAR)

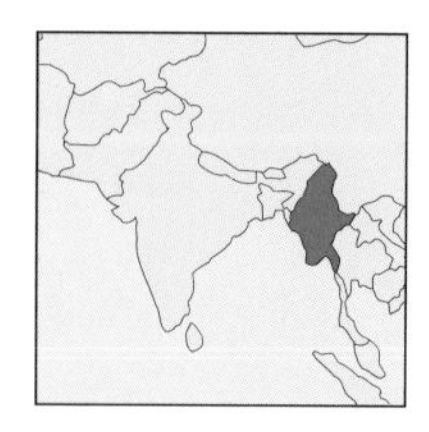

The colors on Burma's flag were adopted in 1948 when the country became independent from Britain. The socialist symbol, added in 1974, includes a ring of 14 stars for the country's states. The gearwheel represents industry and the rice plant agriculture.

The Union of Burma is now officially known as the Union of Myanmar. Its name was changed in 1989, but most people still call the country Burma. Mountains border the country in the east and west, with the highest mountains in the north. Burma's highest mountain is Hkakabo Razi, which is 19,295 ft [5,881 m] high. Between these ranges is central Burma, which contains the fertile valleys of the Irrawaddy and Sittang rivers. The Irrawaddy delta on the Bay of Bengal is one of the world's leading rice-growing areas. Burma also includes the long Tenasserim coast in the southeast.

AREA 261,228 sq miles [676,577 sq km]
POPULATION 41,995,000
CAPITAL (POPULATION) Rangoon (or Yangon, 2,513,000)
GOVERNMENT Military regime
ETHNIC GROUPS Burman 69%, Shan 9%, Karen 6%, Rakhine 5%, Mon 2%, Kachin 1%
LANGUAGES Burmese (official)
RELIGIONS Buddhism 89%, Christianity, Islam
CURRENCY Kyat = 100 pyas

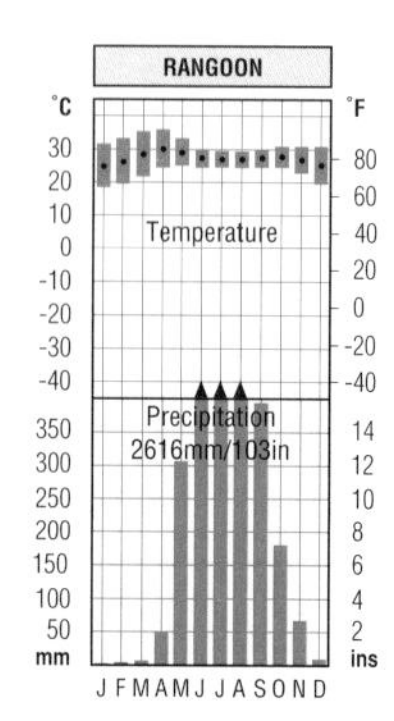

CLIMATE

Burma has a tropical monsoon climate. There are three seasons. The rainy season runs from late May to mid-October. A cool, dry season follows, between late October and the middle part of February. The hot season lasts from late February to mid-May, though temperatures remain high during the humid rainy season. The rainfall varies greatly across the country. The coastal areas are much wetter than the interior plains.

VEGETATION

About 50% of the country is covered by forest. Tropical trees, such as teak, grow on low-lying areas, with mangrove swamps on the coast. Forests of oak and pine grow on mountain slopes, but the lowlands are intensively cultivated.

HISTORY AND POLITICS

Many groups settled in Burma in ancient times. Some, called the hill peoples, live in remote mountain areas where they have retained their own cultures. The ancestors of the main group today, the Burmese, arrived in the 9th century AD.

Britain conquered Burma in the 19th century and made it a province of British India. But, in 1937, the

British granted Burma limited self-government. Japan conquered Burma in 1942, but the Japanese were driven out in 1945. Burma became a fully independent country in 1948.

Revolts by Communists and various hill people led to instability in the 1950s. In 1962, Burma became a military dictatorship and, in 1974, a one-party state. Attempts to control minority liberation movements and the opium trade led to repressive rule. The National League for Democracy led by Aung San Suu Kyi won the elections in 1990. Hovever, the military continued their repressive rule throughout the 1990s, earning Burma the reputation for having one of the world's worst human rights records.

ECONOMY

Burma's internal political problems have also helped to make it one of the world's poorest countries. Agriculture is the main activity, employing 66% of the people. The chief crop is rice. Maize, pulses, oilseeds, and sugarcane are other major products. Forestry is important. Teak and rice together make up about two-thirds of the total value of the exports. Burma has many mineral resources, though they are mostly undeveloped, but the country is famous for its precious stones, especially rubies. Manufacturing is mostly on a small scale.

> **DID YOU KNOW**
> - that Burma produces the world's finest rubies
> - that the Burmese language is related to Tibetan
> - that Burma is the world's leading illicit producer of the opium poppy
> - that the spire of the Buddhist Shwe Dagon pagoda in Rangoon is covered by 8,868 thin sheets of gold; the pagoda is 325 ft [99 m] high

BURUNDI

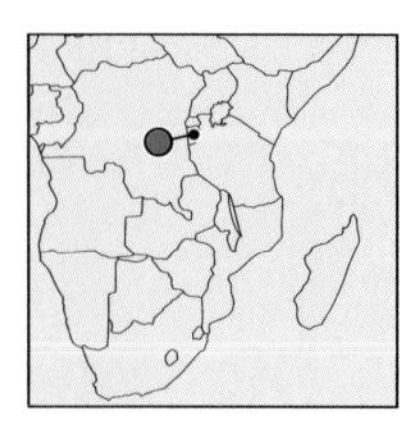

This flag, adopted when Burundi became a republic in 1966, contains three red stars rimmed in green. They symbolize the nation's motto of "Unity, Work, Progress." The green represents hope for the future, the red the struggle for independence, and the white the desire for peace.

The Republic of Burundi is the fifth smallest country on the mainland of Africa. It is also the second most densely populated after its northern neighbor Rwanda. Part of the deep Great Rift Valley, which runs throughout eastern Africa into southwestern Asia, lies in western Burundi. It includes part of Lake Tanganyika.

East of the Rift Valley are high mountains composed partly of volcanic rocks. In central and eastern Burundi, the land descends in a series of steppe-like grassy plateaux.

AREA 10,745 sq miles [27,830 sq km]
POPULATION 6,224,000
CAPITAL (POPULATION) Bujumbura (300,000)
GOVERNMENT Republic
ETHNIC GROUPS Hutu 85%, Tutsi 14%, Twa (pygmy) 1%
LANGUAGES French and Kirundi (both official)
RELIGIONS Roman Catholic 62%, traditional beliefs 23%, Islam 10%
CURRENCY Burundi franc = 100 centimes

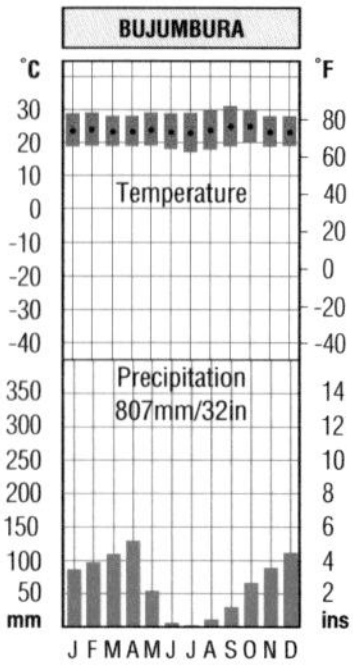

CLIMATE

Bujumbura, the capital of Burundi, lies on the shore of Lake Tanganyika. It has a warm climate. A dry season occurs from June to September, but the other months are fairly rainy. The mountains and plateaux are cooler and wetter, but the rainfall decreases to the east.

VEGETATION

Grasslands cover much of Burundi. The land was formerly mainly forest, but farmers have cleared most of the trees. In some areas, new forests are being planted to protect the soil against the rain and wind. Soil erosion is a serious problem.

HISTORY AND POLITICS

The Twa, a pygmy people, were the first known inhabitants of Burundi. About 1,000 years ago, the Hutu, a people who speak a Bantu language, gradually began to settle the area, pushing the Twa into remote areas.

From the 15th century, the Tutsi, a cattle-owning people from the northeast, gradually took over the country. The Hutu, although greatly outnumbering the Tutsi, were forced to serve the Tutsi overlords.

Germany conquered the area that is now Burundi and Rwanda in the late 1890s. The area, called Ruanda-

Urundi, was taken by Belgium during World War I (1914–18). In 1961, the people of Urundi voted to become a monarchy, while the people of Ruanda voted to become a republic. The two territories became fully independent as Burundi and Rwanda in 1962.

After 1962, the rivalries between the Hutu and Tutsi led to periodic outbreaks of fighting. The Tutsi monarchy was ended in 1966 and Burundi became a republic. Instability continued with coups in 1976, 1987, 1993, and 1996, with periodic massacres as the Tutsi and Hutu fought for power. In the early 2000s, attempts were made to broker peace by introducing a power-sharing government.

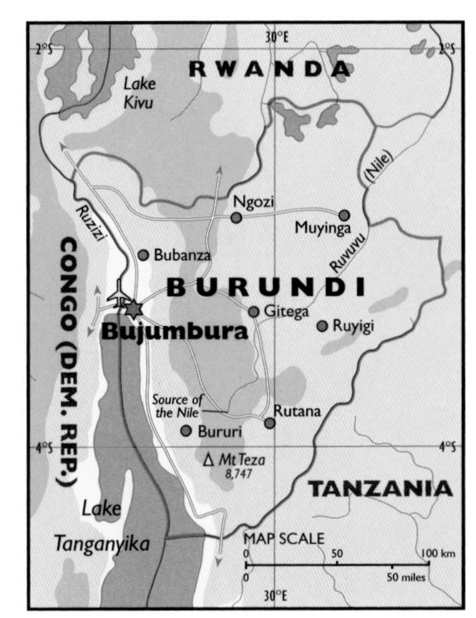

ECONOMY

Burundi is one of the world's ten poorest countries. About 92% of the people are farmers, who mostly grow little more than they need to feed their own families. The main food crops are beans, cassava, maize, and sweet potatoes. Cattle, goats, and sheep are raised. Fishing is also important, but Burundi has to import food.

Major cash crops are coffee (which accounts for 80% of the exports), tea, and cotton. Mining is unimportant and manufacturing is limited. Most factories are in Bujumbura.

This 50-centime stamp, issued in 1964, shows impalas, which are among the most graceful of Africa's antelopes. In 1964, Burundi was a monarchy, as shown by the wording on the stamp – Royaume *means Kingdom. Two years later, Burundi's* mwami *(king) was overthrown and the country became a republic.*

DID YOU KNOW

- that the Tutsi of Burundi, who are also called Watusi, are among the world's tallest people. In some areas, they average 6 ft 5 inches [195.5 cm] in height
- that Burundi also has some of the world's shortest people – the Twa, who are pygmies
- that Burundi only has one doctor for every 21,000 people

CAMBODIA

Red is the traditional color of Cambodia. The blue symbolizes the water resources that are so important to the people, three-quarters of whom depend on farming for a living. The silhouette is the historic temple at Angkor Wat.

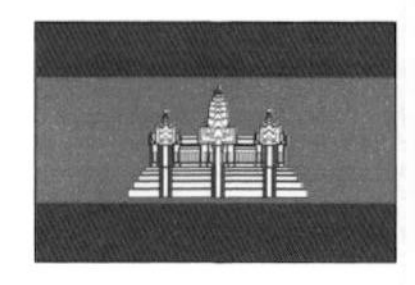

The Kingdom of Cambodia is a country in Southeast Asia. Low mountains border the country except in the southeast. But most of Cambodia consists of plains drained by the River Mekong, which enters Cambodia from Laos in the north and exits through Vietnam in the southeast. The northwest contains Tonlé Sap (or Great Lake). In the dry season, this lake drains into the River Mekong. But in the wet season, the level of the Mekong rises and water flows in the opposite direction from the river into Tonlé Sap – the lake then becomes the largest freshwater lake in Asia.

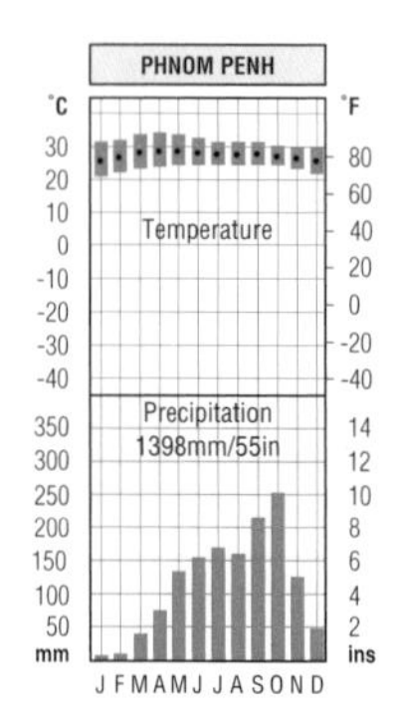

CLIMATE

Cambodia has a tropical monsoon climate, with high temperatures all through the year. The dry season, when winds blow from the north or northeast, runs from November to April. During the rainy season, from May to October, moist winds blow from the south or southeast. The high humidity and heat often make conditions unpleasant. The rainfall is heaviest near the coast, and rather lower inland.

AREA 69,900 sq miles [181,040 sq km]
POPULATION 12,492,000
CAPITAL (POPULATION) Phnom Penh (570,000)
GOVERNMENT Constitutional monarchy
ETHNIC GROUPS Khmer 90%, Vietnamese 5%, Chinese 1%, other 5%
LANGUAGES Khmer (official)
RELIGIONS Buddhism 95%, other 5%
CURRENCY Riel = 100 sen

VEGETATION

Forests cover about three-quarters of Cambodia. The rain forests on the northern mountains contain a tangled undergrowth of bamboos, palms, vines, and other herbaceous plants. Dense mangrove forests line the coast.

HISTORY AND POLITICS

From 802 to 1432, the Khmer people ruled a great empire, which reached its peak in the 12th century. The Khmer capital was at Angkor. The Hindu stone temples built there and at nearby Angkor Wat form the world's largest group of religious buildings.

Cambodia was ruled by France between 1863 and 1954, when the country became an independent monarchy. But King Norodom Sihanouk gave up his throne to become prime

minister. In 1970, his government was overthrown, the monarchy was abolished, and the country became a republic.

In 1970, US and South Vietnamese troops entered Cambodia but left after destroying North Vietnamese Communist camps in the east. The country became involved in the Vietnamese War, and then in a civil war as Cambodian Communists of the Khmer Rouge organization fought for power. The Khmer Rouge took over Cambodia in 1975 and launched a reign of terror in which between 1 million and 2.5 million people were killed. In 1979, Vietnamese and Cambodian troops overthrew the Khmer Rouge government. But fighting continued between several factions. Vietnam withdrew in 1989, and in 1991 Prince Sihanouk was recognized as head of state. Elections were held in May 1993, and in September 1993 the monarchy was restored and Sihanouk became king. In 2001, the government set up a court to try Khmer Rouge leaders.

ECONOMY

Cambodia is a poor country whose economy has been wrecked by war. Agriculture employs about 70% of the people. The main products are rice, rubber, and maize. Apart from rubber processing, manufacturing is small scale, employing only about 6% of the people. A few factories produce items, such as glass, tobacco products, and processed food, for sale in Cambodia.

The numbers of wild animals in Southeast Asia have been greatly reduced as land is cleared. The stamp, issued in 1964, shows a kouprey, a rare wild ox. (Cambodge is the French name for Cambodia.)

DID YOU KNOW

- that alternative names for Cambodia in recent times have been the Khmer Republic and the People's Republic of Kampuchea
- that Tonlé Sap, a lake in central Cambodia, more than doubles in size during the rainy season
- that after the great temple of Angkor Wat was abandoned, it was hidden by dense rain forest until it was rediscovered in 1860 by Albert Henri Mouhot, a French naturalist

CAMEROON

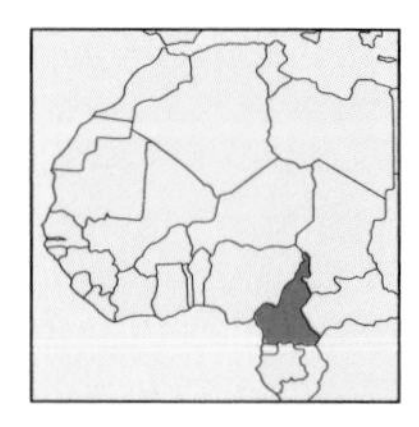

Cameroon uses the colors that appear on the flag of Ethiopia, Africa's oldest independent nation. These colors symbolize African unity. The flag is based on the tricolor adopted in 1957. The design with the yellow liberty star dates from 1975.

The Republic of Cameroon is a country in West Africa. Its name comes from the Portuguese word *camarões*, or prawns. This name was used by Portuguese explorers who fished for prawns along the coast. Behind the narrow coastal plains on the Gulf of Guinea, the land rises to a series of plateaux. In the north, the land slopes down toward the Lake Chad basin. The mountain region in the southwest of the country includes Mount Cameroon, a volcano which erupts from time to time.

AREA 183,567 sq miles [475,440 sq km]

POPULATION 15,803,000

CAPITAL (POPULATION) Yaoundé (800,000)

GOVERNMENT Multiparty republic

ETHNIC GROUPS Fang 20%, Bamileke and Bamum 19%, Duala, Luanda and Basa 15%, Fulani 10%

LANGUAGES French and English (both official), many others

RELIGIONS Christianity 40%, traditional beliefs 40%, Islam 20%

CURRENCY CFA franc = 100 centimes

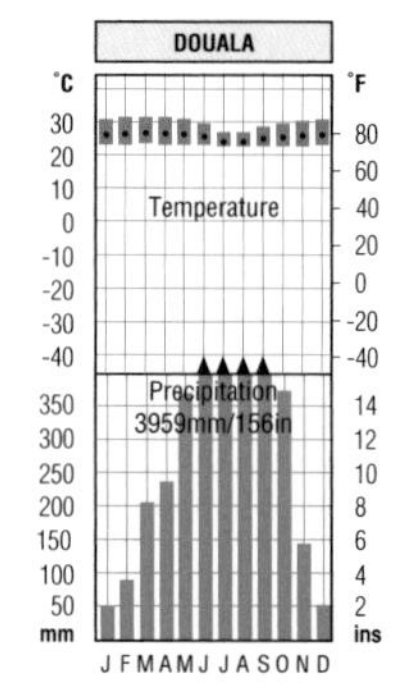

CLIMATE

Douala is a city in the southwest of the country. It lies in a hot and humid region where the rainfall is extremely high, especially on the mountains. The rain is heaviest between June and September. The inland plateaux are cooler. The yearly rainfall decreases to the north. The far north has a hot, dry climate.

VEGETATION

Rain forests flourish in the hot and humid south. Inland, the forests merge into savanna (tropical grassland with scattered woodland). The far north is semidesert. The northern savanna region contains some national parks, where such animals as antelopes, elephants, giraffes, lions, and wart-hogs are protected. The best known national park, the Waza, is north of Maroua.

HISTORY AND POLITICS

Among the early inhabitants of Cameroon were groups of Bantu-speaking people. (There are now more than 160 ethnic groups, each with their own language.) In the late 15th century, Portuguese explorers, who were seeking a sea route to Asia around Africa, reached the Cameroon coast. From the 17th century, south-

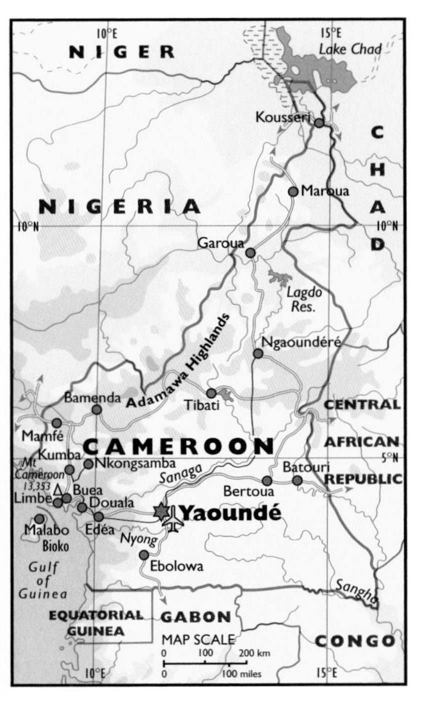

ern Cameroon was a center of the slave trade, but slavery was ended in the early 19th century. In 1884, the area became a German protectorate.

Germany lost Cameroon during World War I (1914–18). The country was then divided into two parts, one ruled by Britain and the other by France. In 1960, French Cameroon became the independent Cameroon Republic. In 1961, after a vote in British Cameroon, part of the territory joined the Cameroon Republic to become the Federal Republic of Cameroon. The other part joined Nigeria.

In 1972, Cameroon became a unitary state called the United Republic of Cameroon. It adopted the name Republic of Cameroon in 1984, but the country had two official languages. In 1995, partly to placate English-speaking people, Cameroon became the 52nd member of the Commonwealth.

ECONOMY

Like most countries in tropical Africa, Cameroon's economy is based on agriculture, which employs 73% of the people. The chief food crops include cassava, maize, millet, sweet potatoes, and yams. The country also has plantations to produce such crops as cocoa and coffee for export.

Cameroon is fortunate in having some oil, the country's chief export, and bauxite. Although Cameroon has few manufacturing and processing industries, its mineral exports and its self-sufficiency in food production make it one of the better-off countries in tropical Africa.

This stamp, which was issued in 1983, shows Cameroon's best known land feature, Mount Cameroon. Mount Cameroon, the country's highest peak, is an active volcano, reaching 13,353 ft [4,070 m] near the coast in southern Cameroon. The ash and lava, which erupted from the volcano, have made the land around the mountain extremely fertile.

CANADA

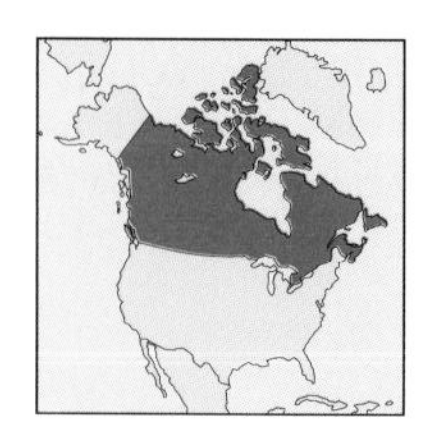

Canada's flag, with its simple 11-pointed maple leaf emblem, was adopted in 1965 after many attempts to find an acceptable design. The old flag, used from 1892, was the British Red Ensign. But this flag became unpopular with Canada's French community.

Canada is the world's second largest country after Russia. But it is thinly populated. Much of the land is too cold or too mountainous for human settlement. Most Canadians live within 186 miles [300 km] of the southern border.

Western Canada is rugged. It includes the Pacific ranges and the mighty Rocky Mountains. East of the Rockies are the interior plains.

The Canadian Shield is a vast region of ancient rocks which covers almost half of the land area, enclosing the Hudson Bay lowlands. In the north are the bleak Arctic islands. But south of the Canadian Shield lie Canada's most populous regions, the lowlands north of lakes Erie and Ontario, and the lowlands in the St Lawrence River valley. The northern-most part of the Appalachian Mountains are in the far southeast.

AREA 3,851,788 sq miles [9,976,140 sq km]

POPULATION 31,593,000

CAPITAL (POPULATION) Ottawa (1,107,000)

GOVERNMENT Federal multiparty constitutional monarchy

ETHNIC GROUPS British 28%, French 23%, other European 15%, Native American (Amerindian/Inuit) 2%, other (mostly Asian, African, Arab and mixed)

LANGUAGES English and French (both official)

RELIGIONS Roman Catholic 46%, Protestant 36%, other (including Judaism, Islam, Hinduism, Sikhism) 18%

CURRENCY Canadian dollar = 100 cents

CLIMATE

Canada has a cold climate. In winter, temperatures fall below freezing point throughout most of Canada. But Vancouver, on the west coast, has a mild climate and average temperatures remain above freezing in the winter months. In the north, along the Arctic Circle, mean temperatures are below freezing for seven months per year.

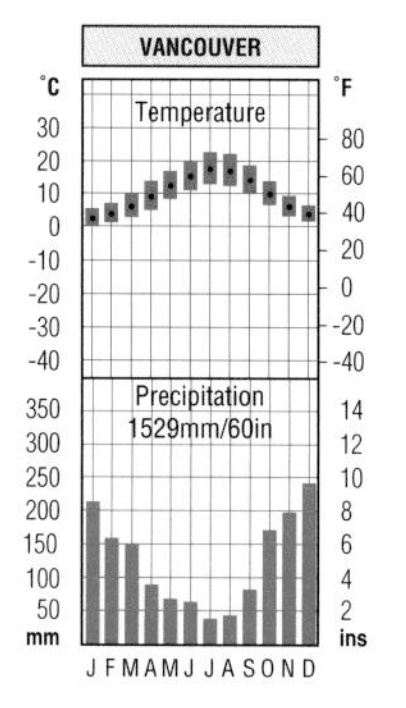

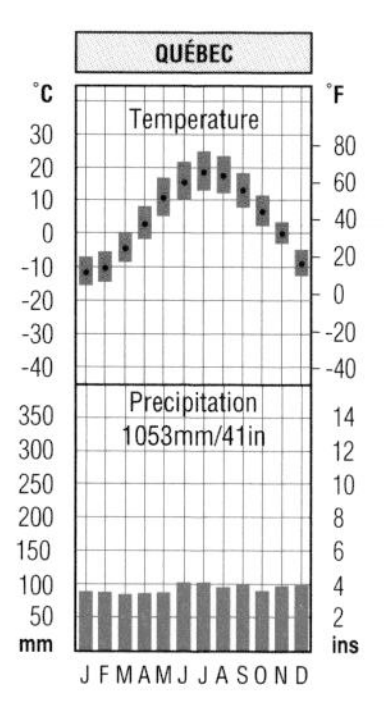

Western Canada has a high rainfall, but the prairies are dry with 10 to 20 inches [250 mm to 500 mm] of rain every year. Southeastern Canada has a moist climate. For example, Québec has an average of about 41 inches [1,040 mm] of rain per year.

VEGETATION

Forests of cedars, hemlocks, and other trees grow on the western mountains, with firs and spruces at the higher levels. The interior plains were once grassy prairies, but today they are used largely for farming and ranching. The far north contains cold, treeless zones called tundra. The southeastern lowlands contain forests of deciduous trees, such as beech, oak, and walnut. The Appalachian region has beautiful mixed forests.

HISTORY

Canada's first people, the ancestors of the Native Americans, or Indians, arrived in North America from Asia around 40,000 years ago. Later arrivals were the Inuit (Eskimos), who also came from Asia. European explorers reached the Canadian coast in 1497 and soon a race began between France and Britain for the riches in this new land.

France gained an initial advantage, and the French founded Québec in 1608. But the British later occupied eastern Canada. In 1867, Britain passed the British North America Act, which set up the Dominion of Canada, which was made up of Québec, Ontario, Nova Scotia, and New Brunswick. Other areas were added, the last being Newfoundland in 1949. Today, Canada is a confederation, consisting of ten provinces and two territories.

Canada fought alongside Britain in both World Wars and many Canadians feel close ties with Britain. Canada is a constitutional monarchy, and the British monarch is Canada's head of state.

POLITICS

Rivalries between French- and English-speaking Canadians continue. In 1995, Québeckers voted against a move to make Québec a sovereign state. The majority was less than 1% and this issue seems unlikely to disappear. Another problem concerns the rights of the aboriginal minorities. In 1999, Canada created a territory called Nunavut for the Inuit people in the north. Nunavut covers about 64% of what was formerly Northwest Territories.

ECONOMY

Canada is a highly developed and prosperous country. Although farmland covers only 8% of the country, Canadian farms are highly productive.

Issued in 1993, this stamp celebrates the bicentenary of Canada's largest metropolitan area, Toronto. The site was selected by John Graves Simcoe in 1793 and named York. But the settlement was renamed Toronto in 1834. Toronto is a Huron Indian word for "meeting place."

CANADA

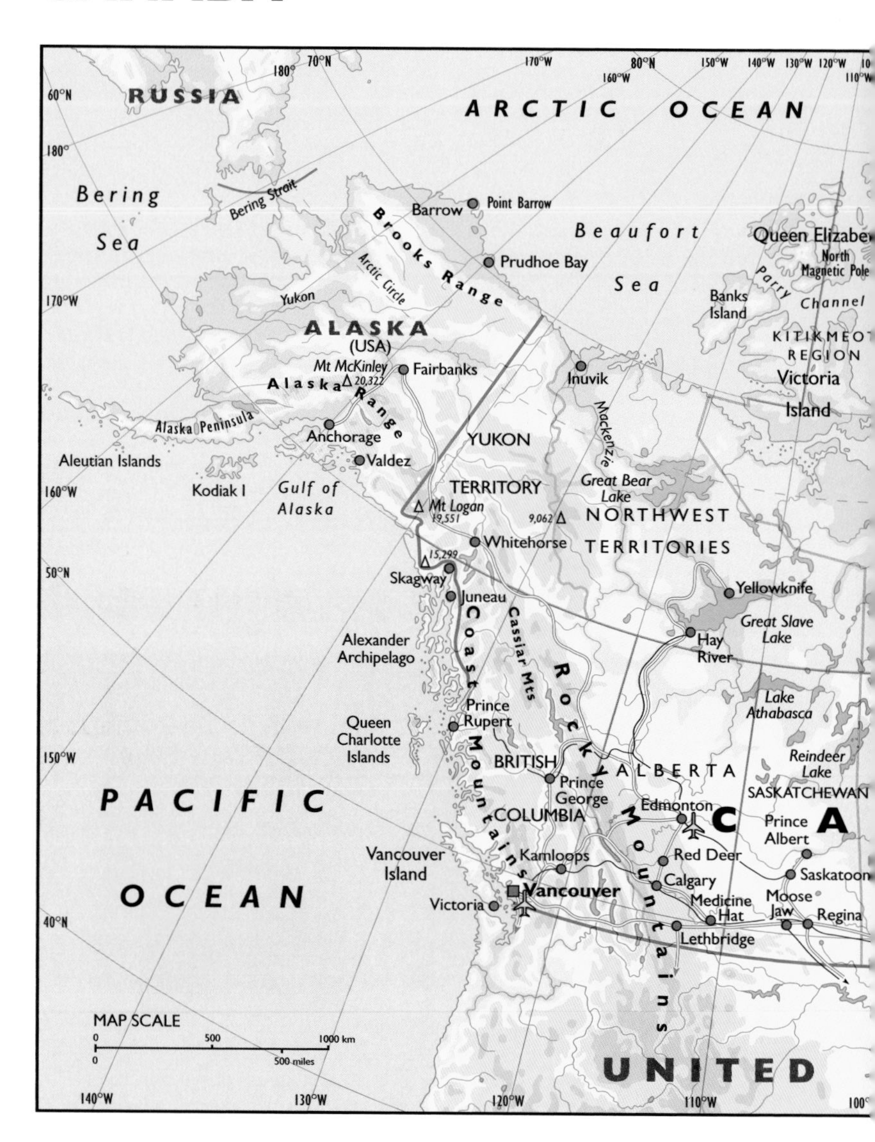

Canada is one of the world's leading producers of barley, wheat, meat, and milk. Forestry and fishing are other important industries. It is rich in natural resources and is a major exporter of minerals. The most important products are oil and natural gas. The country also produces copper, gold, iron

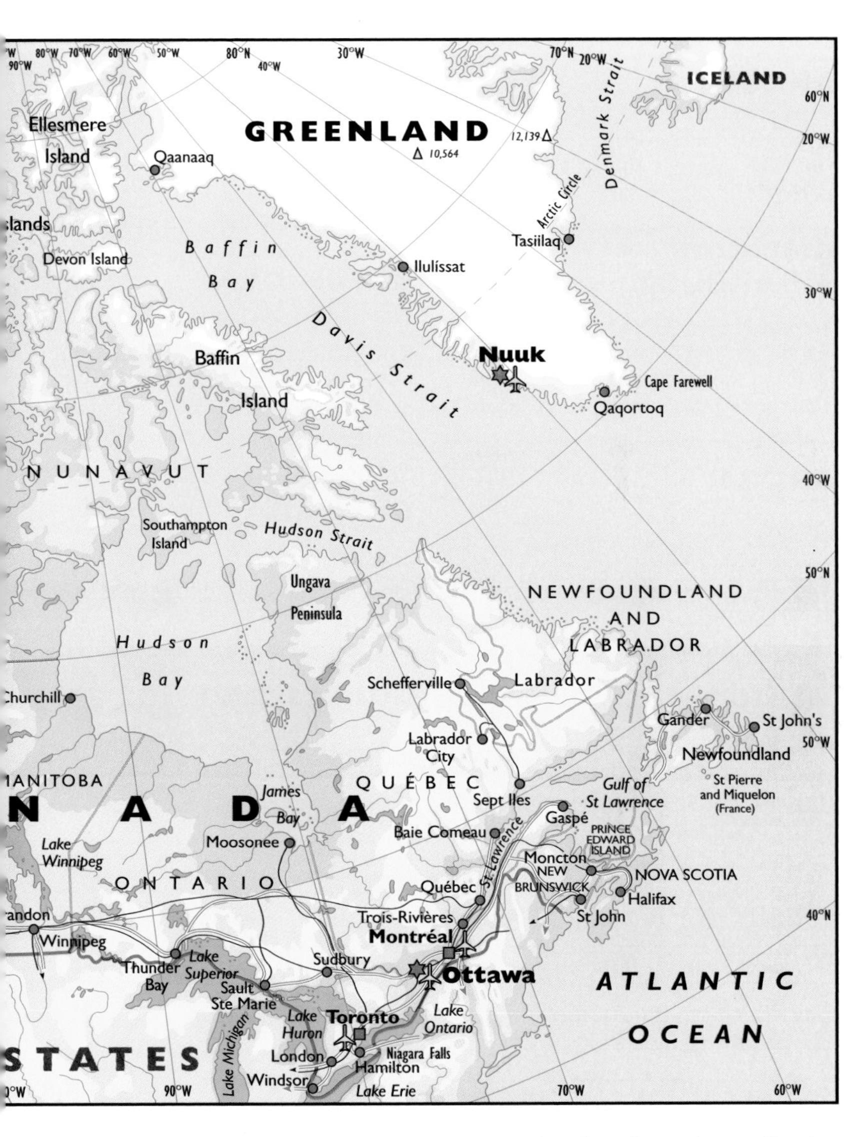

ore, uranium, and zinc. Manufacturing is highly developed, especially in the cities where 78% of the people live. Canada has many factories that process farm and mineral products. It also produces cars, chemicals, electronic goods, machinery, paper, and timber products.

CARIBBEAN SEA

ANGUILLA

A British dependency since 1980. AREA 37 sq miles [96 sq km]; POPULATION 12,000; CAPITAL The Valley.

ANTIGUA & BARBUDA

Antigua became independent from Britain in 1981. AREA 170 sq miles [442 sq km]; POPULATION 67,000; CAPITAL St John's (population 38,000).

ARUBA

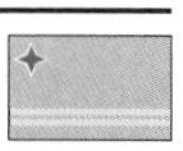

A Dutch island, formerly part of the Netherlands Antilles. AREA 75 sq miles [193 sq km]; POPULATION 70,000; CAPITAL Oranjestad.

BAHAMAS

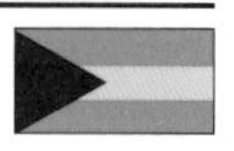

The Bahamas became independent from Britain in 1973. AREA 5,380 sq miles [13,940 sq km]; POPULATION 298,000; CAPITAL Nassau.

BARBADOS

Barbados became independent from Britain in 1960. AREA 166 sq miles [430 sq km]; POPULATION 275,000; CAPITAL Bridgetown.

CAYMAN ISLANDS

A British dependency. AREA 100 sq miles [259 sq km]; POPULATION 36,000; CAPITAL George Town.

DOMINICA

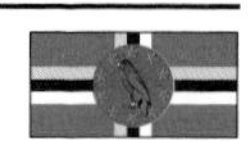

The Commonwealth of Dominica became an independent republic in 1978. AREA 290 sq miles [751 sq km]; POPULATION 71,000; CAPITAL Roseau.

DOMINICAN REPUBLIC

The Dominican Republic occupies eastern Hispaniola. AREA 18,815 sq miles [48,730 sq km]; POPULATION 8,581,000; CAPITAL Santo Domingo.

GRENADA

Grenada became independent in 1974. AREA 131 sq miles [340 sq km]; POPULATION 89,000; CAPITAL St George's.

GUADELOUPE

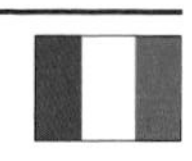

A French overseas department. AREA 658 sq miles [1,706 sq km]; POPULATION 431,000; CAPITAL Basse-Terre.

HAITI

Independent from 1804, Haiti occupies western Hispaniola. AREA 10,714 sq miles [27,750 sq km]; POPULATION 6,965,000; CAPITAL Port-au-Prince.

JAMAICA

Jamaica became independent from Britain in 1962. AREA 4,243 sq miles [10,990 sq km]; POPULATION 2,666,000; CAPITAL Kingston.

MARTINIQUE

A French overseas department. **AREA** 425 sq miles [1,100 sq km]; **POPULATION** 418,000; **CAPITAL** Fort-de-France (population 102,000).

MONTSERRAT

A self-governing British dependency. **AREA** 39 sq miles [102 sq km]; **POPULATION** 12,000; **CAPITAL** Plymouth.

NETHERLANDS ANTILLES

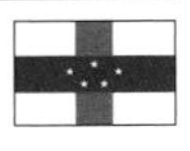

Consisting of Curaçao and Bonaire, off the Venezuelan coast, and Saba, St Eustatius, and the southern part of St Maarten in the Leeward Islands, these islands are a self-governing part of the Netherlands. **AREA** 383 sq miles [993 sq km]; **POPULATION** 212,000; **CAPITAL** Willemstad.

PUERTO RICO

Ceded by Spain to the US in 1898, Puerto Rico became a self-governing Commonwealth in free political association with the United States following a referendum in 1952. **AREA** 3,436 sq miles [8,900 sq km]; **POPULATION** 3,939,000; **CAPITAL** San Juan.

ST KITTS & NEVIS

St Kitts and Nevis became independent from Britain in 1983. **AREA** 139 sq miles [360 sq km]; **POPULATION** 39,000; **CAPITAL** Basseterre.

ST LUCIA

St Lucia became independent from Britain in 1979. **AREA** 236 sq miles [610 sq km]; **POPULATION** 158,000; **CAPITAL** Castries (population 53,000).

ST VINCENT & THE GRENADINES

This country became independent from Britain in 1979. **AREA** 150 sq miles [388 sq km]; **POPULATION** 116,000; **CAPITAL** Kingstown.

TRINIDAD & TOBAGO

This country became independent from Britain in 1962. **AREA** 1,981 sq miles [5,130 sq km]; **POPULATION** 1,170,000; **CAPITAL** Port-of-Spain.

TURKS & CAICOS ISLANDS

A British dependency. **AREA** 166 sq miles [430 sq km]; **POPULATION** 18,000; **CAPITAL** Cockburn Harbour.

VIRGIN ISLANDS, BRITISH

A British dependency. **AREA** 59 sq miles [153 sq km]; **POPULATION** 21,000; **CAPITAL** Road Town.

VIRGIN ISLANDS, US

A US territory from 1917, including the island of St Croix. **AREA** 130 sq miles [340 sq km]; **POPULATION** 122,000; **CAPITAL** Charlotte Amalie.

CARIBBEAN SEA

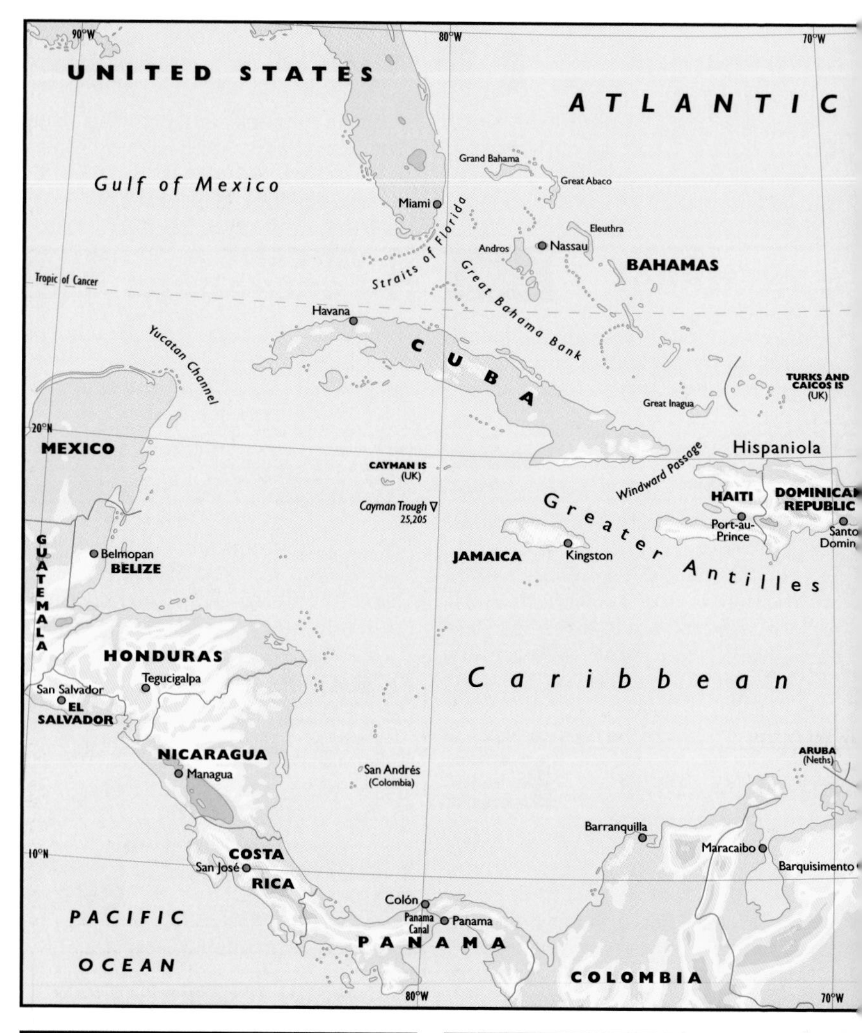

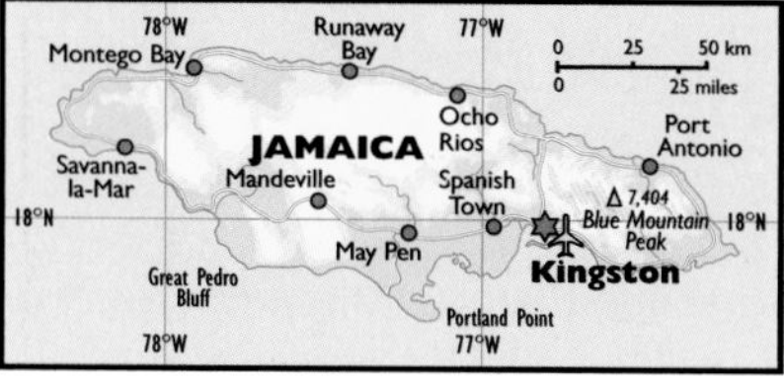

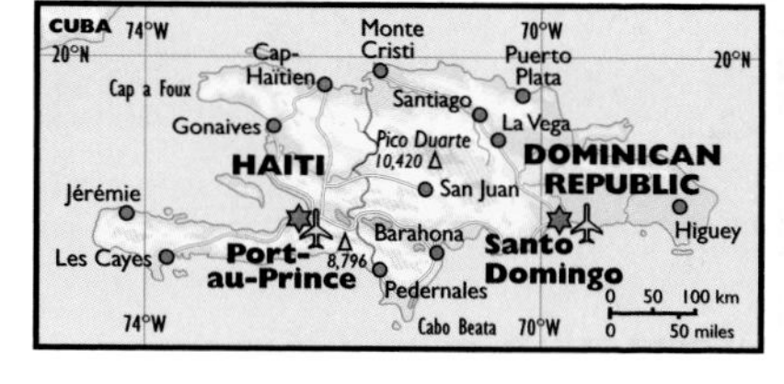

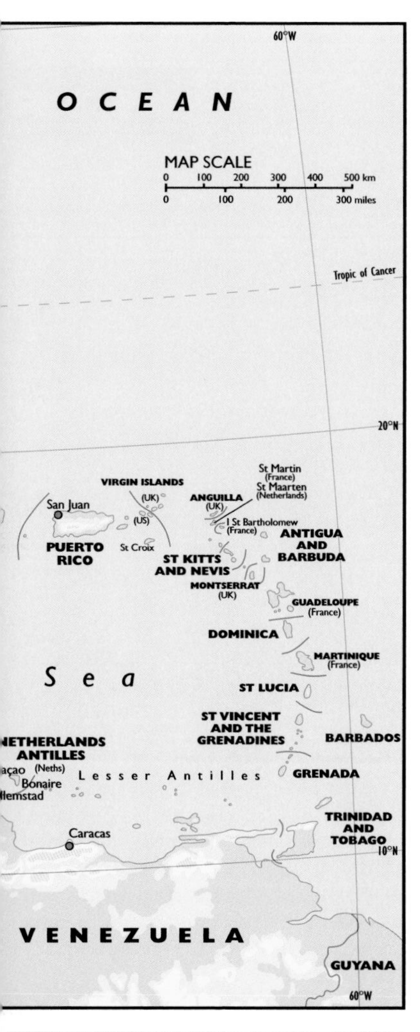
OCEAN
MAP SCALE
0 100 200 300 400 500 km
0 100 200 300 miles
60°W
Tropic of Cancer
20°N
VIRGIN ISLANDS
(UK)
(US)
San Juan
PUERTO RICO
St Croix
ANGUILLA
(UK)
St Martin
(France)
St Maarten
(Netherlands)
I St Bartholomew
(France)
ANTIGUA AND BARBUDA
ST KITTS AND NEVIS
MONTSERRAT
(UK)
GUADELOUPE
(France)
DOMINICA
MARTINIQUE
(France)
Sea
ST LUCIA
ST VINCENT AND THE GRENADINES
BARBADOS
GRENADA
Lesser Antilles
(Neths)
Bonaire
TRINIDAD AND TOBAGO
Caracas
10°N
VENEZUELA
GUYANA
60°W

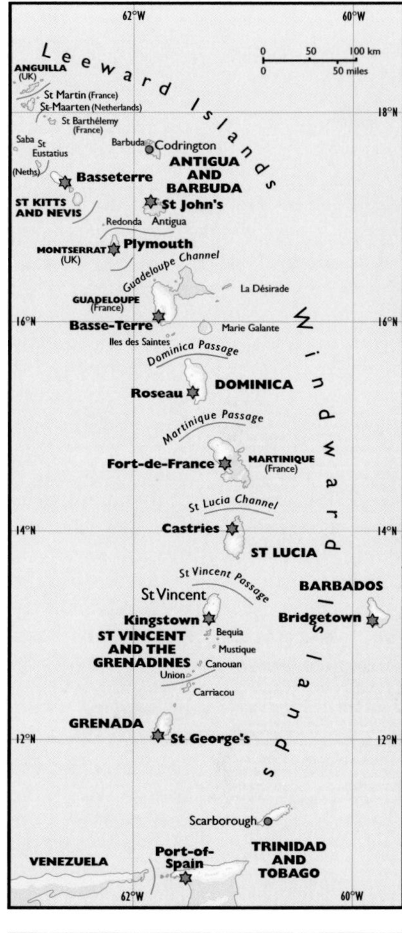
62°W
60°W
0 50 100 km
0 50 miles
Leeward Islands
ANGUILLA
(UK)
St Martin (France)
St Maarten (Netherlands)
St Barthélemy
(France)
18°N
Saba
St Eustatius
(Neths)
Barbuda
Codrington
ANTIGUA AND BARBUDA
Basseterre
ST KITTS AND NEVIS
St John's
Redonda
Antigua
MONTSERRAT
(UK)
Plymouth
Guadeloupe Channel
La Désirade
GUADELOUPE
(France)
16°N
Basse-Terre
Marie Galante
Iles des Saintes
Dominica Passage
Roseau
DOMINICA
Windward Islands
Martinique Passage
Fort-de-France
MARTINIQUE
(France)
St Lucia Channel
14°N
Castries
ST LUCIA
St Vincent Passage
BARBADOS
St Vincent
Kingstown
Bridgetown
ST VINCENT AND THE GRENADINES
Bequia
Mustique
Canouan
Union
Carriacou
GRENADA
12°N
St George's
Scarborough
VENEZUELA
Port-of-Spain
TRINIDAD AND TOBAGO
62°W
60°W

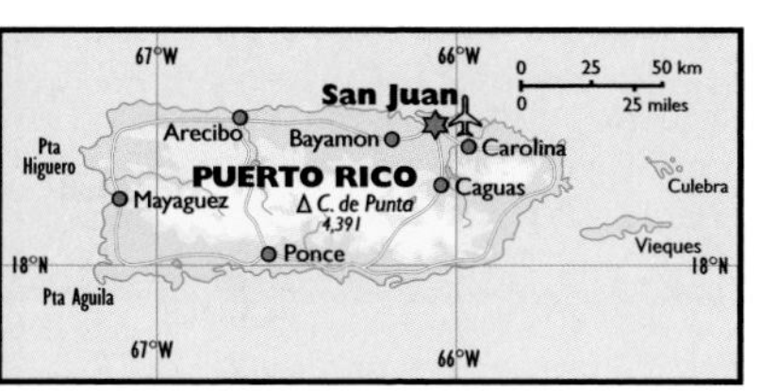
67°W
66°W
0 25 50 km
0 25 miles
San Juan
Arecibo
Bayamon
Carolina
Pta Higuero
PUERTO RICO
Caguas
Culebra
Mayaguez
C. de Punta
4,391
Ponce
Vieques
18°N
Pta Aguila
67°W
66°W

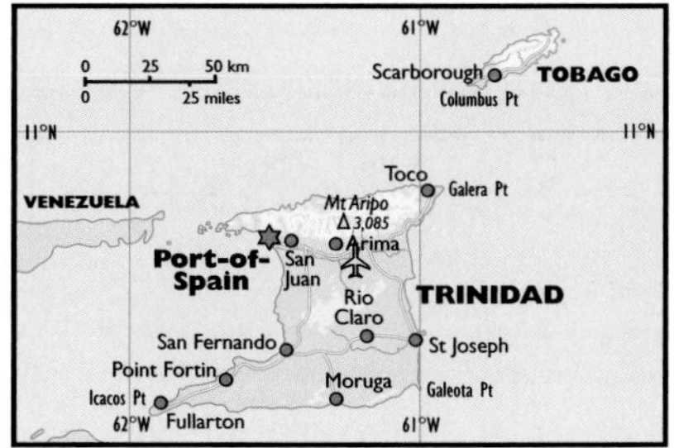
62°W
61°W
0 25 50 km
0 25 miles
Scarborough
TOBAGO
Columbus Pt
11°N
Toco
Galera Pt
VENEZUELA
Mt Aripo
3,085
Port-of-Spain
San Juan
Arima
TRINIDAD
Rio Claro
San Fernando
St Joseph
Point Fortin
Moruga
Icacos Pt
Galeota Pt
Fullarton
62°W
61°W

CENTRAL AFRICAN REPUBLIC

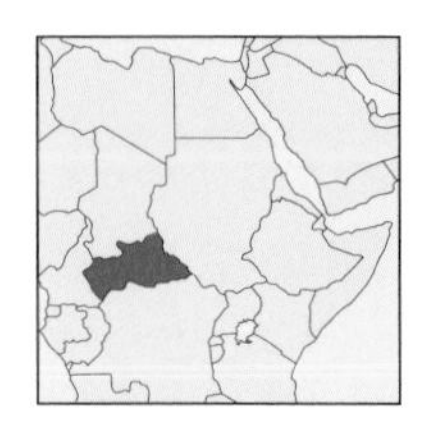

The red, yellow, and green colors on this flag were originally used by Ethiopia, Africa's oldest independent nation. They symbolize African unity. The blue, white, and red recall the flag of France, the country's colonial ruler. This flag was adopted in 1958.

The Central African Republic is a remote, landlocked country enclosed by Chad, Sudan, Congo (Dem. Republic of), Congo, and Cameroon. It lies on a plateau, which lies mostly between 1,970 ft and 2,620 ft [600 m to 800 m] above sea level. This plateau divides the headwaters of two river systems. In the south, the rivers flow into the navigable Ubangi (Oubangi) River. This river itself flows into the River Congo. The Ubangi and the Bomu, its tributary, form much of the country's southern border. In the north, most rivers are headwaters of the Chari (Shari) River. This river flows into Lake Chad to the north.

AREA 240,533 sq miles [622,980 sq km]
POPULATION 3,577,000
CAPITAL (POPULATION) Bangui (553,000)
GOVERNMENT Multiparty republic
ETHNIC GROUPS Baya 34%, Banda 27%, Mandjia 21%, Sara 10%, Mbaka 4%, Mboum 4%
LANGUAGES French (official), Sangho, tribal languages
RELIGIONS Traditional beliefs 57%, Christianity 35%, Islam 8%
CURRENCY CFA franc = 100 centimes

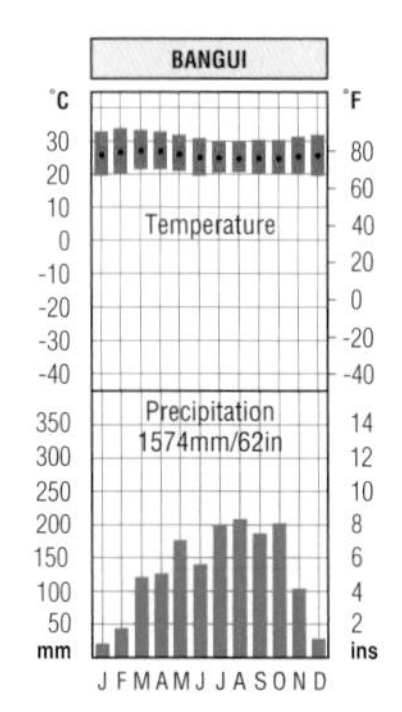

CLIMATE

Bangui, the capital, lies in the southwest of the country on the Ubangi River. The climate is warm throughout the year. The average yearly rainfall totals 62 inches [1,574 mm]. The north is drier, with an average yearly rainfall of about 31 inches [800 mm].

VEGETATION

The southwest has rain forests and tongues of forest extend along the rivers. But wooded savanna covers much of the country, with more open grasslands in the north.

The country has many forest and savanna animals, including buffaloes, elephants, lions, leopards, and many kinds of birds. About 6% of the land is protected in national parks and reserves. But tourism is on a small scale, because of the remoteness of the country.

HISTORY

Little is known of the early history of the area. Between the 16th and 19th centuries, the population was greatly reduced by the slave trade. The country is still thinly populated. Although it is larger than France in area, France has 17 times as many people.

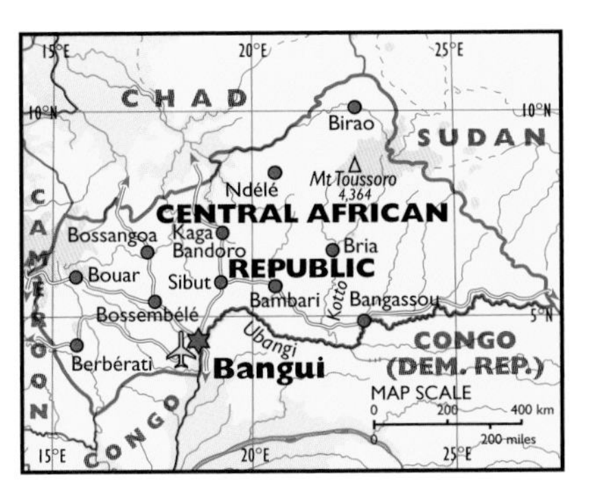

France set up an outpost at Bangui in 1899 and ruled the country as a colony from 1894. Known as Ubangi-Shari, the country was ruled by France for most of the time as part of French Equatorial Africa, a vast area which also included Chad, Congo, and Gabon. The country became fully independent in 1960.

POLITICS

Central African Republic became a one-party state in 1962, but army officers seized power in 1966. The head of the army, Jean-Bedel Bokassa, made himself emperor in 1976. The country was renamed the Central African Empire, but after a brutal and tyrannical reign, Bokassa was overthrown by a military group in 1979. The monarchy was abolished and the country again became a republic.

In March 1981, the country became a multiparty democracy, but the army took over in September and banned all political parties. The country adopted a new, multiparty constitution in 1991. Elections were held in 1993. An army rebellion in 1996 was put down in 1997 with help from French troops. Other failed coups took place in 2001 and 2002.

ECONOMY

The World Bank classifies Central African Republic as a "low-income" developing country. Over 80% of the people are farmers, and most of them produce little more than they need to feed their families. The main crops are bananas, maize, manioc, millet, and yams. Coffee, cotton, timber, and tobacco are produced for export, mainly on commercial plantations.

Diamonds, the only major mineral resource, are the most valuable single export. Manufacturing is on a small scale. Products include beer, cotton fabrics, footwear, leather, soap, and sawn timber. The Central African Republic's development has been impeded by its remote position, its poor transport system and its untrained work force. The country is heavily dependent on aid, especially from France.

Large numbers of beetle species are found in the rain forests of Africa. They are an important part of the ecology of the forests. For example, they eat dead plants and animals, clearing them from the environment. This one-franc timbre taxe *(postage due) stamp, which was issued in 1962, shows two of the beetles found in the Central African Republic. They are the* Phosphorus virescens *and the* Ceroplesis carabarica.

CHAD

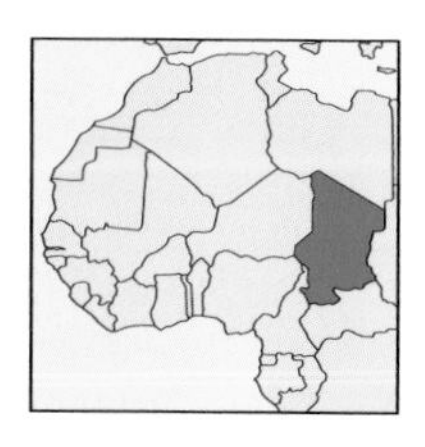

Chad's flag was adopted in 1959 as the country prepared for independence in 1960. The blue represents the sky, the streams in southern Chad, and hope. The yellow symbolizes the sun and the Sahara in the north. The red represents national sacrifice.

The Republic of Chad is a landlocked country in north-central Africa. It is Africa's fifth largest country and is more than twice as big as France, the country which once ruled it as a colony.

Southern Chad is crossed by rivers that flow into Lake Chad, on the western border. Beyond a large depression, northeast of Lake Chad, are the Tibesti Mountains which rise steeply from the sands of the Sahara Desert. The mountains contain Chad's highest peak, Emi Koussi, at 11,204 ft [3,415 m] above sea level.

AREA 495,752 sq miles [1,284,000 sq km]
POPULATION 8,707,000
CAPITAL (POPULATION) Ndjamena (530,000)
GOVERNMENT Multiparty republic
ETHNIC GROUPS In north and center: Arab, Gorane, Zaghawa, Kanembou, Baguirmi, Fulbe, Hausa, Boulala; in south: Sara, Moundang, Moussei
LANGUAGES French and Arabic (both official), many others
RELIGIONS Islam 50%, Christianity 25%, traditional beliefs 25%
CURRENCY CFA franc = 100 centimes

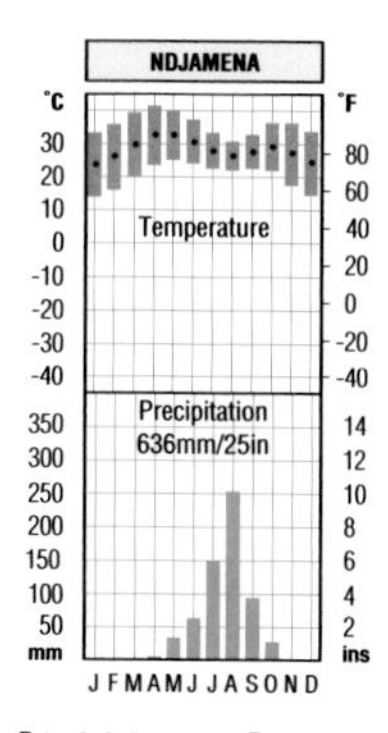

CLIMATE

Ndjamena in central Chad has a hot, tropical climate, with a marked dry season from November to April. The south of the country is wetter, with an average yearly rainfall of around 39 inches [1,000 mm]. The burning-hot desert in the north has an average yearly rainfall of less than 5 inches [130 mm].

VEGETATION

The far south contains forests. Central Chad is a region of savanna, which merges into the dry grasslands of a region called the Sahel. In the north, plants are rare in the bleak desert. Droughts are common in north-central Chad. Long droughts, overgrazing by livestock, and the cutting down of bushes for firewood have exposed the soil of the Sahel to the wind. Wind erosion is turning the land into desert, making the Sahara move southward. This process is called "desertification."

HISTORY AND POLITICS

Chad straddles two worlds. The north is populated by Muslim Arab and Berber peoples, while black Africans, who follow traditional beliefs or who have converted to Christianity, live in the south.

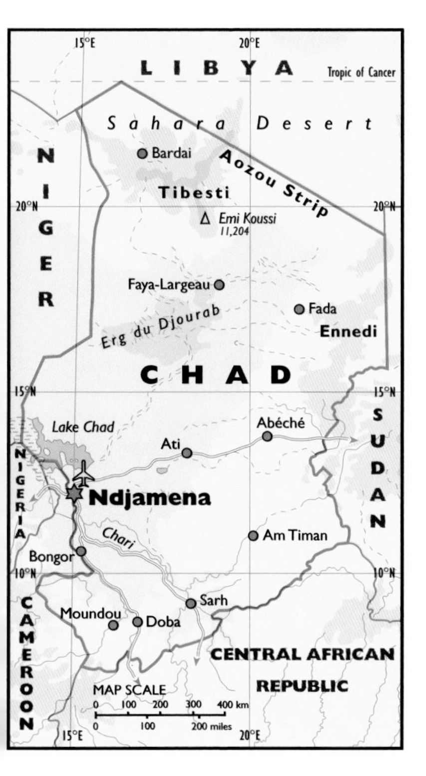

Southern Chad was part of the Kanem empire, which was founded in about AD 700. Other smaller kingdoms developed around Kanem. French explorers were active in the area in the late 19th century. France finally made Chad a colony in 1902.

Chad's progress since it became independent in 1960 has been held back by conflict between various rival ethnic groups. In the 1960s, a group of Muslim northerners mounted a rebellion. In the 1970s, the rebels were supported by Libya, which occupied part of northern Chad, namely a region called the Aozou Strip. Civil conflict also occurred in the south. The 1970s were marked by civil war and coups.

Chad and Libya agreed a truce in 1987 and, in 1994, the International Court of Justice ruled against Libya's claim on the Aozou Strip. Chad enjoyed more stable conditions in the early 1990s. However, more fighting occurred between 1998 and 2002 when a rebel group in the north launched an uprising.

ECONOMY

Hit by drought and civil war, Chad is one of the world's poorest countries. Farming, fishing, and livestock raising employ 83% of the people, but many farmers produce little more than they need to feed their families. Millet, rice, groundnuts, and sorghum are major food crops in the south, while cotton accounts for half of the exports. Chad has few resources and very few manufacturing industries.

Combating contagious tropical diseases is one of the priorities for governments of developing countries. This stamp commemorates the campaign against trachoma, an eye disease spread by a bacteria in warm countries where it is a leading cause of blindness. However, trachoma is treatable with antibiotics.

CHILE

Chile's flag was adopted in 1817. It was designed in that year by an American serving in the Chilean army who was inspired by the US Stars and Stripes. The white represents the snow-capped Andes, the blue the sky, and the red the blood of the nation's patriots.

The Republic of Chile stretches about 2,650 miles [4,260 km] from north to south, although the maximum east–west distance is only about 267 miles [430 km]. The high Andes Mountains form Chile's eastern borders with Argentina and Bolivia. To the west are basins and valleys, with coastal uplands overlooking the shore. In the south, the land has been worn by glaciers and the coastal uplands are islands, while the inland valleys are arms of the sea. Most people live in the central valley, which contains the capital, Santiago.

AREA 292,258 sq miles [756,950 sq km]
POPULATION 15,328,000
CAPITAL (POPULATION) Santiago (4,691,000)
GOVERNMENT Multiparty republic
ETHNIC GROUPS Mestizo 95%, Amerindian 3%
LANGUAGES Spanish (official)
RELIGIONS Roman Catholic 89%, Protestant 11%
CURRENCY Peso = 100 centavos

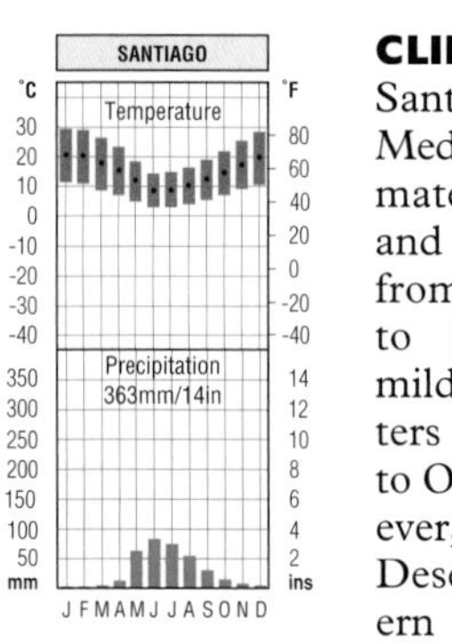

CLIMATE

Santiago has a Mediterranean climate, with hot and dry summers from November to March, and mild, moist winters from April to October. However, the Atacama Desert in northern Chile is one of the world's driest places, while southern Chile is cold and stormy.

VEGETATION

The Atacama Desert stretches about 994 miles [1,600 km] south from the Peruvian border. This region's plants include varieties of cactus and shrubs. Central Chile has mixed forests of beech and laurel. The wet south is a region of thick forests, glaciers, scenic lakes, and windswept, rocky slopes.

HISTORY AND POLITICS

Amerindian people reached the southern tip of South America at least 8,000 years ago. In 1520, the Portuguese navigator Ferdinand Magellan became the first European to sight Chile, but the country became a Spanish colony in the 1540s. Chile became independent in 1818 and, during a war (1879–83), it gained mineral-rich areas from Peru and Bolivia.

In 1970, Salvador Allende became the first Communist leader ever to be elected democratically. He was overthrown in 1973 by army officers, who were supported by the United States

Central Intelligence Agency (CIA). General Augusto Pinochet then ruled as a dictator. A new constitution was introduced in 1981 and elections were held in 1989, though Pinochet remained important because he served as commander-in-chief of the armed forces. In 2000, a socialist, Ricardo Lagos, was elected president.

ECONOMY

The World Bank classifies Chile as a "lower-middle-income" developing country. Mining is important, especially copper production. Minerals dominate Chile's exports. But the most valuable activity is manufacturing; products include processed foods, metals, iron and steel, wood products, transport equipment, and textiles.

Agriculture employs 14% of the people. The chief crop is wheat. Also important are beans, fruits, maize, and livestock products. Chile's fishing industry is one of the world's largest.

Chile's shape and its inhospitable terrain have hampered the development of a good transport system. Air transport is therefore important in linking Chile's cities, as shown by this airmail stamp.

CHINA

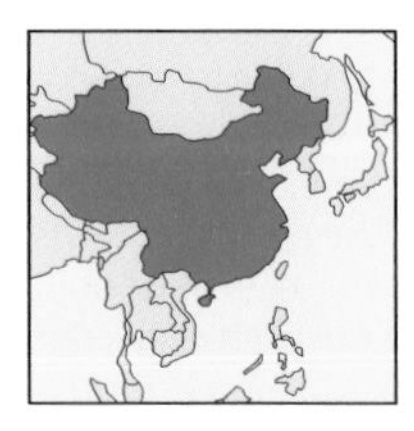

China's flag was adopted in 1949, when the country became a Communist People's Republic. Red is the traditional color of both China and Communism. The large star represents the Communist Party program. The smaller stars symbolize the four main social classes.

The People's Republic of China is the world's third largest country. It is also one of the few countries with more than 1,000 million people. Most people live in the east – on the coastal plains, in the highlands, or in the fertile valleys of such rivers as the Huang He (Hwang Ho), the Chang Jiang (Yangtze Kiang), which is Asia's longest at 3,960 miles [6,380 km], and the Xi Jiang (Si Kiang).

Western China is thinly populated. It includes the bleak Tibetan plateau which is bounded by the Himalayas, the world's highest mountain range. Other ranges include the Kunlun Shan, the Altun Shan, and the Tian Shan. China also has deserts, including the Gobi Desert on the border with Mongolia.

AREA 3,705,386 sq miles [9,596,960 sq km]

POPULATION 1,273,111,000

CAPITAL (POPULATION) Beijing (or Peking, 12,362,000)

GOVERNMENT Single-party Communist republic

ETHNIC GROUPS Han Chinese 92%, 55 minority groups

LANGUAGES Mandarin Chinese (official)

RELIGIONS Majority of population say that they are atheists; Buddhism, Taoism, Islam, Christianity

CURRENCY Renminbi (yuan) = 10 jiao = 100 fen

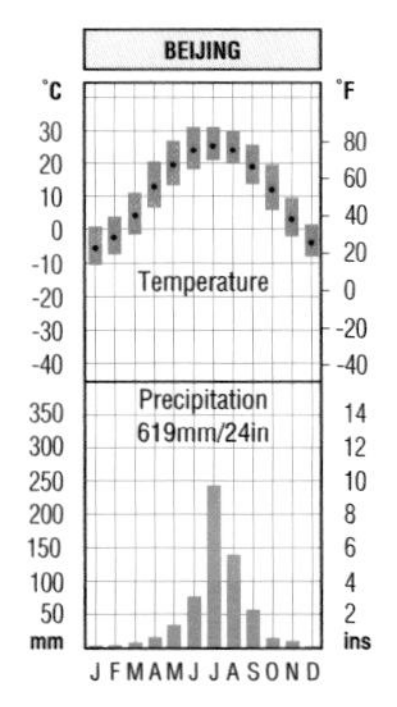

CLIMATE

Beijing in northeastern China has cold winters and warm summers, with a moderate rainfall. Shanghai, in the east-central region of China, has milder winters and more rain. The southeast has a wet, subtropical climate. In the west, the climate is severe. Lhasa has very cold winters and a low rainfall.

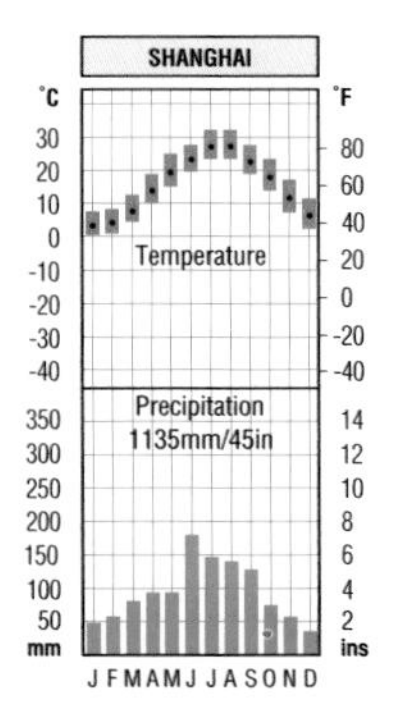

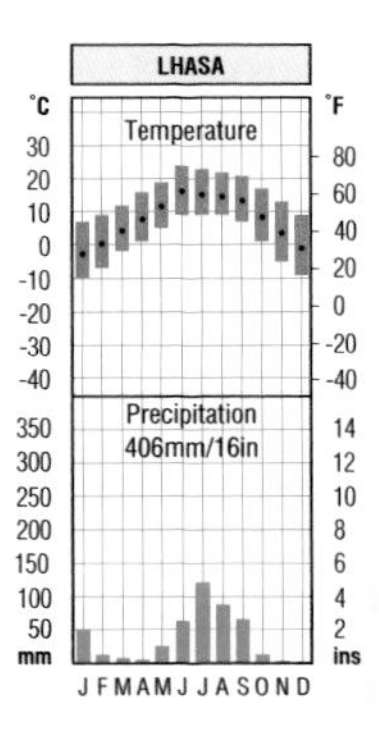

VEGETATION

Large areas in the west are covered

by sparse grasses or desert. The most luxuriant forests, including the bamboo forest home of the rare giant panda, are in the southeast.

HISTORY

China is one of the world's oldest civilizations, going back 3,500 years. Under the Han dynasty (202 BC to AD 220), the Chinese empire was as large as the Roman empire. Mongols conquered China in the 13th century, but Chinese rule was restored in 1368. The Manchu people of Mongolia ruled the country from 1644 to 1912, when the country became a republic.

War with Japan (1937–45) was followed by civil war between the nationalists and the Communists. The Communists triumphed in 1949, setting up the People's Republic of China.

The Nationalists, who had fought the Communists, set up a government in the island of Taiwan.

POLITICS

In the early 1950s, the Communists brought stability to China and living standards for the peasants rose. By 1957, virtually all economic activities were under the control of the government. However, between 1959 and 1961, a government plan called the "Great Leap Forward" caused a fall in industrial production and food shortages, which led to a famine.

Perhaps as many as 20 million people died during this period.

In 1966, a period known as the Cultural Revolution began in which radical students, known as the Red Guards, attacked moderates. Violence broke out and China's leader, Mao Zedong, had to use troops to restore order. Mao died in 1976 and a political struggle began, in which the moderates finally prevailed.

China began to introduce economic reforms, encouraging private enterprise and foreign investment, in special economic zones in eastern China, policies forbidden under Mao. Between 1989 and 2002, the economy grew, on average, by an astonishing 9.3%. In 1997, the former British colony of Hong Kong, which had achieved great economic success, was restored to China, followed by the formerly Portuguese territory of Macau in 1999. Both territories became Special Administrative Regions of China, and Hong Kong, in particular, is making a major contribution to China's fast-developing economy.

However, despite all the economic reforms, the Communist government did not allow political freedoms, and China continued to have one of the world's worst records on human rights.

ECONOMY

The expansion of China's economy, which is one of the world's largest, suggests that China will soon become a major superpower. In the late 1990s, the percentage of people employed in agriculture fell from about 70% to 48%, mainly as a result of people flocking to the fast-developing industrial cities in the east.

Only about 10% of the land is used for crops. Major products include rice, sweet potatoes, tea, and wheat, together with many fruits and vegetables. Livestock farming is also important. Pork is a popular meat

CHINA

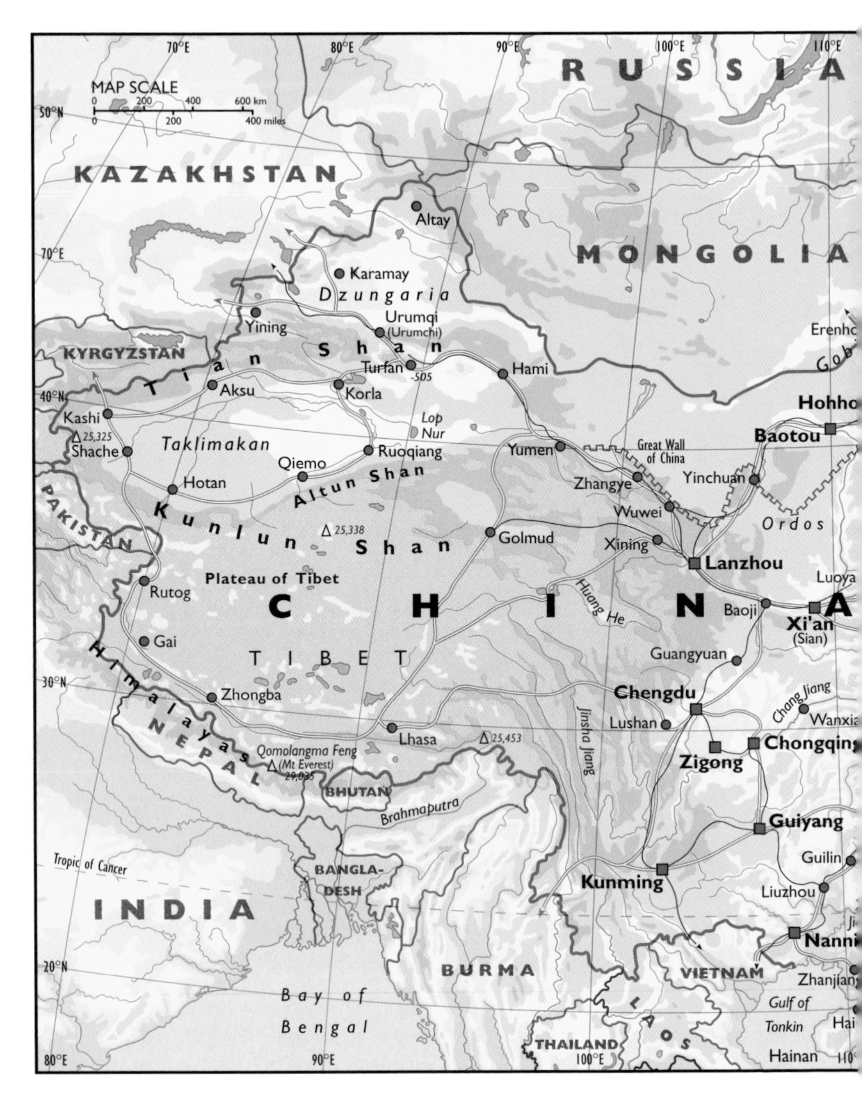

and China has more than a third of the world's pigs.

China's resources include coal, oil, iron ore, and other metals. China has huge steel industries. Manufactures include cement, chemicals, fertilizers, machinery, telecommunications, recording equipment, and textiles. Consumer goods include high-quality but cheap air-conditioners, cameras,

computer monitors and hard disk drives, television sets, and washing machines. These products are now important exports.

TAIWAN

Taiwan (formerly called Formosa) is officially called the Republic of China. High mountains run down the length of the island and are covered in thick forests of evergreen trees and conifers. It was established as a country separate from China when about 2 million nationalists, defeated by the Communists, settled on the island.

At first, Taiwan occupied China's seat in the United Nations but, in 1971, the UN admitted the People's Republic of China instead of Taiwan. In the 1980s, as Taiwan's economy grew quickly, China's relations with Taiwan improved. China regarded Taiwan as a break-away province and proposed that China and Taiwan should be regarded as one nation with two equal governments. But problems arose in 1999 when the Taiwanese president declared that relations between China and Taiwan should be on a nation-by-nation basis. Tension between the two continued into the 21st century, but the United States reaffirmed its policy of defending Taiwan if it was attacked.

With help from the United States, Taiwan is now a prosperous industrial country. The main products include electronic items, clothing, and color television sets, which are produced mainly for export.

AREA 13,900 sq miles [36,000 sq km]
POPULATION 22,370,000
CAPITAL (POPULATION) Taipei (2,596,000)

COLOMBIA

The yellow on Colombia's flag depicts the land, which is separated from the tyranny of Spain by the blue, symbolizing the Atlantic Ocean. The red symbolizes the blood of the people who fought to make the country independent. The flag has been used since 1806.

The Republic of Colombia, in northeastern South America, is the only country in the continent to have coastlines on both the Pacific and the Caribbean Sea. Colombia contains the northernmost ranges of the Andes Mountains. The fertile valleys between the ranges contain about three-quarters of Colombia's people. East of the Andes are plains drained by headwaters of the Amazon and Orinoco rivers; this area covers about two-thirds of the country but less than 2% of the population live here. West of the mountains lie the Caribbean lowlands in the north and the Pacific lowlands in the west.

AREA 439,733 sq miles [1,138,910 sq km]
POPULATION 40,349,000
CAPITAL (POPULATION) Bogotá (6,005,000)
GOVERNMENT Multiparty republic
ETHNIC GROUPS Mestizo 58%, White 20%, Mulatto 14%, Black 4%, mixed Black-Amerindian 3%, Amerindian 1%
LANGUAGES Spanish (official)
RELIGIONS Roman Catholic 90%
CURRENCY Peso = 100 centavos

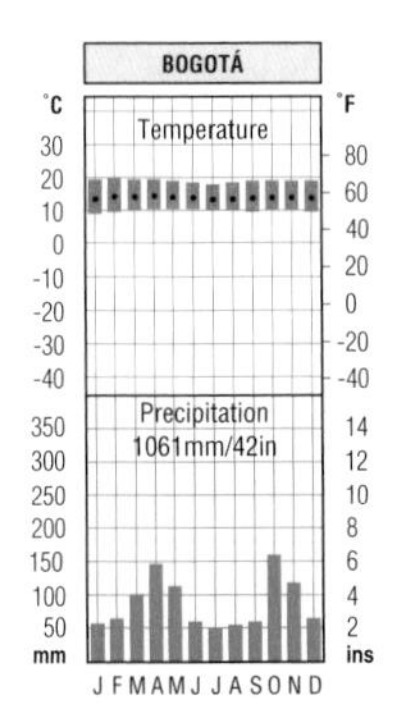

CLIMATE

There is a tropical climate in the lowlands. But the altitude greatly affects the climate of the Andes. The capital, Bogotá, which stands on a plateau in the eastern Andes at about 9,200 ft [2,800 m] above sea level, has mild temperatures throughout the year. The rainfall is heavy, especially on the Pacific coast, though the Caribbean lowlands and the Magdalena valley, which cuts through the Andes, have dry seasons.

VEGETATION

Plant life in the Andes varies with the altitude, with snow-capped peaks and tundra at the highest levels, and grassy meadows and forest at lower levels. The western lowlands have some dense forests, with mangrove swamps on the coast, but the original forests on the Caribbean lowlands have been largely cleared. The northeastern plains are covered by tropical grassland, called *llanos*. Rain forest grows in the southeast.

HISTORY

Amerindian people have lived in Colombia for thousands of years. But today, only a small proportion of the people are of unmixed Amerindian ancestry. Mestizos (people of mixed white and Amerindian ancestry) form the largest group, followed by whites and mulattos (people of mixed European and African ancestry).

Spaniards opened up the area in the early 16th century and they set up a territory known as the Viceroyalty of the New Kingdom of Granada, which included Colombia, Ecuador, Panama, and Venezuela. In 1819, the area became independent, but Ecuador and Venezuela soon split away, followed by Panama in 1903.

POLITICS

Colombia's recent history has been very unstable. Rivalries between the main political parties led to civil wars in 1899–1902 and 1949–57, when the parties agreed to form a coalition. The coalition government ended in 1986 when the Liberal Party was elected. Colombia faces many economic problems. Its government is also trying to contain a large illicit drug industry run by violent dealers, while combating left-wing guerrillas.

In 1998, Andrés Pastrana was elected president. His attempts to end the guerrilla war ended when peace talks collapsed in 2002.

DID YOU KNOW

- that Colombia was named after Christopher Columbus
- that Colombia is the world's second largest coffee producer and the largest producer of emeralds
- that the eruption of the volcano Nevado del Ruiz, south of Medellín, caused a *lahar* (mud avalanche) that buried the town of Armero, killing most of its inhabitants
- that Colombia is notorious for its drug traffic in cocaine; drug dealers have declared war on anyone trying to stop their trade

ECONOMY

The World Bank classifies Colombia as a "lower-middle-income" developing country. Agriculture is important and coffee is the leading export crop. Other crops include bananas, cocoa, maize, tobacco, and vegetables. Dairy and beef cattle are raised. But mining and manufacturing are becoming more important in the economy. Colombia exports coal and oil, and it also produces emeralds and gold. Manufacturing is based in the main cities and products include cement, chemicals, processed foods, and textiles.

CONGO

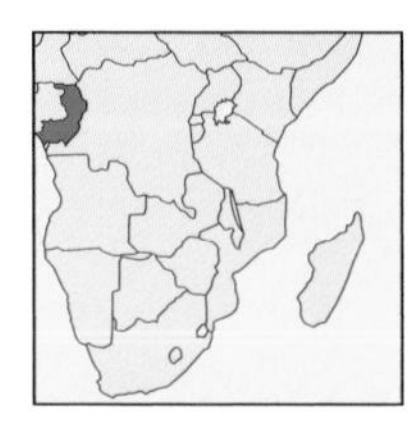

Congo's red flag, with the national emblem of a crossed hoe and mattock (a kind of pick-axe), was dropped in 1990, when the country officially abandoned the Communist policies it had followed since 1970. This new flag was adopted in its place.

The Republic of Congo is a country on the River Congo in west-central Africa. The Equator runs through the center of the country. Congo has a narrow coastal plain on which its main port, Pointe Noire, stands. Behind the plain are uplands through which the River Niari has carved a fertile valley. Central Congo consists of high plains. The north contains large swampy areas in the valleys of rivers that flow into the River Congo and its large tributary, the Oubangi.

AREA 132,046 sq miles [342,000 sq km]
POPULATION 2,894,000
CAPITAL (POPULATION) Brazzaville (938,000)
GOVERNMENT Military regime
ETHNIC GROUPS Kongo 48%, Sangha 20%, Teke 17%, M'bochi 12%
LANGUAGES French (official), many others
RELIGIONS Christianity 50%, Animist 48%, Islam 2%
CURRENCY CFA franc = 100 centimes

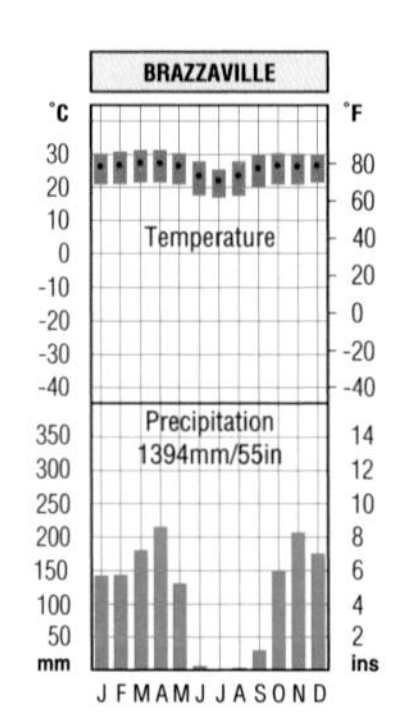

CLIMATE

Congo has a hot, wet equatorial climate. Brazzaville has a dry season between June and September. The coastal plain is drier and cooler than the rest of the country. This is because a cold ocean current, the Benguela Current, flows northward along the coast.

VEGETATION

The mostly treeless coastal plains are backed by forested highlands. Around Brazzaville most of the land has been cleared by farmers. Central Congo contains areas of savanna (tropical grassland) with forests in the valleys. The north is mainly forest. The trees include the valuable okoumé and mahogany.

HISTORY

Between the 15th and 18th centuries, part of Congo probably belonged to the huge Kongo kingdom, whose center lay to the south. But the Congo coast became a center of the European slave trade. The European exploration of the interior took place in the late 19th century. The area came under French protection in 1880 and it was later governed as part of a larger region called French Equatorial Africa. The country remained under French control until 1960.

POLITICS

Congo became a one-party state in 1964 and a military group took over the government in 1968. In 1970, Congo declared itself a Communist country. The country abandoned its Communist policies in 1990. Multi-party elections were held in 1992, when Pascal Lissouba was elected president. In 1997, Denis Sassou-Nguesso, a former military dictator, overthrew Lissouba. After peace was restored, Sassou-Nguesso was elected president in 2002.

ECONOMY

The World Bank classifies Congo as a "lower-middle-income" developing country. Agriculture is the most important activity, employing more than 50% of the people. But many farmers live at subsistence level. Food crops include bananas, cassava, maize, and rice. The main cash crops are coffee and cocoa. The main exports are oil (which makes up 90% of the total) and timber. Manufacturing is small scale.

DID YOU KNOW

- that Congo was once known as Congo (Brazzaville) to distinguish it from the other Congo which was renamed Zaïre in 1971
- that once again, Africa has another country called Congo since Zaïre was renamed Congo (Democratic Republic of the) in 1998
- that Congo declared itself Africa's first Communist state in 1970
- that Brazzaville was named after the explorer Pierre Savorgnan de Brazza, who founded it in 1880

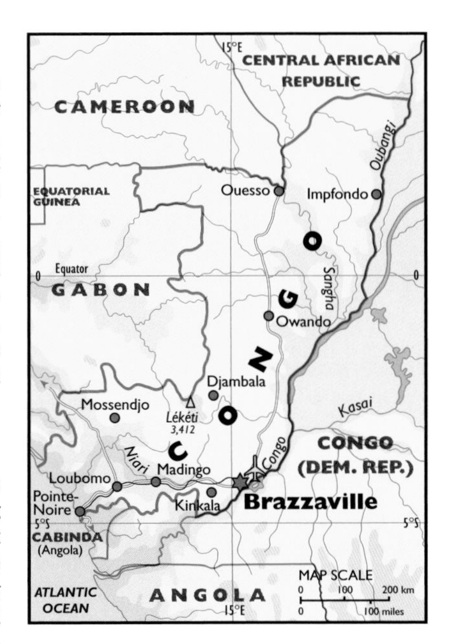

Congo's forests are the home for many birds, including parrots. The African grey parrot, shown on this stamp, issued in 1990, is a popular pet because of its ability to mimic human speech.

CONGO (DEMOCRATIC REPUBLIC OF THE)

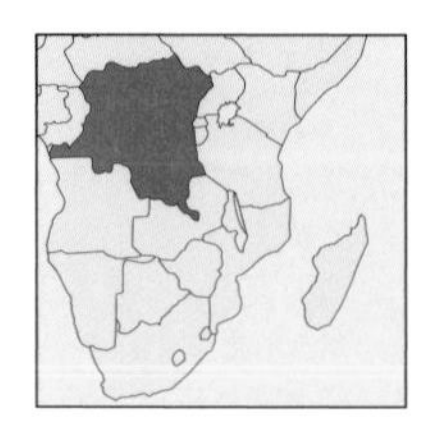

The Democratic Republic of the Congo adopted a new flag in 1997 after Laurent Kabila rose to power. The blue represents the UN's role in securing independence for the country, and the six stars represent the original provinces of the independent state.

AREA 905,365 sq miles [2,344,885 sq km]

POPULATION 53,625,000

CAPITAL (POPULATION) Kinshasa (2,664,000)

GOVERNMENT Single-party republic

ETHNIC GROUPS Over 200 groups, of which the majority are Bantu; the four largest are Mongo, Luba, Kongo (all Bantu), and the Mangbetu-Azande (Hamitic)

LANGUAGES French (official), tribal languages

RELIGIONS Roman Catholic 50%, Protestant 20%, Kimbanguist 10%, Islam 10%, others 10%

CURRENCY Congolese franc = 100 centimes

The Democratic Republic of the Congo (formerly Zaïre) is the world's 12th largest country. Much of the land lies within the drainage basin of the River Congo, which was also called the River Zaïre, and reaches the sea along the country's short coastline, about 25 miles [40 km] long.

Mountains rise in the east where Congo's borders run through lakes Tanganyika, Kivu, Edward, and Albert. The lakes lie on the floor of an arm of the Great Rift Valley, which runs through East Africa. Southern Congo is also mountainous.

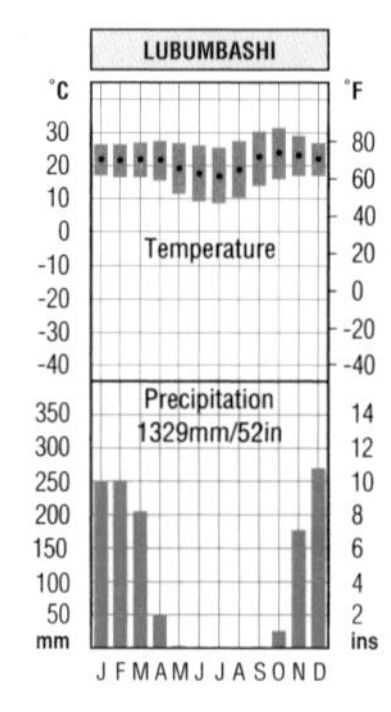

CLIMATE

The line of the Equator passes through northern Congo. The Equatorial region has very high temperatures and heavy rainfall throughout the year. In the subtropical south, where the town of Lubumbashi is situated, there is a marked wet and dry season.

VEGETATION

Dense rain forests grow in northern Congo, but to the south, the forests merge into savanna (tropical grassland with scattered trees). The mountains have zones of vegetation which vary according to the altitude.

HISTORY AND POLITICS

Pygmies were the first inhabitants of Congo. Later on, people speaking Bantu languages set up large kingdoms, including those of the Kongo, Luba, and Lunda. Portuguese navigators reached the coast in 1482, but the interior was not explored until the late 19th century. In 1885, the country, called Congo Free State, became the personal property of King Léopold II of Belgium. In 1908, the country became a Belgian colony.

The Belgian Congo became independent in 1960 and was renamed Zaïre in 1971. Conflict between ethnic groups continued until 1965 when a one-party state was established under President Mobutu. In 1996, fighting broke out, as the Tutsi-Hutu conflict spilled over in the east. In 1997, an army led by Laurent Kabila overthrew Mobutu. Zaïre was renamed Congo. Civil war broke out in 1998 and Kabila was assassinated in 2001.

ECONOMY

The World Bank classifies Congo as a "low-income" developing country, despite its reserves of copper, the main export, and other minerals. Agriculture employs 65% of the people, though many farmers grow barely enough to feed their own families. Cocoa, coffee, cotton, and tea are the major cash crops. Food crops include bananas, cassava, maize, and rice.

COSTA RICA

Costa Rica's flag is based on the blue-white-blue pattern used by the Central American Federation (1823–39). This Federation consisted of Costa Rica, El Salvador, Guatemala, Honduras and Nicaragua. The red stripe, which was adopted in 1848, reflects the colors of France.

The Republic of Costa Rica in Central America has coastlines on both the Pacific Ocean and also on the Caribbean Sea. Central Costa Rica consists of mountain ranges and plateaux with many volcanoes. The Meseta Central, where the capital San José is situated, and the Valle del General in the southeast, have rich volcanic soils and are the most thickly populated parts of Costa Rica. The highlands descend to the Caribbean lowlands and the Pacific Coast region, with its low mountain ranges. The main rivers are the Grande (which drains into the Pacific Ocean), the Reventazon, and the General.

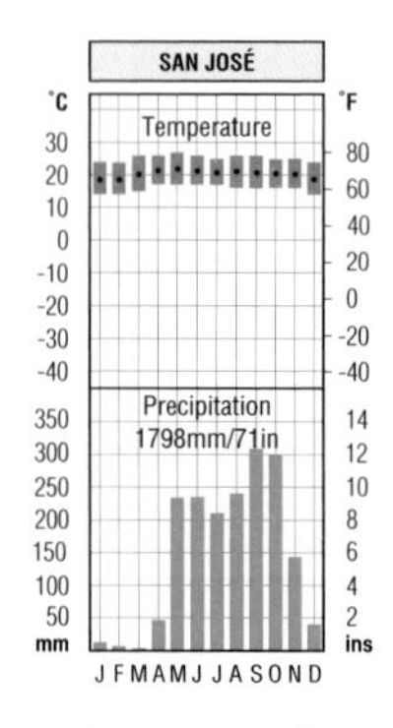

CLIMATE

San José stands at about 3,840 ft [1,170 m] above sea level and has a pleasant climate. Inland, the average annual temperature is 68°F [20°C], compared with more than 81°F [27°C] on the coasts. The coolest months are December and January. The northeast trade winds bring heavy rain to the Caribbean coast. There is less rainfall in the highlands and on the Pacific coastlands.

AREA 19,730 sq miles [51,100 sq km]
POPULATION 3,773,000
CAPITAL (POPULATION) San José (1,220,000)
GOVERNMENT Multiparty republic
ETHNIC GROUPS White 85%, Mestizo 8%, Black and Mulatto 3%, East Asian (mostly Chinese) 1%
LANGUAGES Spanish (official)
RELIGIONS Roman Catholic 76%, Evangelical 14%
CURRENCY Colón = 100 céntimos

VEGETATION

Evergreen forests with such trees as mahogany and tropical cedar cover about half of Costa Rica. Oaks grow in the highlands, with scrub at the higher levels where trees cannot grow. Palm trees grow along the Caribbean coast. Mangrove swamps are common on the Pacific coast.

HISTORY AND POLITICS

Christopher Columbus reached the Caribbean coast in 1502 and rumors of treasure soon attracted many Spaniards to settle in the country. Spain ruled the country until 1821, when Spain's Central American colonies broke away to join Mexico in 1822. In 1823, the Central American states broke with Mexico and set up the Central American Federation.

Later, this large union broke up and Costa Rica became fully independent in 1838.

From the late 19th century, Costa Rica experienced a number of revolutions, with periods of dictatorship and periods of democracy. In 1948, following a revolt, the armed forces were abolished. Since 1948, Costa Rica has enjoyed a long period of stable democracy, which many in Latin America admire and envy.

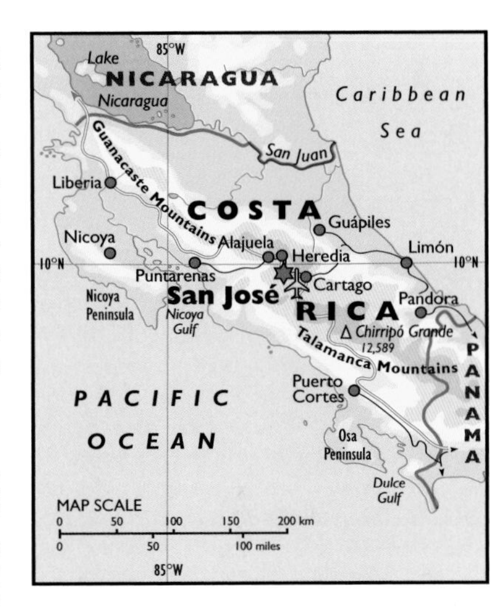

ECONOMY

Costa Rica is classified by the World Bank as a "lower-middle-income" developing country and one of the most prosperous countries in Central America. There are high educational standards and a high life expectancy (to an average of 77.5 years). Agriculture employs 19% of the people. Major crops include coffee grown in the highlands, bananas, and sugar, all of which are exported. Other crops include beans, citrus fruits, cocoa, maize, potatoes, and vegetables. Cattle ranching is another important activity, particularly in the far northwest of the country.

The country's resources include its forests, but it lacks minerals apart from some bauxite and manganese. Manufacturing is increasing. Products include cement, clothing, fertilizers, machinery, furniture, processed food, and textiles. The United States is Costa Rica's chief trading partner. Tourism is a fast-growing industry.

DID YOU KNOW

- that Costa Rica means "rich coast," because early Spanish explorers mistakenly believed that the land was rich in gold and other minerals
- that, following a civil war in 1948, Costa Rica abolished its armed forces to preserve the peace
- that about 12% of the country is protected in national parks and other areas
- that 93% of Costa Ricans can read and write; this is the highest literacy level in Central America

Costa Rica's scenic central highlands are dotted with volcanoes, which periodically erupt. This stamp, showing a plane flying over one of the active volcanoes, Volcán Poas, commemorates Costa Rica's First Annual Fair in 1937.

CROATIA

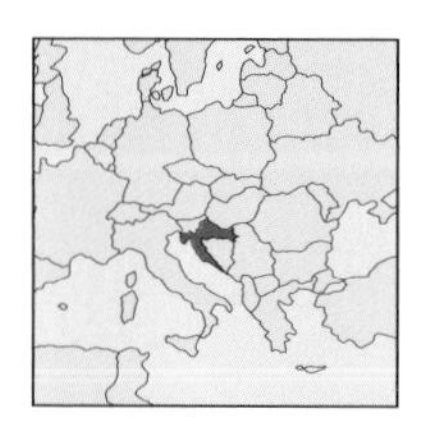

Croatia adopted a red, white, and blue flag in 1848. Under Communist rule, a red star appeared at the center. In 1990, the red star was replaced by the present coat of arms, which symbolizes the various parts of the country.

The Republic of Croatia was one of the six republics that made up the former Communist country of Yugoslavia until it became independent in 1991. The region bordering the Adriatic Sea is called Dalmatia. It includes the coastal ranges, which contain large areas of bare limestone. They reach 6,276 ft [1,913 m] at Mount Troglav. Other highlands lie in the northeast. Most of the rest of the country consists of the fertile Pannonian plains, which are drained by Croatia's two main rivers, the Drava and the Sava.

AREA 21,824 sq miles [56,538 sq km]

POPULATION 4,334,000

CAPITAL (POPULATION) Zagreb (868,000)

GOVERNMENT Multiparty republic

ETHNIC GROUPS Croat 78%, Serb 12%, Bosnian

LANGUAGES Serbo-Croatian

RELIGIONS Roman Catholic 77%, Eastern Orthodox 11%, Islam 1%

CURRENCY Kuna = 100 lipas

CLIMATE

The coastal area has a typical Mediterranean climate, with hot, dry summers and mild, moist winters. Inland, the climate becomes more continental. Winters are often bitterly cold, while temperatures often soar to 100°F [38°C] in the summer months.

VEGETATION

Pasture and farmland cover 70% of Croatia, with woodland occupying only 15%. Sparse Mediterranean scrub is the chief vegetation in Dalmatia.

HISTORY

Slav people settled in the area around 1,400 years ago. In 803, Croatia became part of the Holy Roman Empire and the Croats soon adopted Christianity. Croatia was an independent kingdom in the 10th and 11th centuries. But in 1102, the king of Hungary also became king of Croatia, creating a union that lasted 800 years.

In 1526, part of Croatia and Hungary came under the Turkish Ottoman empire, while the rest of Croatia came under the Austrian Habsburgs. In 1699, all of Croatia came under Habsburg rule. In 1867, the Habsburg empire became the dual monarchy of Austria-Hungary.

After Austria-Hungary was defeated in World War I (1914–18), Croatia became part of the new

Kingdom of the Serbs, Croats, and Slovenes. This kingdom was renamed Yugoslavia in 1929. Germany occupied Yugoslavia during World War II (1939–45). Croatia was proclaimed independent, but it was really ruled by the invaders.

After the war, Communists took power with Josip Broz Tito as the country's leader. Despite ethnic differences between the people, Tito held Yugoslavia together until his death in 1980. In the 1980s, economic and ethnic problems, including a deterioration in relations with Serbia, threatened stability. In the 1990s, Yugoslavia split into five nations, one of which was Croatia, which declared itself independent in 1991.

POLITICS

After Serbia supplied arms to Serbs living in Croatia, war broke out between the two republics, causing great damage. Croatia lost more than 30% of its territory. But in 1992, the United Nations sent a peacekeeping force to Croatia, which effectively ended the war with Serbia.

In 1992, when war broke out in Bosnia-Herzegovina, Bosnian Croats occupied parts of the country. But in 1994, Croatia helped to end Croat-Muslim conflict in Bosnia-Herzegovina and, in 1995, after retaking some areas occupied by Serbs, it helped to draw up the Dayton Peace Accord which ended the civil war there.

ECONOMY

The wars of the early 1990s disrupted Croatia's economy, which had been quite prosperous before the disturbances. Tourism on the Dalmatian coast had been a major industry. Croatia also had major manufacturing industries, and manufactures remain the chief exports. Manufactures include cement, chemicals, refined oil and oil products, ships, steel, and wood products. Major farm products include fruits, livestock, maize, soya beans, sugar beet, and wheat.

Josip Broz Tito (1892–1980), the Communist leader of Yugoslavia between 1945 and 1980, was born in Croatia. This stamp is one of a set of 28 issued in 1967 on the occasion of Tito's 75th birthday.

CUBA

Cuba's flag, the "Lone Star" banner, was designed in 1849, but it was not adopted as the national flag until 1901, after Spain had withdrawn from the country. The red triangle represents the Cuban people's bloody struggle for independence.

The Republic of Cuba is the largest island country in the Caribbean Sea. It consists of one large island, Cuba, the Isle of Youth (Isla de la Juventud) and about 1,600 small islets.

Mountains and hills cover about a quarter of Cuba. The highest mountain range, the Sierra Maestra in the southeast, reaches 6,562 ft [2,000 m] above sea level. The rest of the land consists of gently rolling country or coastal plains, crossed by fertile valleys carved by the short, mostly shallow and narrow rivers.

AREA 42,803 sq miles [110,860 sq km]
POPULATION 11,184,000
CAPITAL (POPULATION) Havana (2,204,000)
GOVERNMENT Socialist republic
ETHNIC GROUPS Mulatto 51%, White 37%, Black 11%
LANGUAGES Spanish (official)
RELIGIONS Roman Catholic 40%, Protestant 3%
CURRENCY Cuban peso = 100 centavos

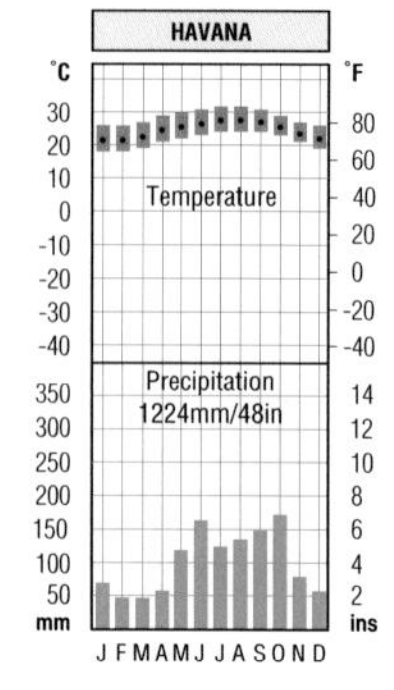

CLIMATE

Cuba lies in the tropics. But sea breezes moderate the temperature, warming the land in winter and cooling it in summer. The dry season runs from November to April, while May to October is the rainy season. Hurricanes sometimes strike between August and October.

VEGETATION

Farmland covers about half of Cuba. Forests of pine still grow, especially in the southeast. Other trees include ebony and mahogany. Parts of the coast are lined by mangrove swamps.

HISTORY

Christopher Columbus discovered the island in 1492 and Spaniards began to settle there from 1511. Spanish rule ended in 1898, when the United States defeated Spain in the Spanish-American War. American influence in Cuba remained strong until 1959, when revolutionary forces under Fidel Castro overthrew the dictatorial government of Fulgencio Batista.

POLITICS

The United States opposed Castro's policies, when he turned to the Soviet Union for assistance. In 1961, Cuban exiles attempting an invasion were defeated. In 1962, the US learned that nuclear missile bases armed by the Soviet Union had been established in Cuba. The US ordered the Soviet

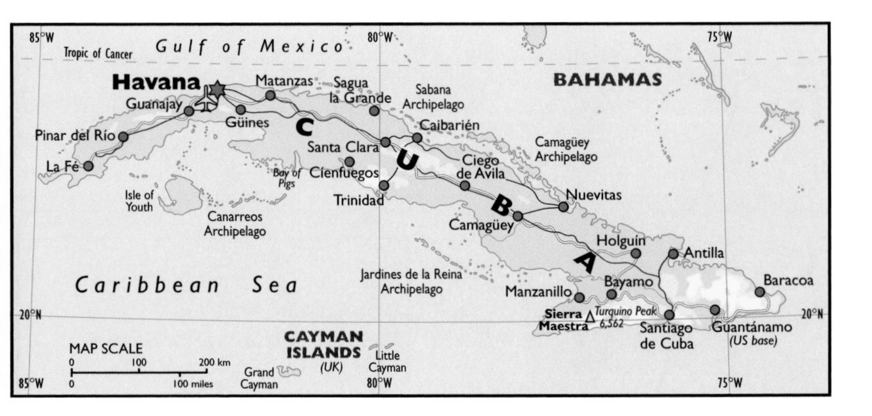

Union to remove the missiles and bases and, after a few days, when many people feared that a world war might break out, the Soviet Union agreed to the American demands.

Cuba's relations with the Soviet Union remained strong until 1991, when the Soviet Union was broken up. The loss of Soviet aid greatly damaged Cuba's economy, but Castro continued the country's left-wing policies. Attempts to thaw US-Cuban relations were scuttled in 2002 when President George W. Bush stated that Cuba was part of an "axis of evil."

ECONOMY

The government runs Cuba's economy and owns 70% of the farmland. Sugar is the chief export, followed by refined nickel ore. Other exports include cigars, citrus fruits, fish, medical products, and rum.

Before 1959, US companies owned most of Cuba's manufacturing industries. But under Fidel Castro, they became government property. Among the chief manufactures are processed foods, textiles, chemicals, and metals. After the collapse of Communist governments in the Soviet Union and its allies, Cuba worked to increase its trade with Latin America and China.

Ché Guevara, a Latin American revolutionary leader born in Argentina, served in Cuba's government in the 1960s. Portraits of him are shown on this stamp issued in 1977 to commemorate the tenth anniversary of Guerrillas Heroes Day. Guevara was killed in 1967 while leading a revolt against the government in Bolivia.

CYPRUS

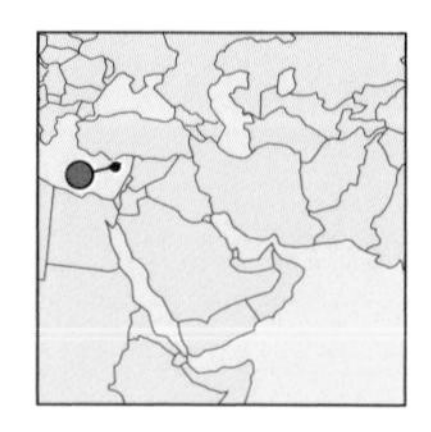

This flag became the official flag when the country became independent from Britain in 1960. It shows an outline map of the island, with two olive branches. Since Cyprus was divided, the separate communities have flown the Greek and Turkish flags.

The Republic of Cyprus is an island nation in the northeastern Mediterranean Sea. Geographers regard it as part of Asia, but it resembles southern Europe in many ways.

Cyprus has scenic mountain ranges, including the Kyrenia range in the north and the Troodos Mountains in the south, which rise to 6,401 ft [1,951 m] at Mount Olympus. The island also contains several fertile lowlands, including the broad Mesaoria plain between the Kyrenia and Troodos mountains.

AREA 3,571 sq miles [9,250 sq km]
POPULATION 763,000
CAPITAL (POPULATION) Nicosia (189,000)
GOVERNMENT Multiparty republic
ETHNIC GROUPS Greek Cypriot 81%, Turkish Cypriot 19%
LANGUAGES Greek and Turkish (both official)
RELIGIONS Greek Orthodox, Islam
CURRENCY Cyprus pound = 100 cents

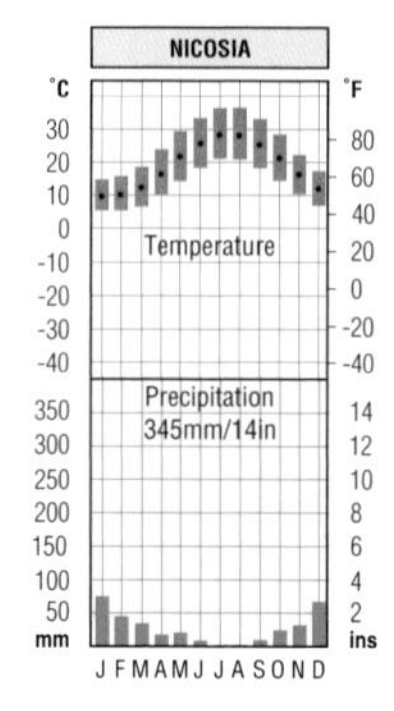

CLIMATE

Cyprus has a Mediterranean climate, with hot, dry summers and mild, moist winters. But the summers are hotter than in the western Mediterranean lands; this is because Cyprus lies close to the hot mainland of southwestern Asia. The mountains are cooler and wetter than the coast.

VEGETATION

Large pine forests grow on the slopes of the Kyrenia and Troodos mountains. The plains are largely farmed, but groves of citrus and olive trees are common. The land is usually parched by late summer, but wildflowers grow in abundance during the sunny winter. Some areas are covered by low shrubs and bushes, with patches of acacia, cypress, and eucalyptus trees.

HISTORY

Greeks settled on Cyprus around 3,200 years ago. From AD 330, the island was part of the Byzantine empire. In the 1570s, Cyprus became part of the Turkish Ottoman empire. Turkish rule continued until 1878 when Cyprus was leased to Britain. Britain annexed the island in 1914 and proclaimed it a colony in 1925.

In the 1950s, Greek Cypriots, who made up four-fifths of the population, began a campaign for *enosis* (union) with Greece. Their leader was the

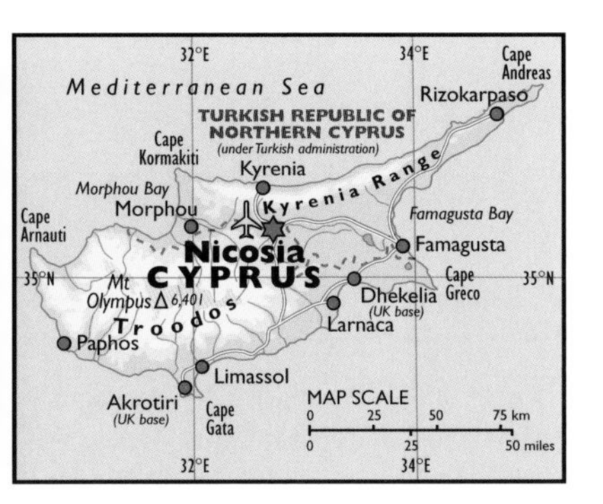

Greek Orthodox Archbishop Makarios. A secret guerrilla force called EOKA attacked the British, who exiled Makarios. Cyprus became an independent country in 1960, although Britain retained two military bases. Archbishop Makarios became the first president of Cyprus.

POLITICS

Independent Cyprus had a constitution which provided for power-sharing between the Greek and Turkish Cypriots. But the constitution proved unworkable and fighting broke out between the two communities. In 1964, the United Nations sent in a peacekeeping force. Communal clashes recurred in 1967.

In 1974, Cypriot forces led by Greek officers overthrew Makarios. This led Turkey to invade northern Cyprus, a territory occupying about 40% of the island. Many Greek Cypriots fled from the north which, in 1979, was proclaimed an independent state called the Turkish Republic of Northern Cyprus. But the United Nations still regarded Cyprus as a single nation under the Greek-Cypriot government in the south. Efforts to unite the island were given urgency in 2002 when Cyprus was invited to become a member of the European Union in May 2004.

ECONOMY

Cyprus got its name from the Greek word *kypros*, meaning copper. But little copper remains and the chief minerals today are asbestos and chromium. Industry employs 37% of the work force and manufactures include cement, clothes, footwear, tiles, and wine. Farming employs 14% of workers, and crops include barley, citrus fruits, grapes, olives, potatoes, and wheat. However, the most valuable activity in Cyprus is tourism.

In the early 1990s, the United Nations reclassified Cyprus as a developed rather than a developing country. But the economy of the Turkish-Cypriot north lags behind that of the Greek-Cypriot south.

The leading map-maker of the 16th century was a Flemish geographer, Gerardus Mercator. His world maps were ideal for navigation, although the accuracy of the maps was limited by the amount of information available at that time, as this reproduction of his map of Cyprus reveals. Mercator's map dates back to 1554. This stamp was issued in 1969.

CZECH REPUBLIC

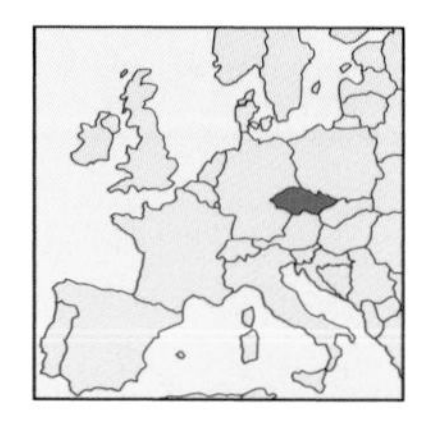

After independence, on January 1, 1993, the Czech Republic adopted the former flag of Czechoslovakia. It features the red and white of Bohemia in the west, together with the blue of Moravia and Slovakia. Red, white, and blue are the colors of Pan-Slavic liberation.

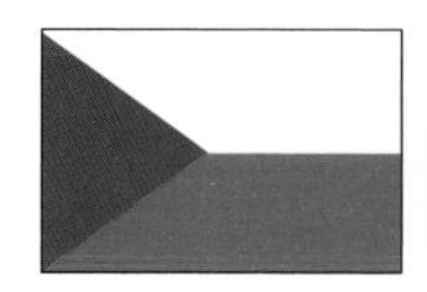

The Czech Republic is the western three-fifths of the former country of Czechoslovakia. It contains two regions: Bohemia in the west, and Moravia in the east. Mountains border much of the country in the west. The Bohemian basin in the north-center is a fertile lowland region, with Prague, the capital city, as its main center. Highlands cover much of the center of the country, with lowlands in the southeast. Some rivers, such as the Elbe (Labe) and Oder (Odra), flow north into Germany and Poland. In the south, rivers flow into the Danube basin.

AREA 30,449 sq miles [78,864 sq km]
POPULATION 10,264,000
CAPITAL (POPULATION) Prague (or Praha, 1,203,000)
GOVERNMENT Multiparty republic
ETHNIC GROUPS Czech 81%, Moravian 13%, Slovak 3%, Polish, German, Silesian, Gypsy, Hungarian, Ukrainian
LANGUAGES Czech (official)
RELIGIONS Atheist 40%, Roman Catholic 39%, Protestant 4%
CURRENCY Czech koruna = 100 haler

CLIMATE

The climate is influenced by its landlocked position in east-central Europe. It is sometimes affected by air streams from Atlantic Europe and sometimes by easterly winds that come from Russia, bringing bitterly cold weather in winter. Prague has warm, sunny summers and cold winters. The average rainfall is moderate, with 20 to 30 inches [500 mm to 750 mm] every year in lowland areas. Thunderstorms often occur on summer afternoons.

VEGETATION

Many of the country's forests have been cut down to create farmland, but forests of oak and spruce still survive. Acid rain is now damaging trees in the north.

HISTORY AND POLITICS

The ancestors of the Czech people began to settle in what is now the Czech Republic around 1,500 years ago. Bohemia, in the west, became important in the 10th century as a kingdom within the Holy Roman Empire. In 1526, the Austrian Habsburgs began to rule Austria. But a Czech rebellion in 1618 led to the Thirty Years' War. German culture dominated the area until the late 18th century. Although Austria continued to rule Bohemia and Moravia, Czech nationalism continued to grow throughout the 19th century.

After World War I (1914–18), Czechoslovakia was created. Germany seized the country in World War II (1939–45). In 1948, Communist leaders took power and Czechoslovakia was allied to the Soviet Union. When democratic reforms were introduced in the Soviet Union in the late 1980s, the Czechs also demanded reforms. Free elections were held in 1990, but differences between the Czechs and Slovaks led the government to agree in 1992 to the partitioning (dividing) of the country on January 1, 1993. The break was peaceful and the nations kept many ties. In 2002, the Czech Republic and Slovakia were invited to join the European Union in May 2004.

ECONOMY

Under Communist rule the Czech Republic became one of the most industrialized parts of Eastern Europe. The country has deposits of coal, uranium, iron ore, magnesite, tin, and zinc. The country produces chemicals, iron and steel, and machinery, while light industries make glassware and textiles for export. Manufacturing employs about 28% of the work force.

Farming remains important. The main crops include barley, fruit, hops for beer-making, maize, potatoes, sugar beet, vegetables, and wheat. Cattle and other livestock are raised. Under Communist rule, the land was owned by the government. But the private ownership of the land is now being restored.

DID YOU KNOW

- that Prague is often called "the city of a hundred spires" because of its many beautiful churches
- that Antonin Dvorak, Leos Janacek, and Bedrich Smetana were all Czech composers
- that the "Prague Spring" (1968) was a time when the Communist leaders of Czechoslovakia tried to introduce reforms; troops from the Soviet Union and other Communist countries invaded Czechoslovakia and replaced the government

"Animals in Heraldry" was the subject of a set of stamps issued in Czechoslovakia in 1979. This stamp displays the coat of arms of Jesenik, a town in the northeastern Czech Republic. The bear and the eagle were formerly common creatures throughout Central Europe, but the number of bears has declined sharply.

DENMARK

Denmark's flag is called the Dannebrog, or "the spirit of Denmark." It may be the oldest national flag in continuous use. It represents a vision thought to have been seen by the Danish King Waldemar II before the Battle of Lyndanisse, which took place in Estonia in 1219.

AREA 16,629 sq miles [43,070 sq km]
POPULATION 5,353,000
CAPITAL (POPULATION) Copenhagen (1,362,000)
GOVERNMENT Parliamentary monarchy
ETHNIC GROUPS Danish 97%
LANGUAGES Danish (official)
RELIGIONS Lutheran 95%, Roman Catholic 1%
CURRENCY Krone = 100 øre

The Kingdom of Denmark is the smallest country in Scandinavia. It consists of a peninsula, called Jutland (or Jylland), which is joined to Germany, and more than 400 islands, 89 of which are inhabited. The capital city, Copenhagen, is on the largest island, Sjælland. It faces Sweden across a narrow strait, called The Sound, which leads from the Baltic Sea to the Kattegat and North Sea.

The land is flat and mostly covered by rocks dropped there by huge ice sheets during the last Ice Age. The highest point in Denmark is on Jutland. It is only 568 ft [173 m] above sea level.

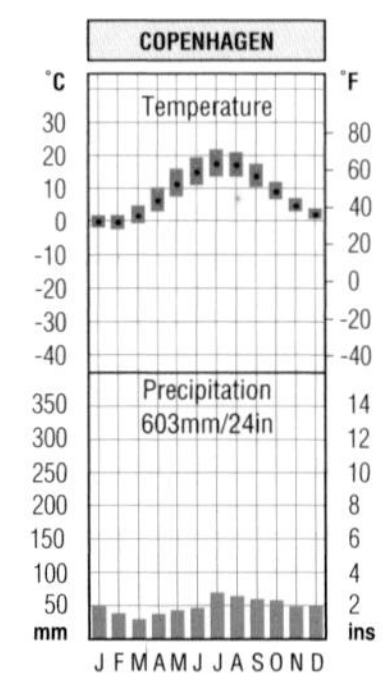

CLIMATE

Denmark has a cool but pleasant climate, except during cold spells in the winter when The Sound between Sjælland and Sweden may freeze over. Summers are warm. Rainfall occurs all through the year. The wettest seasons are summer and fall. Atlantic storms sometimes hit Denmark.

VEGETATION

Much of Denmark is a patchwork of green fields, lakes, sandy beaches, and drifting sand dunes. Forests of deciduous trees, such as oak and elm, once covered the land. But most of these forests have been cut down. Today, planted belts of beech, pine, and spruce help to break the force of the strong westerly winds.

HISTORY

Danish Vikings terrorized much of Western Europe for about 300 years after AD 800. Danish kings ruled England in the 11th century. In the late 14th century, Denmark formed a union with Norway and Sweden (which included Finland). Sweden broke away in 1523, while Denmark lost Norway to Sweden in 1814.

In the late 19th century, the Danes developed their economy. They set up cooperatives and improved farming techniques. Many countries copied their work. Denmark was neutral in World War I (1914–18). In 1944, it made its colony of Iceland independent. Germany occupied Denmark in World War II (1939–45).

POLITICS

After 1945, Denmark played an important part in European affairs, becoming a member of the North Atlantic Treaty Organization (NATO). In 1973, Denmark joined the European Union. The Danes now enjoy high standards of living. Among the country's problems are the high costs of the extensive social welfare services and combating pollution.

ECONOMY

Denmark has few natural resources apart from some oil and gas from wells deep under the North Sea. But the economy is highly developed. Manufacturing industries, which employ about 16% of all workers, produce a wide variety of products, including furniture, processed food, machinery, television sets, and textiles.

Farms cover about three-quarters of the land. Farming employs only 4% of the workers, but it is highly scientific and productive. Meat and dairy farming are the chief activities. Fishing is also important, but service industries – including tourism – employ 76% of the work force.

DID YOU KNOW

- that Danes believe that storks bring good luck
- that Denmark is the world's leading exporter of insulin, which is used to treat diabetes
- that Legoland, which produces the world's best-selling construction toy, is in eastern Jutland, to the northwest of Vejle
- that Helsingør (or Elsinore) on the Danish island of Sjælland is the setting for Shakespeare's famous play *Hamlet, Prince of Denmark*

DANISH POSSESSIONS

Two self-governing territories are closely linked with Denmark. One is the world's largest island, **Greenland** (*see Atlantic Ocean, page 22*).

The other territory, the Faroe Islands, has its own parliament, but it also sends two representatives to the Danish parliament. The 18 islands cover 541 sq miles [1,400 sq km] between Iceland and the Shetland Islands. The islands are mostly made up of volcanic material and are home to many millions of seabirds. The population is 46,000. The capital is Tórshavn on Streymoy island, and the main activities of the people are sheep rearing and fishing.

DJIBOUTI

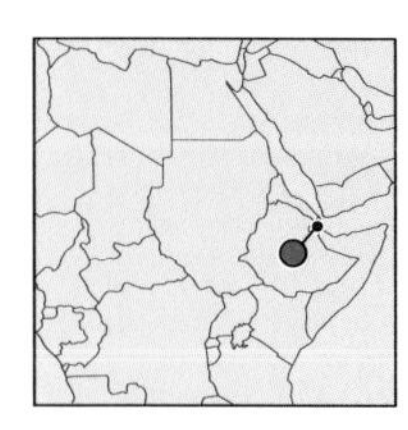

Djibouti's flag was adopted in 1977, when the country became independent from France. It was the flag used by the independence movement since 1972. The colors represent the country's two main peoples: the Issas (blue) and the Afars (green).

The Republic of Djibouti is a small country in eastern Africa. It occupies a strategic position where the Red Sea meets the Gulf of Aden, which leads into the Indian Ocean. Djibouti lies around an inlet of the Gulf of Aden, called the Gulf of Tadjoura. Behind the coastal plain on the northern side of the Gulf of Tadjoura is a highland region, the Mabla Mountains, rising to 5,850 ft [1,783 m] above sea level. Djibouti also contains Lake Assal, the lowest point on land in Africa.

AREA 8,958 sq miles [23,200 sq km]
POPULATION 461,000
CAPITAL (POPULATION) Djibouti (383,000)
GOVERNMENT Multiparty republic
ETHNIC GROUPS Somali (Issa) 60%, Afar 35%
LANGUAGES Arabic and French (both official)
RELIGIONS Islam 94%, Christianity 6%
CURRENCY Djibouti franc = 100 centimes

CLIMATE

Djibouti has one of the world's hottest and driest climates. Summer days are very hot with recorded temperatures of more than 100°F [42°C]. The sparse rainfall on the coast is most likely to occur in winter. On average, it rains on only 26 days every year. Inland, where the rainfall is slightly higher on highland areas, rain is more likely between April and October.

VEGETATION

A few areas of woodland occur in the mountains north of the Gulf of Tadjoura, where the average annual rainfall reaches about 20 inches [500 mm]. But nearly 90% of the land in Djibouti is desert, covered by some scrub vegetation or dry grassland. The shortage of pasture and surface water makes life difficult for the people, most of whom are nomads who wander around, searching for pasture.

HISTORY AND POLITICS

Islam was introduced into the area which is now Djibouti in the 9th century AD. The conversion of the Afars led to conflict between them and the Christian Ethiopians who lived in the interior. By the 19th century, the Issas, who are Somalis, had moved north and occupied much of the traditional grazing land of the Afars. France gained influence in the area in the second half of the 19th century and, in 1888, they set up a territory called French Somaliland. The capital of the territory, Djibouti, became important when the Ethiopian emperor, Menelik II, decided to build a railroad to it from Addis Ababa. This made Djibouti the main port handling Ethiopian trade, though Djibouti lost trade when Eritrea was federated with Ethiopia in 1952, giving Ethiopia another port, Assab, on the Red Sea.

In 1967, the people voted to retain their links with France, though most of the Issas were opposed and favored independence. The country was renamed the French Territory of the Afars and Issas, but it was named Djibouti when it became fully independent in 1977.

Djibouti became a one-party state in 1981, but a new constitution was introduced in 1992, permitting four parties which must maintain a balance between the ethnic groups in the country. Tensions flared up between the Afars and the Issas in 1992 and 1993, and government troops had to put down an Afar rebellion. A peace agreement was signed in 1994.

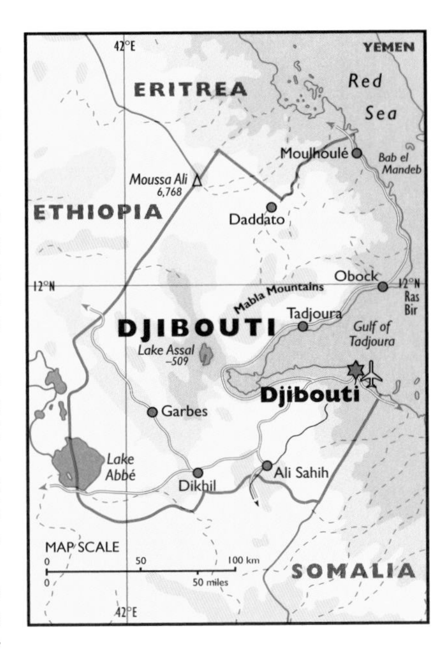

DID YOU KNOW

- that Djibouti was once known as French Somaliland and, for a time, the French Territory of the Afars and Issas
- that the Issas are a Somali clan; the Afars (or Danakils) live both in Djibouti and also in Ethiopia
- that a railroad links the port of Djibouti to Addis Ababa, capital of neighboring Ethiopia; much of Ethiopia's trade passes through Djibouti's harbor
- that Djibouti contains the lowest point on land in Africa – Lake Assal, whose shoreline is 509 ft [155 m] below sea level

ECONOMY

Djibouti is a poor country. Its economy is based mainly on money it gets for use of its port and the railroad that links it to Addis Ababa. Djibouti is a free trade zone and it also has an international airport. The country has no major resources and manufacturing is on a small scale – with some soft drinks and electrical goods produced. The only important activity is livestock raising (of goats, sheep, camels, and cattle) and half of the people are pastoral nomads. Most of the food the country needs has to be imported.

ECUADOR

Ecuador's flag was created by a patriot, Francisco de Miranda, in 1806. The armies of Simón Bolívar, the South American general whose armies won victories over Spain, flew this flag. At the center is Ecuador's coat of arms, showing a condor over Mount Chimborazo.

The Republic of Ecuador straddles the Equator on the west coast of South America. Three ranges of the high Andes Mountains form the backbone of the country. Between the towering, snow-capped peaks of the mountains, some of which are volcanoes, lie a series of high plateaux, or basins. Nearly half of Ecuador's population lives on these plateaux.

West of the Andes lie the flat coastal lowlands, which border the Pacific Ocean and average 60 miles [100 km] wide. The eastern lowlands, often called the Oriente, are drained by headwaters of the River Amazon.

AREA 109,483 sq miles [283,560 sq km]
POPULATION 13,184,000
CAPITAL (POPULATION) Quito (1,574,000)
GOVERNMENT Multiparty republic
ETHNIC GROUPS Mestizo (mixed White and Amerindian) 40%, Amerindian 40%, White 15%, Black 5%
LANGUAGES Spanish (official), Quechua
RELIGIONS Roman Catholic 92%
CURRENCY US dollar = 100 cents

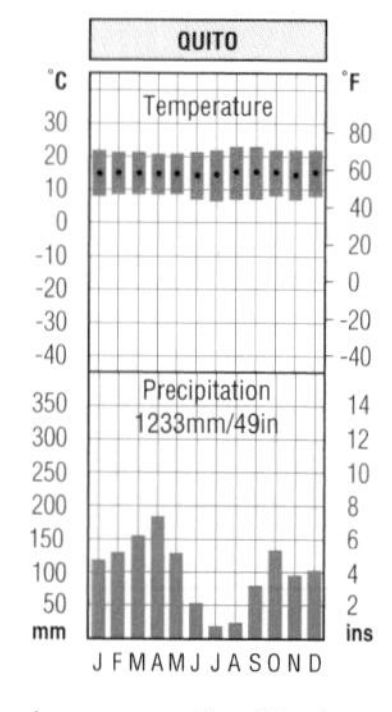

CLIMATE

The climate in Ecuador depends on the height above sea level. Though the coastline is cooled by the cold Peruvian Current, temperatures are between 73°F and 77°F [23°C to 25°C] all through the year. In Quito, at 8,200 ft [2,500 m] above sea level, temperatures are 57°F to 59°F [14°C to 15°C], though the city is just south of the Equator. The rainfall is low in the southwest. Other areas have abundant rainfall. The east is hot and wet.

VEGETATION

The vegetation in the Andes varies from snowfields at the highest levels to grassy meadows and forests lower down. The northern coastal lowlands contain large tropical forests, with deciduous woodland in the central coastal lowlands. Palm trees are common and the fiber from one species, the *Carludovica palmata*, is used to make Panama hats. Balsa trees grow in the Guayas valley north of Guayaquil. The southern coast, bordering Peru, is desert. Dense rain forests cover the eastern lowlands.

HISTORY AND POLITICS

The Inca people of Peru conquered much of what is now Ecuador in the late 15th century. They introduced their language, Quechua, which is widely spoken today. Spanish forces defeated the Incas in 1533 and took control of Ecuador. The country became independent in 1822, following the defeat of a Spanish force in a battle near Quito. Ecuador became part of Gran Colombia, a confederation which also included Colombia and Venezuela. But Ecuador became a separate nation in 1830.

In the 19th and 20th centuries, Ecuador suffered from political instability, while successive governments failed to tackle social and economic problems. A war with Peru in 1941 and in 1995 led to a loss of territory, but a border agreement was signed in 1998. Civilian governments have ruled Ecuador since 1979, but not without occasional crises.

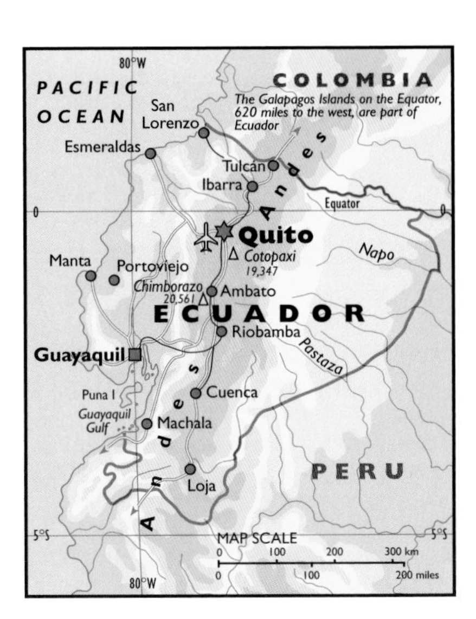

DID YOU KNOW

- that Ecuador is the Spanish name for the Equator, which runs through the country
- that Ecuador is the world's leading producer of balsa wood and the fourth largest producer of bananas
- that the Galapagos Islands, 620 miles [1,000 km] off the coast, belong to Ecuador; these islands contain unusual animals and plants which were studied by Charles Darwin
- that Panama hats are made in Ecuador, not Panama; they are called Panama hats because Panama was a major exporter of hats in the 19th century
- that Ecuador contains more than 30 active volcanoes, including Chimborazo, Ecuador's highest peak, and Cotopaxi, one of the world's highest active volcanoes

ECONOMY

The World Bank classifies Ecuador as a "lower-middle-income" developing country. Agriculture employs 15% of the people and bananas, cocoa, and coffee are all important export crops. Other products in the hot coastal lowlands include citrus fruits, rice, and sugarcane, while beans, maize, and wheat are important in the highlands. Cattle are raised for dairy products and meat, while fishing is important in the coastal waters. Seafood, including shrimps, is exported. Forestry is a major activity, and Ecuador produces balsa wood and such hardwoods as mahogany. Mining has become increasingly important and oil and oil products are Ecuador's leading exports. Manufactures include cement, Panama hats, paper products, processed food, and textiles.

EGYPT

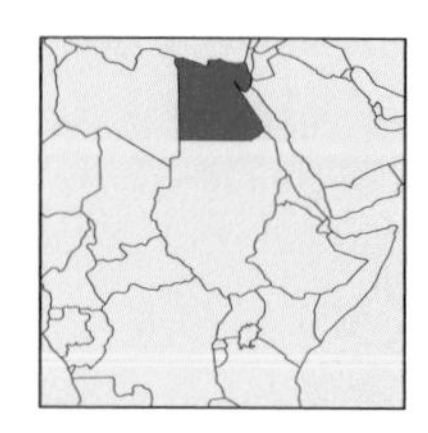

A flag consisting of three bands of red, white, and black, the colors of the Pan-Arab movement, was adopted in 1958. The present design, adopted in 1958, has a gold eagle in the center. This symbolizes Saladin, the warrior who led the Arabs in the 12th century.

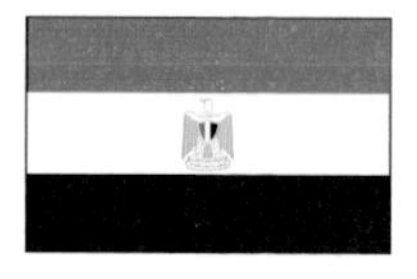

The Arab Republic of Egypt is Africa's second largest country by population after Nigeria, though it ranks 13th in area. Most of Egypt is desert. Almost all the people live either in the Nile Valley and its fertile delta or along the Suez Canal, the artificial waterway between the Mediterranean and Red seas. This canal shortens the sea journey between the United Kingdom and India by 6,027 miles [9,700 km].

Apart from the Nile Valley, Egypt has three other main regions. The Western and Eastern deserts are parts of the Sahara. The Sinai peninsula (Es Sina), to the east of the Suez Canal, is very mountainous and contains Egypt's highest peak, Gebel Katherina (8,652 ft [2,637 m]); few people live in this area.

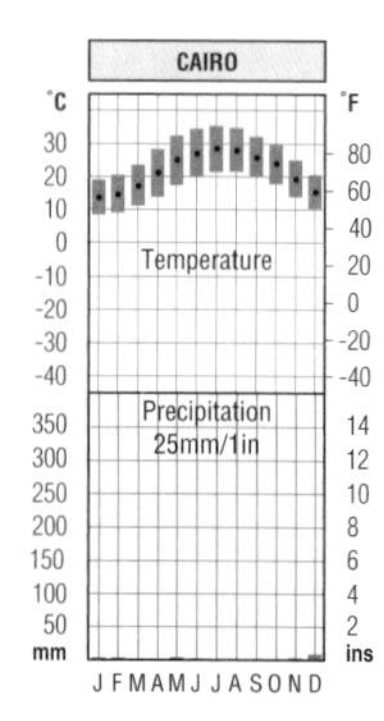

CLIMATE

Egypt is a dry country. The low rainfall occurs, if at all, in winter. This country is one of the sunniest places on Earth. Winters are mild but summers are hot. Conditions become unpleasant when hot and dusty winds blow from the deserts into the Nile Valley.

AREA 386,660 sq miles [1,001,450 sq km]
POPULATION 69,537,000
CAPITAL (POPULATION) Cairo (or El Qahira, 6,800,000)
GOVERNMENT Republic
ETHNIC GROUPS Egyptian 99%
LANGUAGES Arabic (official), French, English
RELIGIONS Islam (Sunni Muslim 94%), Christianity (mainly Coptic Christian 6%)
CURRENCY Pound = 100 piastres

VEGETATION

The Nile Valley forms a long, green ribbon of farmland, dotted with palm trees. But dry landscapes, with sand dunes, plains of loose gravel, and areas of bare rock, cover 90% of the country. The largest of these areas is the Western Desert, which covers about three-quarters of Egypt; within the desert are oases which provide water.

HISTORY

Ancient Egypt, which was founded about 5,000 years ago, was one of the great early civilizations. Throughout the country, pyramids, temples, and richly decorated tombs are memorials to its great achievements.

After Ancient Egypt declined, the country came under successive foreign rulers. Arabs occupied Egypt in AD 639–42. They introduced the Arabic language and Islam. Their influence was so great that most Egyptians now regard themselves as Arabs.

Egypt came under British rule in 1882, but it gained partial independence in 1922, becoming a monarchy. The monarchy was abolished in 1952, when Egypt became a republic. The creation of Israel in 1948 led Egypt into a series of wars in 1948–9, 1956, 1967, and 1973.

POLITICS

Since the late 1970s, Egypt has sought for peace. In 1979, Egypt signed a peace treaty with Israel and regained the Sinai region which it had lost in a war in 1967. Extremists opposed contacts with Israel and, in 1981, President Sadat, who had signed the treaty, was assassinated.

Egypt has also been concerned in international diplomacy. During the 1991 Gulf War, Egyptian troops were members of the allied force which drove Iraqi troops out of Kuwait, which Iraq had invaded in 1990.

> **DID YOU KNOW**
> - that the ancient Greek historian Herodotus called Egypt "the gift of the Nile"
> - that the peasants of Egypt, called *fellahin*, wear long, shirt-like garments called *galabiyahs*
> - that *fool*, or *ful*, which is made from beans, is the national dish of Egypt
> - that the Egyptian writer Naguib Mafouz was the first Arabic-language writer to win the Nobel Prize for Literature

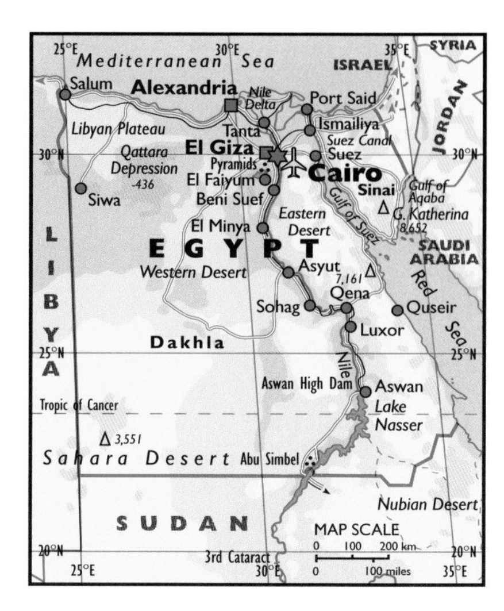

Most people in Egypt are poor. Some people dislike Western influences on their way of life and favor a return to the fundamental principles of Islam. During the 1990s, attacks on foreign visitors by Islamic terrorists led to a decline in tourism. Tourism was again badly hit by the terrorist attacks on the USA on September 11, 2001.

ECONOMY

Egypt is Africa's second most industrialized country after South Africa, but it remains a developing country. The majority of the people are poor. Farming employs 29% of the workers. Most *fellahin* (peasants) grow such food crops as beans, maize, rice, sugarcane, and wheat, but the main cash crop is cotton. Textiles form the second most valuable export after oil.

Oil power stations produce much electricity, but electricity is also generated at the Aswan High Dam. Egypt has important manufacturing industries in the northern cities.

EL SALVADOR

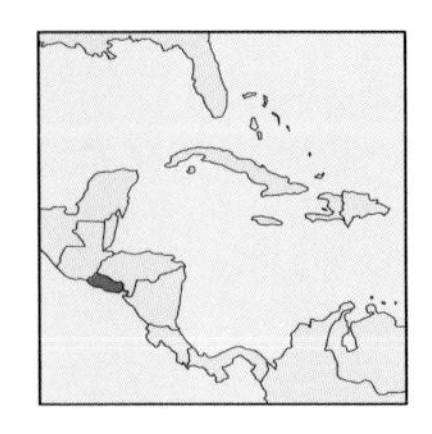

This flag was adopted in 1912, replacing the earlier "Stars and Stripes." The blue and white stripes are featured on the flags of several Central American countries which gained their independence from Spain at the same time in 1821.

The Republic of El Salvador is the only country in Central America which does not have a coast on the Caribbean Sea. El Salvador has a narrow coastal plain along the Pacific Ocean. Behind the coastal plain, the coastal range is a zone of rugged mountains, including volcanoes, which overlooks a thickly populated inland plateau. The lower slopes of the coastal range and the plateau are El Salvador's most thickly populated region. Beyond the plateau, the land rises to the thinly populated interior highlands.

AREA 8,124 sq miles [21,040 sq km]
POPULATION 6,238,000
CAPITAL (POPULATION) San Salvador (1,522,000)
GOVERNMENT Republic
ETHNIC GROUPS Mestizo (mixed White and Amerindian) 89%, White 10%, Amerindian 1%
LANGUAGES Spanish (official)
RELIGIONS Roman Catholic 86%
CURRENCY Colón = 100 centavos; US dollar = 100 cents

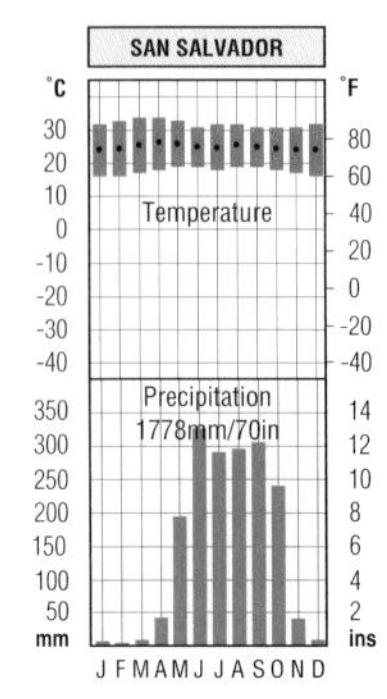

CLIMATE

The coast has a hot, tropical climate. Inland, the climate is moderated by the altitude. San Salvador in the central region has similar temperatures by day, but nights are cooler. Rain falls on practically every afternoon between May and October.

VEGETATION

Grasslands and patches of the original oak and pine forests grow on the interior highlands. The central plateau and valleys have areas of grass and deciduous woodland. Low-lying regions contain savanna (tropical grassland with scattered trees) or forest. The balsam tree, whose scented resin is used to make antiseptics, and cedar and mahogany are among the valuable trees in El Salvador.

HISTORY

Amerindians have lived in El Salvador for thousands of years. The ruins of Mayan pyramids built between AD 100 and 1000 are still found in the western part of the country. Spanish soldiers conquered the area in 1524 and 1525, and Spain ruled until 1821. In 1823, all the Central American countries, except for Panama, set up a Central American Federation. But El Salvador withdrew in 1840 and declared its independence in 1841.

El Salvador suffered from instability throughout the 19th century. The 20th century saw more stable government, but from 1931 military dictatorships alternated with elected governments. In 1969, El Salvador and Honduras fought in the brief "Soccer War." This war was sparked off by the ill-treatment of fans during a World Cup soccer series between the countries. But the real cause of the war was the alleged mistreatment of El Salvadorean immigrants in Honduras.

POLITICS

In the 1970s, El Salvador was plagued by conflict as protesters demanded that the government introduce reforms to help the poor. Kidnappings and murders committed by left- and right-wing groups caused instability.

A civil war broke out in 1979 between the US-backed, right-wing government forces and left-wing guerrillas in the FMLN (Farabundo Marti National Liberation Front). In the 12 years that followed, more than 75,000 people were killed and hundreds of thousands were made homeless. A ceasefire came into effect in 1992, and presidential elections were held in 1993 and 1999.

DID YOU KNOW

- that El Salvador has more than 20 volcanoes; some are still active
- that El Salvador is the most densely populated country in Central America
- that earthquakes are common in El Salvador; an earthquake destroyed San Salvador in 1854; another quake in 2001 made about one-sixth of the population homeless
- that San Salvador was named after the Roman Catholic festival of *San Salvador del Mundo* (Holy Savior of the World); the country was later named after the city

ECONOMY

According to the World Bank, El Salvador is a "lower-middle-income" developing country. Farmland and pasture cover about three-quarters of the country. Coffee, grown in the highlands, is the main export, followed by sugar and cotton, which grow on the coastal lowlands. Fishing for lobsters and shrimps is important, but manufacturing is on a small scale.

Francisco Antonio Gavidia, the El Salvadorean philosopher and humanist, was celebrated by a set of six stamps issued in 1965.

EQUATORIAL GUINEA

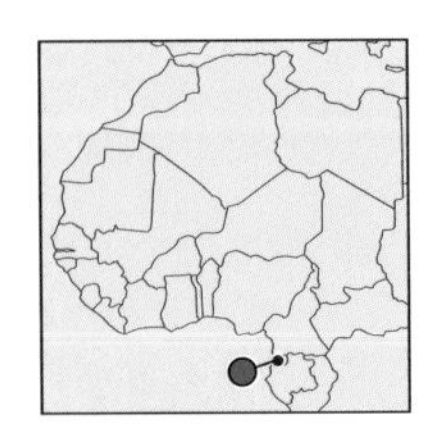

Equatorial Guinea's flag was adopted in 1968, when the country became independent from Spain. The green represents the country's natural resources, the blue represents the sea, the red symbolizes the struggle for independence, and the white stands for peace.

AREA 10,830 sq miles [28,050 sq km]
POPULATION 486,000
CAPITAL (POPULATION) Malabo (35,000)
GOVERNMENT Multiparty republic (transitional)
ETHNIC GROUPS Fang 83%, Bubi 10%, Ndowe 4%
LANGUAGES Spanish and French (both official)
RELIGIONS Roman Catholic 89%
CURRENCY CFA franc = 100 centimes

The Republic of Equatorial Guinea is a small republic in west-central Africa. It consists of a mainland territory which makes up 90% of the land area, called Mbini (or Rio Muni), between Cameroon and Gabon, and five offshore islands in the Bight of Bonny, the largest of which is Bioko. The island of Annobon lies 350 miles [560 km] southwest of Mbini.

Mbini consists mainly of hills and plateaux behind the coastal plains. Its main river, the Lolo (formerly Mbini), rises in Gabon and flows across Mbini. Bioko is a volcanic island with fertile soils. Malabo's harbor is part of a submerged volcano.

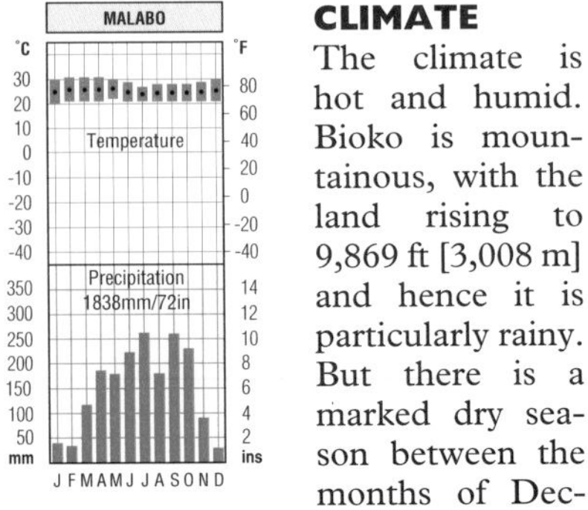

CLIMATE

The climate is hot and humid. Bioko is mountainous, with the land rising to 9,869 ft [3,008 m] and hence it is particularly rainy. But there is a marked dry season between the months of December and February. Mainland Mbini has a similar climate, though the rainfall diminishes inland.

VEGETATION

Dense rain forests cover most of Equatorial Guinea. The forests contain valuable trees, including mahogany, okoumé, and African walnut. Mangrove forests border much of the coast. Bioko is more varied, with grasslands at higher levels.

HISTORY

Portuguese navigators reached the area in 1471. In 1778, Portugal granted Bioko, together with rights over Mbini, to Spain. In return, Spain agreed to allow Portugal to advance the western frontiers of Brazil beyond longitude 50°W. But the Spaniards who settled on Bioko were hit by yellow fever and they withdrew from the island in 1781. No one settled on mainland Mbini.

In 1827, Spain leased bases on Bioko to Britain, which was working to suppress the slave trade. The British settled some freed slaves on Bioko. Their descendants are called Fernandinos. The Spanish returned to the area in the mid-19th century. Toward the end of the century, Spain began to develop plantations on the fertile island of Bioko.

In 1959, Spain made Bioko and Mbini provinces of overseas Spain and, in 1963, it gave the provinces a degree of self-government. Equatorial Guinea became independent in 1968.

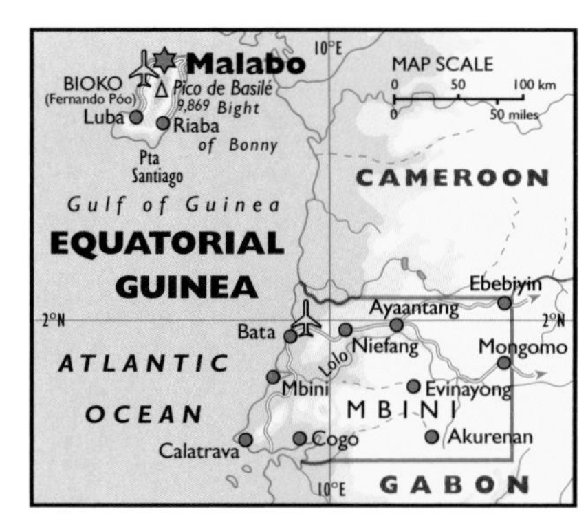

POLITICS

The first president of Equatorial Guinea, Francisco Macias Nguema, proved to be a tyrant. He was overthrown in 1979 and a group of officers, led by Lt.-Col. Teodoro Obiang Nguema Mbasogo, set up a Supreme Military Council to rule the country. In 1991, people voted for a multiparty democracy. Elections were held in the 1990s and early 2000s. But Teodoro Obiang Nguema Mbasogo continued to rule in a semidictatorial manner.

This stamp, one of a series on wildlife protection, shows the endangered bare-headed rock fowl.

ECONOMY

Equatorial Guinea is a poor country. Agriculture employs around 58% of the people. Food crops include sweet potatoes, bananas, and cassava, but the chief cash crop is cocoa. Forestry is important and oil is produced off Bioko. The main exports are timber, oil, and cocoa. Industry is small scale.

DID YOU KNOW

- that the name Guinea comes from an ancient African kingdom
- that the name Guinea was once used for the entire coast of West Africa, from Senegal to Angola; it now survives in three country names – Equatorial Guinea, Guinea, and Guinea-Bissau
- that the island of Bioko was called Fernando Póo in colonial times; it was named after a Portuguese explorer, Ferñao de Po, who discovered the island in 1471–2

ESTONIA

Estonia's flag was used between 1918 and 1940, when the country was an independent republic. It was readopted in June 1988. The blue is said to symbolize the sky, the black Estonia's black soil, and the white the snows that blanket the land in winter.

AREA 17,300 sq miles [44,700 sq km]
POPULATION 1,423,000
CAPITAL (POPULATION) Tallinn (435,000)
GOVERNMENT Multiparty republic
ETHNIC GROUPS Estonian 65%, Russian 28%, Ukrainian 3%, Belarussian 2%, Finnish 1%
LANGUAGES Estonian (official), Russian
RELIGIONS Lutheran, Russian and Estonian Orthodox, Methodist, Baptist
CURRENCY Kroon = 100 sents

The Republic of Estonia is the smallest of the three states on the Baltic Sea, which were formerly part of the Soviet Union, but which became independent in the early 1990s. Estonia consists of a generally flat plain, rising to a maximum height of only 1,043 ft [318 m] at Munamagi, in the southeast. The area was covered by ice sheets during the Ice Age and the land is strewn with moraine (rocks deposited by the ice).

The country is dotted with more than 1,500 small lakes, and water, including the large Lake Peipus (Chudskoye Ozero) and the River Narva, makes up much of Estonia's eastern border with Russia. Estonia has more than 800 islands, which together make up about a tenth of the country. The largest island is Saaremaa (Sarema).

CLIMATE

Despite its northerly position, Estonia has a fairly mild climate because of its nearness to the sea. This is because sea winds tend to warm the land in winter and cool it in summer. Average temperatures fall below freezing point in winter, but the summers are warm. The average annual rainfall varies between about 19 and 23 inches [480 mm to 580 mm].

VEGETATION

Farmland and pasture cover more than one-third of the land. Swamps and lakes cover about 15% and forests about 31%. Common trees include aspen, birch, fir, and pine.

HISTORY

The ancestors of the Estonians, who are related to the Finns, settled in the area several thousand years ago. German crusaders, known as the Teutonic Knights, introduced Christianity in the early 13th century. By the 16th century, German noblemen owned much of the land in Estonia.

In 1561, Sweden took the northern

part of the country and Poland the south. From 1625, Sweden controlled the entire country until Sweden handed it over to Russia in 1721.

Estonian nationalists campaigned for their independence from around the mid-19th century. Finally, Estonia was proclaimed independent in 1918. In 1919, the government began to break up the large estates and distribute land among the peasants.

In 1939, Germany and the Soviet Union agreed to take over parts of Eastern Europe. In 1940, Soviet forces occupied Estonia, but they were driven out by the Germans in 1941. Soviet troops returned in 1944 and Estonia became one of the 15 Soviet Socialist Republics of the Soviet Union. The Estonians strongly opposed Soviet rule. Many of them were deported to Siberia.

Political changes in the Soviet Union in the late 1980s led to renewed demands for freedom. In 1990, the Estonian government declared the

country independent and, finally, the Soviet Union recognized this act in September 1991, shortly before the Soviet Union was dissolved. Estonia adopted a new constitution in 1992, when multiparty elections were held for a new national assembly. In 2002, Estonia was invited to join the European Union in May 2004.

DID YOU KNOW

- that the country's official name in Estonian is *Eesti Vabariik*, meaning Republic of Estonia
- that the people of Estonia were not converted to Christianity until the 13th century
- that music plays an important part in Estonian culture; a major song festival, held in Tallinn every five years, attracts many singers and thousands of visitors
- that the Estonian language is not an Indo-European language; it belongs instead to the Finno-Ugric family, together with Finnish and Hungarian

ECONOMY

Under Soviet rule, Estonia was the most prosperous of the three Baltic states. It has deposits of oil shale, which is used to fuel power plants and is also used in the petrochemical industry. Estonia's forests are its other main resource. Other manufactures include fertilizers, machinery, processed food, and textiles.

Agriculture and fishing are also important. Dairy farming and pig rearing are valuable activities, while barley, potatoes, and oats are major crops, since they are suited to the cool climate and good soils. Since 1988, Estonia has begun to change its government-dominated economy to one based on private enterprise, and the country has started to strengthen its links with the rest of Europe.

ETHIOPIA

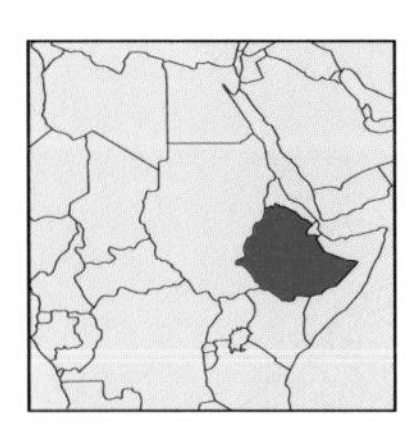

The tricolor flag of Ethiopia was first flown as three separate pennants.The red, yellow, and green combination dates from the late 19th century. It appeared in flag form in 1897. The central pentangle was introduced in 1996, and represents the common will of the country's 68 ethnic groups.

Ethiopia is a landlocked country in northeastern Africa. The land is mainly mountainous, though there are extensive plains in the east, bordering southern Eritrea, and in the south, bordering Somalia. The highlands are divided into two blocks by an arm of the Great Rift Valley, which runs throughout eastern Africa. North of the Rift Valley, the land is especially rugged, rising to 15,157 ft [4,620 m] at Ras Dashen. Southeast of Ras Dashen is Lake Tana, source of the River Abay (Blue Nile).

AREA 435,521 sq miles [1,128,000 sq km]

POPULATION 65,892,000

CAPITAL (POPULATION) Addis Ababa 2,316,000)

GOVERNMENT Federation of nine provinces

ETHNIC GROUPS Oromo 40%, Amharic 32%, Sidamo 9%, Shankella 6%, Somali 6%

LANGUAGES Amharic (official), 280 others

RELIGIONS Islam 47%, Eastern Orthodox 40%, traditional beliefs 11%

CURRENCY Birr = 100 cents

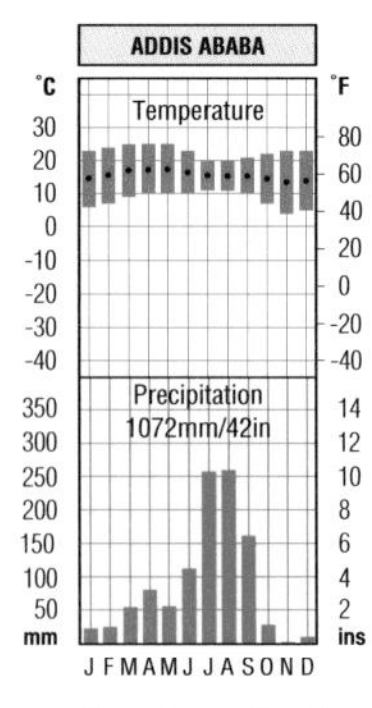

CLIMATE

The climate in Ethiopia is greatly affected by the altitude. Addis Ababa, at 8,000 ft [2,450 m], has an average yearly temperature of 68°F [20°C]. The rainfall is generally more than 39 inches [1,000 mm]. But the lowlands, including the Eritrean coast, are hot and dry.

VEGETATION

Grassland, farmland, and scattered trees cover much of the highlands. Semidesert and savanna (tropical grassland with such trees as acacias) cover parts of the lowlands. Dense rain forests grow in the southwest of the country.

HISTORY AND POLITICS

Ethiopia was the home of an ancient monarchy, which became Christian in the 4th century. In the 7th century, Muslims gained control of the lowlands, but Christianity survived in the highlands. In the 19th century, Ethiopia resisted attempts to colonize it. Italy invaded Ethiopia in 1935, but Ethiopian and British troops defeated the Italians in 1941.

In 1952, Eritrea, on the Red Sea coast, was federated with Ethiopia. But in 1961, Eritrean nationalists demanded their freedom and began

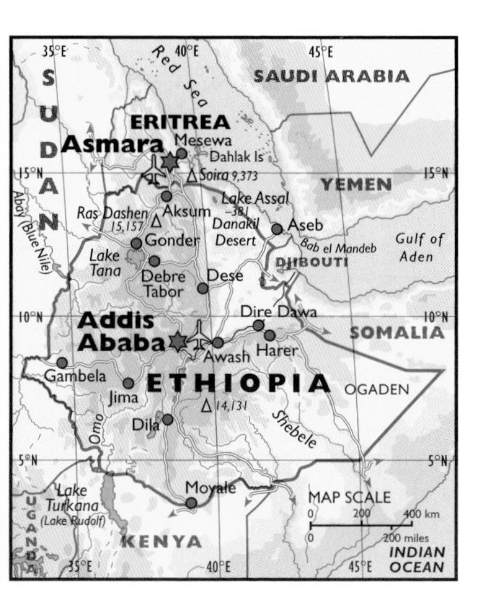

a struggle that ended in their independence in 1993. Relations with Eritrea gradually soured and border clashes occurred in 1998 and 1999.

The monarchy was abolished in 1974 and a socialist military group took control. Civil war broke out and the military regime collapsed in 1991. Border conflict between Ethiopia and Eritrea broke out in 1998. Peace was restored in 2000.

ECONOMY

Ethiopia is one of the world's poorest countries, particularly in the 1970s and 1980s when it was plagued by civil war and famine caused partly by long droughts. Many richer countries have sent aid (money and food) to help the Ethiopian people.

Agriculture is the main activity. Various grains (such as maize, barley, and wheat) are grown for food, but coffee is the main commercial crop and export. Many of the exports, including hides and skins, and oilseeds, are shipped abroad through the port of Djibouti.

ERITREA

Eritrea, which was an Italian colony from the 1880s, was part of Ethiopia from 1952 until 1993, when it became a fully independent nation. Its flag, hoisted in May 1993, is a variation on the flag of the Eritrean People's Liberation Front, which led the struggle against Ethiopian forces. Eritrea contains a hot, dry coastal plain facing the Red Sea, with highlands in the center, where most people live. Farming and nomadic livestock rearing are the main activities in this poor, war-ravaged territory. Eritrea has a few manufacturing industries, based mainly in Asmara.

AREA 36,293 sq miles [94,000 sq km]
POPULATION 4,298,000
CAPITAL (POPULATION) Asmara (367,500)

The camel is a beast of burden in the dry lands of Eritrea. This stamp was issued in 1933 by the Italian authorities who controlled Eritrea at that time.

FINLAND

The flag of Finland was adopted in 1918, after the country had become an independent republic in 1917, following a century of Russian rule. The blue represents Finland's many lakes. The white symbolizes the blanket of snow which masks the land in winter.

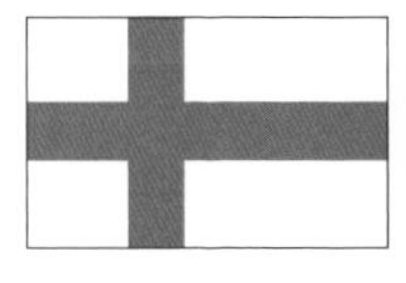

The Republic of Finland is a beautiful country in northern Europe. Part of the country lies north of the Arctic Circle, in the "Land of the Midnight Sun." Here the sun shines for 24 hours a day for periods of time in summer, especially in June.

In the south, behind the coastal lowlands where most Finns live, lies the Lake District, a region of sparkling lakes worn out by ice sheets in the Ice Age. The thinly populated northern uplands cover about two-fifths of the country.

HELSINKI
Temperature
Precipitation 688mm/27in
J F M A M J J A S O N D

CLIMATE

Helsinki, the capital city, has warm summers, but the average temperatures between the months of December and March are below freezing point. Snow covers the land in winter. The north has less precipitation (rain and snow) than the south, but it is much colder.

VEGETATION

Forests of birch, pine, and spruce cover much of Finland. These trees can survive the cold winters when most of the country's lakes freeze over. The vegetation becomes more and more sparse to the north. The far north is tundra, a treeless plain, where mosses, lichens, and some flowering plants grow in summer.

AREA 130,552 sq miles [338,130 sq km]

POPULATION 5,176,000

CAPITAL (POPULATION) Helsinki (532,000)

GOVERNMENT Multiparty republic

ETHNIC GROUPS Finnish 93%, Swedish 6%

LANGUAGES Finnish and Swedish (both official); small Lapp- and Russian-speaking minorities

RELIGIONS Evangelical Lutheran 89%, Russian Orthodox 1%

CURRENCY Euro = 100 cents

HISTORY

Between 1150 and 1809, Finland was under Swedish rule. The close links between the countries continue today. Swedish remains an official language in Finland, and many towns have Swedish as well as Finnish names.

In 1809, Finland became a grand duchy of the Russian empire. It finally declared itself independent in 1917, after the Russian Revolution and the collapse of the Russian empire. But during World War II (1939–45), the Soviet Union declared war on Finland and took part of Finland's territory.

Finland allied itself with Germany, but it lost more land to the Soviet Union at the end of the war.

POLITICS

After World War II, Finland became a neutral country and negotiated peace treaties with the Soviet Union. Finland also strengthened its relations with other northern European countries and became an associate member of the European Free Trade Association (EFTA) in 1961. Finland became a full member of EFTA in 1986, and in 1992 it applied for membership of the European Union, which it finally achieved on January 1, 1995. On January 1, 1999, Finland adopted the euro, the single European currency.

ECONOMY

Forests are Finland's most valuable resource, and forestry accounts for about 28% of the country's exports. The chief manufactures are wood products, pulp, and paper. Since World War II, Finland has set up many other industries, producing such things as machinery and transport equipment. Its economy has expanded rapidly, but there has been a large increase in the number of unemployed people.

Farmland covers less than a tenth of the land and farming employs only 5% of the people. Livestock and dairy farming are the chief activities. Crops include barley, oats, potatoes, and sugar beet.

DID YOU KNOW

- that Finland has about 60,000 lakes; they cover about a tenth of the country
- that Finns drink more coffee per person than any other people in the world
- that sauna baths are a feature of Finnish life. This tradition goes back more than 1,000 years
- that about 6,000 Sami (also called Lapps) live in northern Finland. They were once nomads who followed the migration of herds of reindeer, but most of them now live in permanent homes

FRANCE

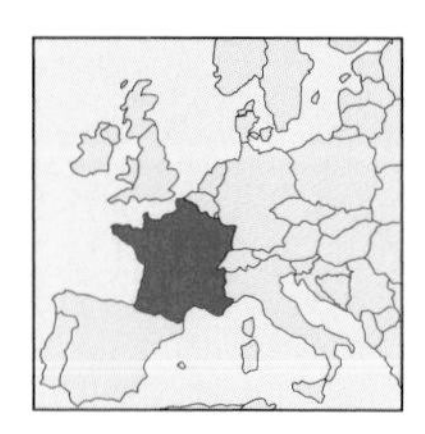

The colors of this flag originated during the French Revolution of 1789. The red and blue are said to represent Paris, while the white represented the monarchy. The present design was adopted in 1794. It is meant to symbolize republican principles.

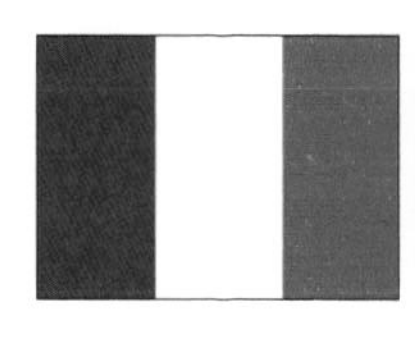

The Republic of France is the largest country in Western Europe. The scenery is extremely varied. The Vosges Mountains overlook the Rhine valley in the northeast, the Jura Mountains and the Alps form the borders with Switzerland and Italy in the southeast, while the Pyrenees straddle France's border with Spain. The only large highland area entirely within France is the Massif Central in southern France.

Brittany (Bretagne) and Normandy (Normande) form a scenic hill region. Fertile lowlands cover most of northern France, including the densely populated Paris basin. Another major lowland area, the Aquitanian basin, is in the southwest, while the Rhône-Saône valley and the Mediterranean lowlands are in the southeast.

AREA 212,934 sq miles [551,500 sq km]

POPULATION 59,551,000

CAPITAL (POPULATION) Paris (11,175,000)

GOVERNMENT Multiparty republic

ETHNIC GROUPS Celtic, Latin, Arab, Teutonic, Slavic

LANGUAGES French (official), Breton, Occitan

RELIGIONS Roman Catholic 90%, Islam 3%

CURRENCY Euro = 100 cents

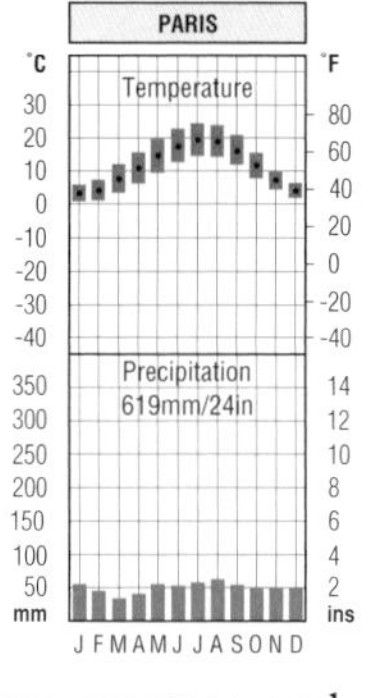

CLIMATE

The climate of France varies from west to east and from north to south. The west comes under the moderating influence of the Atlantic Ocean, giving generally mild weather. To the east, summers are warmer and winters colder. The climate also becomes warmer as one travels from north to south. The Mediterranean Sea coast has hot, dry summers and mild, moist winters. The Alps, Jura, and Pyrenees mountains have snowy winters. Winter sports centers are found in all three areas. Large glaciers occupy high valleys in the Alps.

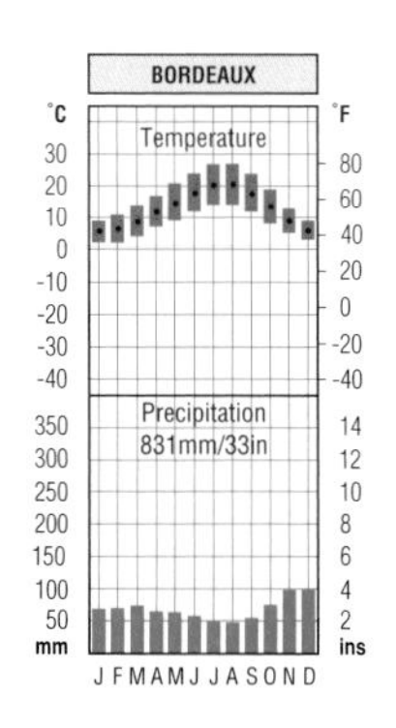

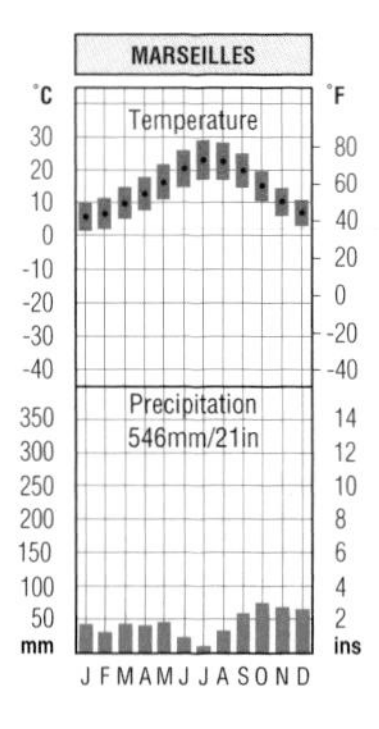

VEGETATION

A patchwork of fields and grassy meadows covers about three-fifths of the land in France. Forests occupy about 27%, with beech and oak being especially common in the north, while birch, pine, and poplar are found in the center, and olive trees and thorny plants in the Mediterranean south.

HISTORY

The Romans conquered France (then called Gaul) in the 50s BC. Roman rule began to decline in the fifth century AD and, in 486, the Frankish realm (as France was called) became independent under a Christian king, Clovis. In 800, Charlemagne, who had been king since 768, became emperor of the Romans. He extended France's boundaries, but, in 843, his empire was divided into three parts and the area of France contracted. After the Norman invasion of England in 1066, large areas of France came under English rule, but English rule was finally ended in 1453.

France later became a powerful monarchy. But the French Revolution

France has many historic towns, such as the fortified town of St-Emilion, in the lower Dordogne valley, east of Bordeaux. The famous red wine called St-Emilion is produced in nearby vineyards.

DID YOU KNOW

- that Bastille Day on July 14, France's national holiday, commemorates the capture of the Bastille prison during the French Revolution in 1789
- that the world's fastest passenger train runs between Paris and Lyons; the journey takes two hours and the average speed is 132 mph [212.5 km/h]
- that France is the world's second largest wine producer after Italy
- that nuclear power plants produce more than half of France's electricity
- that France is among the world's five top producers of apples, barley, butter, cars, cheese, maize, sugar beet, sunflower seed, and wheat

(1789–99) ended absolute rule by French kings. In 1799, Napoleon Bonaparte took power and fought a series of brilliant military campaigns before his final defeat in 1815. The monarchy was restored until 1848, when the Second Republic was founded. In 1852, Napoleon's nephew became Napoleon III, but the Third Republic was established in 1875. France was the scene of much fighting during World War I (1914–18) and World War II (1939–45), causing great loss of life and much damage to the economy.

In 1946, France adopted a new constitution, establishing the Fourth Republic. But political instability and costly colonial wars slowed France's postwar recovery. In 1958, Charles de Gaulle was elected president and he introduced a new constitution, giving the president extra powers and inaugurating the Fifth Republic.

FRANCE

POLITICS

Since the 1960s, France has made rapid economic progress, becoming one of the most prosperous nations in the European Union. But France's government faced a number of problems, including unemployment, pollution, and the growing number of elderly people, who find it difficult to live when inflation rates are high. One social problem concerns the presence in France of large numbers of immigrants from Africa and southern Europe, many of whom live in poor areas.

ECONOMY

France is one of the world's most developed countries. Its natural resources include its fertile soil, together with deposits of bauxite, coal, iron ore, oil and natural gas, and potash. France is one of the world's top manufacturing nations. Paris is a world center of fashion industries, but France has many other industrial towns and cities. Major manufactures include aircraft, cars, chemicals, electronic products, machinery, metal products, processed food, steel, and textiles.

Agriculture employs about 4% of the people, but France is the largest producer of farm products in Western Europe, producing most of the food it needs. Wheat is the leading crop and livestock farming is of major importance. Fishing and forestry are leading industries, while tourism is a major activity.

MONACO

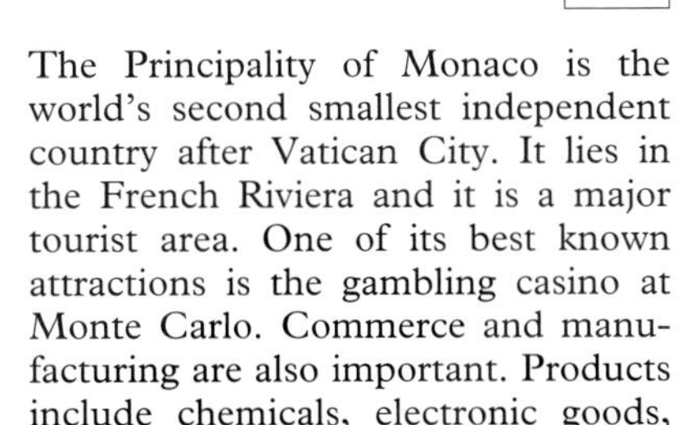

The Principality of Monaco is the world's second smallest independent country after Vatican City. It lies in the French Riviera and it is a major tourist area. One of its best known attractions is the gambling casino at Monte Carlo. Commerce and manufacturing are also important. Products include chemicals, electronic goods, paper, and plastics.

AREA 0.6 sq miles [1.5 sq km]
POPULATION 32,000
CAPITAL (POPULATION) Monaco (32,000)
GOVERNMENT Constitutional monarchy under French protection
ETHNIC GROUPS French 47%, Monégasque 16%, Italian 16%, other 21%
LANGUAGES French (official), English, Italian, Monégasque
RELIGIONS Roman Catholic 90%
CURRENCY Euro = 100 cents

Monaco has been independent since AD 980 and it has been ruled by the Grimaldi family since 1297. Its flag, adopted in 1881, uses colors from the Prince of Monaco's coat of arms, which dates back to medieval times.

Among Monaco's tourist attractions are its sporting events. The top stamp, illustrating a 1931 Bugatti racing car, was issued in 1967 as one of a set of 14 stamps commemorating the 25th Monaco Grand Prix. The lower stamp, issued in 1964, commemorated the 50th anniversary of the first Monte Carlo Aerial Rally.

FRENCH GUIANA

The French tricolor is the flag used by French Guiana, or Guyane as it is called in the French language. This is because the territory is an overseas department of France and is treated as part of mainland France. Its citizens send representatives to the French parliament.

French Guiana is the smallest country on the mainland of South America. The coastal plain is swampy in places, though there are also dry areas which are cultivated. Inland lies a low plateau, with the low Tumachumac Mountains in the south. This range reaches a height of 2,264 ft [690 m] in the southwestern corner of French Guiana. Most of the rivers, including the Maroni, which forms part of the western border, and the Oyapock, which flows along the eastern border with Brazil, run northward to the Atlantic Ocean.

AREA 34,749 sq miles [90,000 sq km]
POPULATION 178,000
CAPITAL (POPULATION) Cayenne (42,000)
GOVERNMENT Overseas department of France
ETHNIC GROUPS Mulatto 66%, Chinese and Amerindian 12%, White 10%
LANGUAGES French (official)
RELIGIONS Roman Catholic 80%, Protestant 4%
CURRENCY Euro = 100 cents

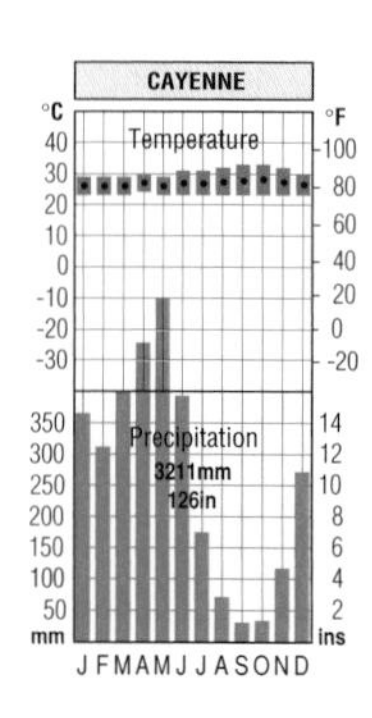

CLIMATE

French Guiana has a hot, equatorial climate, with high temperatures throughout the year. The rainfall is heavy, especially between December and June, but it is dry between August and October. The northeast trade winds blow constantly across the country.

VEGETATION

Rain forests containing many valuable hardwood species cover nearly 90% of the land. Mangrove swamps line parts of the coast, while other areas are covered by savanna (tropical grassland with scattered trees).

HISTORY AND POLITICS

The first people to live in what is now French Guiana were Amerindians. Today, only a few of them survive in the interior. The first Europeans to explore the coast arrived in 1500, and they were followed by adventurers seeking El Dorado, the mythical city of gold. The French were the first people to settle in the area in 1604, and Cayenne was founded in 1637 by a group of French merchants. The first attempt at colonization failed, and the area changed hands several times before it finally became a French colony in the late 17th century.

The colony, whose plantation economy depended on African slaves, remained French except for a brief period in the early 19th century. Slavery was abolished in 1848 and Asian laborers were introduced to work the land. France used the colony as a penal settlement for political prisoners from the times of the French Revolution in the 1790s. From the 1850s to 1945, the country became notorious as a place where prisoners were harshly treated. Many of them died, unable to survive in the tropical conditions.

In 1946, French Guiana became an overseas department of France, and in 1974 it also became an administrative region. An independence movement developed in the 1980s, but most people want to retain their links with France and continue to obtain financial aid to develop their territory.

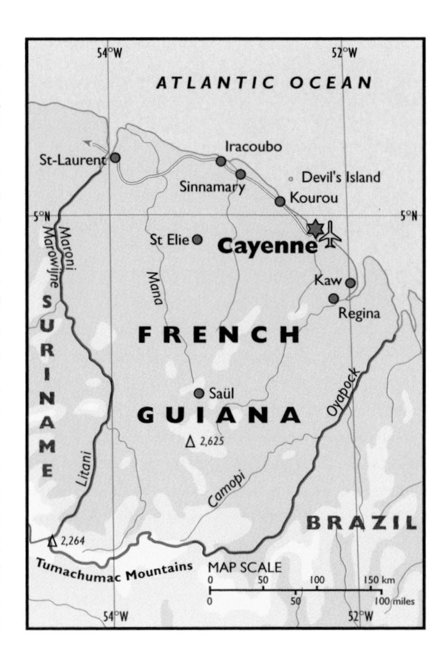

ECONOMY

Although it has rich forest and mineral resources, such as bauxite (aluminum ore), French Guiana is a developing country. It depends greatly on France for money to run its services and the government is the country's biggest employer. Since 1968, Kourou in French Guiana, the European Space Agency's rocket-launching site, has earned money for France by sending communications satellites into space.

Other activities include fishing (mostly for shrimps), forestry, gold mining, and agriculture, which is confined to coastal areas. Crops include bananas, cassava, rice, and sugarcane, but French Guiana depends on food imports to feed its people, and petroleum to produce energy. Exports include shrimps, valuable timber from the large forests, rum, and rosewood essence. Unemployment is high.

DID YOU KNOW

- that Kourou, northwest of Cayenne, is the site of a space research and rocket-launching station
- that the people of French Guiana have full French citizenship; they hold French passports
- that Devil's Island off the coast was a notorious penal colony for French criminals and political prisoners from 1852 until 1945
- that the people of French Guiana elect two deputies to France's National Assembly, and one senator to the Senate

GABON

Gabon's flag was adopted in 1960 when the country became independent from France. The central yellow stripe symbolizes the Equator which runs through Gabon. The green stands for the country's forests. The blue symbolizes the sea.

The Gabonese Republic lies on the Equator in west-central Africa. In area, it is a little larger than the United Kingdom, with a coastline 500 miles [800 km] long. Behind the narrow, partly lagoon-lined coastal plain, the land rises to hills, plateaux, and mountains divided by deep valleys carved by the River Ogooué and its tributaries. The Ogooué, which rises in Congo, reaches the sea near Port Gentil. This seaport lies on an island in the river's estuary near Cape Lopez, Gabon's most westerly point.

AREA 103,347 sq miles [267,670 sq km]

POPULATION 1,221,000

CAPITAL (POPULATION) Libreville (418,000)

GOVERNMENT Multiparty republic

ETHNIC GROUPS Four major Bantu tribes: Fang, Eshira, Bapounou and Bateke

LANGUAGES French (official), Bantu languages

RELIGIONS Roman Catholic 65%, Protestant 19%, African churches 12%, traditional beliefs 3%, Islam 2%

CURRENCY CFA franc = 100 centimes

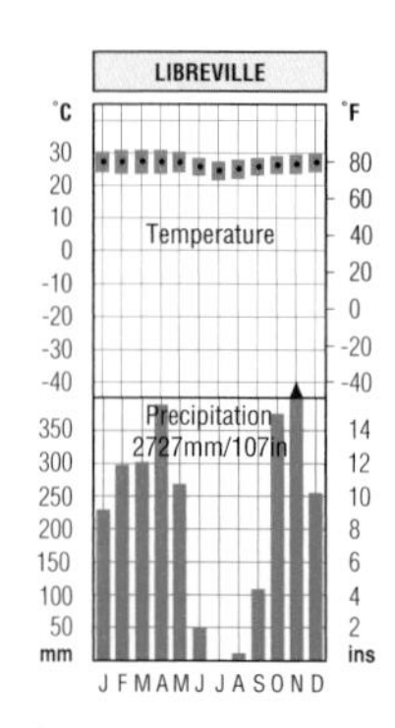

CLIMATE

Most of Gabon has an equatorial climate, with high temperatures and humidity throughout the year. The rainfall is heavy and the skies are often cloudy. The capital Libreville has a dry season between June and August, when winds blow outward from the land to the sea.

VEGETATION

Dense rain forest covers about three-quarters of Gabon. In the east and south, the rain forest gives way to savanna (tropical grassland).

The forests contain abundant animals, with antelopes, including the rare bongo and sitatonga, buffaloes, chimpanzees, elephants, and gorillas. Gabon has several national parks and wildlife reserves.

HISTORY AND POLITICS

Portuguese explorers reached the Gabon coast in the 1470s. The area later became a source of slaves. France established a settlement in Gabon in 1839. In 1849, after the French landed a group of slaves at the site, the settlement was named Libreville. Gabon became a French colony in the 1880s, but achieved full independence in 1960. In 1964, an attempted coup was put down when French troops intervened and crushed the revolt. In 1967, Bernard-Albert Bongo, who later renamed himself El Hadj Omar Bongo, became president. He declared Gabon a one-party state in 1968. Opposition parties were legalized in 1991, but Bongo was re-elected president in 1993 and 1998.

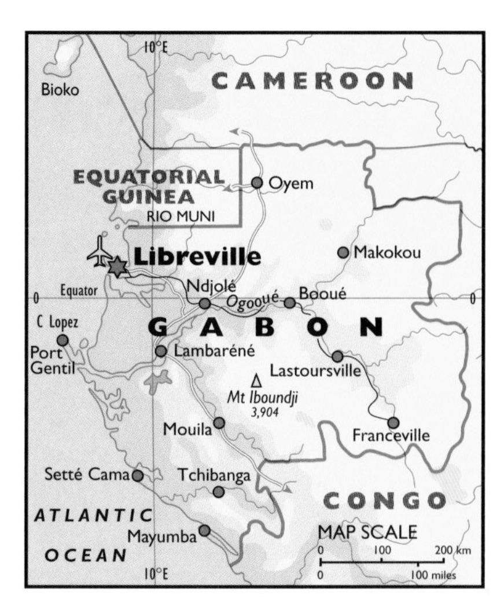

ECONOMY

Gabon's abundant natural resources include its forests, oil and gas deposits near Port Gentil, together with manganese and uranium. These mineral deposits make Gabon one of Africa's better-off countries.

However, agriculture still employs about 42% of the people and many of the farmers are poor, producing little more than they need to feed their own families. Crops include bananas, cassava, maize, and sugarcane. Cocoa and coffee are grown for export. The chief exports are oil, manganese, timber and wood products, and uranium.

DID YOU KNOW

- that the capital Libreville, a word meaning "free town," was founded in 1849 as a settlement for slaves freed from illegal slaving ships
- that the striking, heart-shaped masks made by the Fang people of Gabon influenced several European artists, including Pablo Picasso
- that Lambaréné is the site of a hospital and leper colony built by Albert Schweitzer, who won the Nobel Peace Prize for his humanitarian work
- that Gabon has the world's largest known deposit of the metal manganese
- that Portuguese explorers in the 1470s thought that the Gabon estuary, on which Libreville now stands, had the shape of a hooded cloak, or *gabão* in Portuguese; as a result, the country later became known as Gabon

The stamp, one of a set of two issued in 1978 depicting "Views of Gabon," shows oceangoing ships moored at the port of Owendo on the Gabon estuary just south of the capital Libreville. Owendo has a deep-water port. It is now a suburb of Libreville and its main shipping point, exporting such products as hardwoods, palm products, and rubber.

GAMBIA, THE

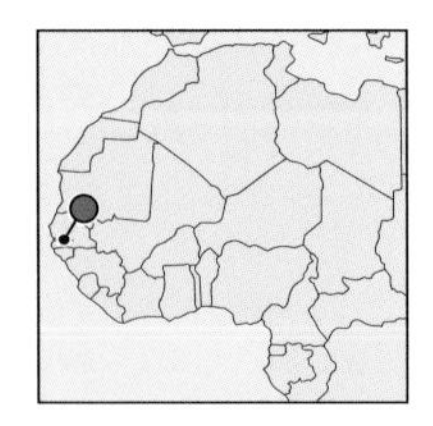

The Gambia's flag was adopted when the country became independent from Britain in 1965. The blue stripe represents the River Gambia which flows from east to west through the country. The red stands for the sun and the green for the land.

The Republic of The Gambia is the smallest country in mainland Africa. It consists of a narrow strip of land bordering the River Gambia. The Gambia is almost entirely enclosed by Senegal, except along the short Atlantic coastline. The land is low-lying. Near the sea, the land is flat and the soils are salty. The middle part of the river is bordered by terraces, called *banto faros*, which are flooded after heavy rains. The upper river flows through a deep valley which the river has cut into a sandstone plateau.

AREA 4,363 sq miles [11,300 sq km]
POPULATION 1,411,000
CAPITAL (POPULATION) Banjul (171,000)
GOVERNMENT Military regime
ETHNIC GROUPS Mandinka 42%, Fulani 18%, Wolof 16%, Jola 10%, Serahuli 9%
LANGUAGES English (official), Mandinka, Wolof, Fula
RELIGIONS Islam 90%, Christianity 9%, traditional beliefs 1%
CURRENCY Dalasi = 100 butut

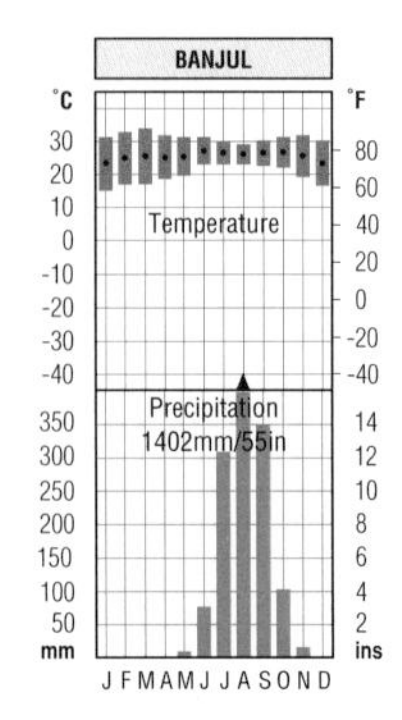

CLIMATE

The Gambia has hot and humid summers, but the winter temperatures (November to May) drop to around 61°F [16°C]. In winter, dry northeasterly winds blow over the country. In the summer, moist southwesterlies bring rain, which is heaviest on the coast.

VEGETATION

Mangrove swamps line the river banks from the coast through to the center of the country. But savanna used to cover most of the land until it was largely cleared for farming. The Gambia is rich in birdlife. Crocodiles and hippopotamuses live in the River Gambia.

HISTORY AND POLITICS

Portuguese mariners reached The Gambia's coast in 1455 when the area was part of the large Mali empire. In the 16th century, Portuguese and English slave traders operated in the area.

English traders bought rights to trade on the River Gambia in 1588, and in 1664 the English established a settlement on an island in the river estuary. In 1765, the British founded a colony called Senegambia, which included parts of The Gambia and Senegal. In 1783, Britain handed this colony over to France.

In 1816, Britain founded Bathurst (now Banjul) as a base for its anti-

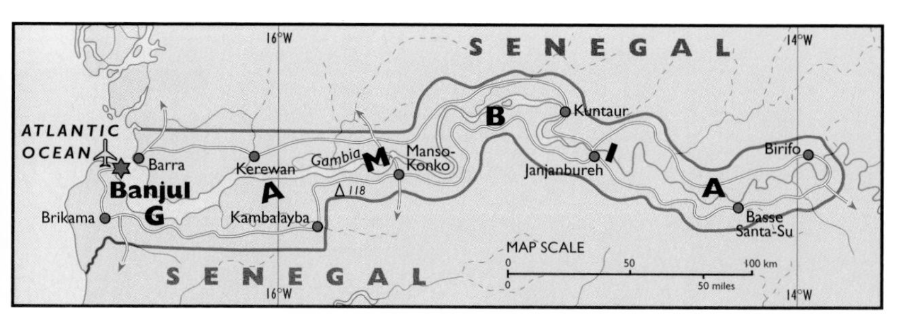

slavery operations. But control of inland areas was made difficult by wars between local Muslims and non-Muslims. In the 1860s and 1870s, Britain and France discussed the exchange of The Gambia for some other French territory. But no agreement was reached and Britain made The Gambia a British colony in 1888. It remained under British rule until it achieved full independence in 1965. In 1970, The Gambia became a republic. But in July 1994, a military group overthrew the president, Sir Dawda Jawara, who fled into exile. Captain Yahya Jammeh, who took power, was elected president in 1996 and 2001.

DID YOU KNOW

- that, in 1765, parts of what are now The Gambia became parts of Britain's first African colony, which was called Senegambia
- that the African American writer Alex Haley, author of the bestseller *Roots* which became a major TV series, traced his ancestors back to a Gambian village called Juffure
- that the number of tourists visiting The Gambia rose from a few hundred in 1965 to more than 100,000 in 1990–91

One major political issue concerns relations between the English-speaking Gambians and their French-speaking Senegalese neighbors. In 1981, an attempted coup in The Gambia was defeated with the help of Senegalese troops. In 1982, The Gambia and Senegal set up a defense alliance, the Confederation of Senegambia. But this alliance was dissolved in 1989.

ECONOMY

Agriculture employs more than 80% of the people. The main food crops include cassava, millet, and sorghum, but groundnuts and groundnut products are the leading exports. Tourism is growing in importance.

The Gambia's rich birdlife was the subject of a set of 13 stamps issued in 1966, when the country's unit of currency was still the pound. The dalasi was not introduced until 1971. This stamp shows the red-throated bee-eater.

GEORGIA

Georgia's flag was first used between 1917 and 1921. It was readopted when Georgia became independent. The wine-red color represents the good times of the past and the future. The black symbolizes Russian rule and the white represents hope for peace.

Georgia is a country on the borders of Europe and Asia, facing the Black Sea. The land is rugged with the Caucasus Mountains forming its northern border. The highest mountain in this range, Mount Elbrus (18,481 ft [5,633 m]), lies over the border in Russia.

Lower ranges run through southern Georgia, through which pass the borders with Turkey and Armenia. The fertile Black Sea coastal plains are in the west. In the east a low plateau extends into Azerbaijan. The main river in the east is the River Kura, on which the capital Tbilisi stands.

AREA 26,910 sq miles [69,700 sq km]
POPULATION 4,989,000
CAPITAL (POPULATION) Tbilisi (1,253,000)
GOVERNMENT Multiparty republic
ETHNIC GROUPS Georgian 70%, Armenian 8%, Russian 6%, Azeri 6%, Ossetian 3%, Greek 2%, Abkhaz 2%, others 3%
LANGUAGES Georgian (official), Russian
RELIGIONS Georgian Orthodox 65%, Islam 11%, Russian Orthodox 10%, Armenian Apostolic 8%
CURRENCY Lari = 100 tetri

CLIMATE

The Black Sea plains have hot summers and mild winters, when the temperatures seldom drop below freezing point. The rainfall is heavy, but inland Tbilisi has moderate rainfall, with the heaviest rains in the spring and early summer. Summers are hot. Winters are colder than in the west. Snow covers the highest mountains throughout the year.

VEGETATION

Forest and shrub cover about half of the country. Common trees in the west include alder, beech, chestnut, and oak. Evergreen forests flank the higher mountain slopes, with alpine meadows above the tree line. The most luxuriant vegetation is found on the subtropical Black Sea lowlands. Orchards of apples and pears, together with orange groves, are quite common.

HISTORY

The first Georgian state was set up nearly 2,500 years ago. But for much of its history, the area was ruled by various conquerors. Christianity was introduced in AD 330.

Georgia freed itself of foreign rule in the 11th and 12th centuries, but

Mongol armies attacked in the 13th century. From the 16th to the 18th centuries, Iran and the Turkish Ottoman empire struggled for control of the area, and in the late 18th century Georgia sought the protection of Russia, its northern neighbor. By the early 19th century, Georgia was part of the Russian empire.

After the Russian Revolution of 1917, Georgia declared itself an independent state, and in 1921 it was recognized by the League of Nations (the predecessor to the United Nations). But Russian troops invaded and made Georgia part of the Soviet regime. The Russians then combined Georgia, Armenia, and Azerbaijan in the Transcaucasian Soviet Federated Socialist Republic, but this federation was broken up in 1936. Georgia then became a separate Soviet Socialist Republic within the Soviet Union.

In 1991, following reforms in the Soviet Union, Georgia declared itself independent. It became a separate country when the Soviet Union was dissolved in December 1991.

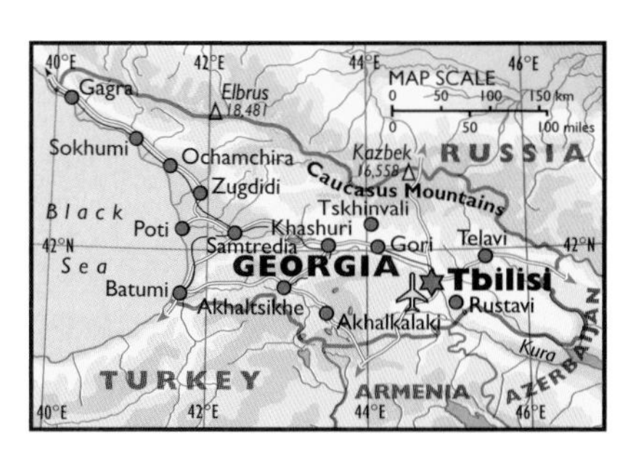

DID YOU KNOW

- that Josef Stalin, the dictator of the Soviet Union between 1929 and 1953, was a Georgian, born in Gori, 40 miles [65 km] northwest of Tbilisi
- that Georgia is famous for the longevity of its people; claims have been made that one in every 2,500 people is more than 100 years old
- that the Georgian language has its own alphabet and a rich literary tradition
- that Georgia was the "Land of the Golden Fleece" in ancient Greek mythology

POLITICS

Georgia contains three regions containing minority peoples: Abkhazia in the northwest, South Ossetia in north-central Georgia, and Adjaria (also spelled Adzharia) in the southwest. Communal conflict in the early 1990s led to outbreaks of civil war in South Ossetia and Abkhazia, where the people expressed the wish to set up their own independent countries. In 2002, Islamic terrorists were reported to be taking refuge in Pankisi gorge in eastern Georgia.

ECONOMY

Georgia is a developing country. Farm products include barley, citrus fruits, grapes for wine-making, maize, tea, tobacco, and vegetables. Food processing, and silk- and perfume-making are also important. Sheep and cattle are reared.

Barite (barium ore), coal, copper, and manganese are mined, and tourism is a major industry on the Black Sea coast. Georgia has huge hydroelectric potential, but most of the country's electricity is generated in Russia or Ukraine.

GERMANY

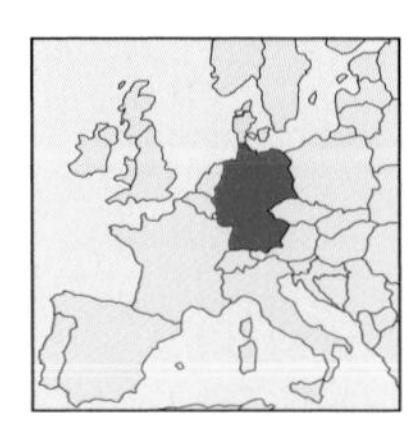

This flag, adopted by the Federal Republic of Germany (West Germany) in 1949, became the flag of the reunified Germany in 1990. The red, black, and gold colors date back to the Holy Roman Empire. They are associated with the struggle for a united Germany from the 1830s.

The Federal Republic of Germany is the fourth largest country in Western Europe, after France, Spain, and Sweden. The North German plain borders the North Sea in the northwest and the Baltic Sea in the northeast. The plain is covered by moraine (rocks deposited by glaciers during the Ice Age). Major rivers draining the plain include the Weser, Elbe, and Oder.

The central highlands contain plateaux and highlands, including the Harz Mountains, the Thuringian Forest (Thüringer Wald), the Ore Mountains (Erzgebirge), and the Bohemian Forest (Böhmerwald) on the border with the Czech Republic.

AREA 137,803 sq miles [356,910 sq km]
POPULATION 83,030,000
CAPITAL (POPULATION) Berlin (3,426,000)
GOVERNMENT Federal multiparty republic
ETHNIC GROUPS German 93%, Turkish 2%, Serbo-Croat 1%, Italian 1%, Greek, Polish, Spanish
LANGUAGES German (official)
RELIGIONS Protestant (mainly Lutheran) 38%, Roman Catholic 34%, Islam 2%
CURRENCY Euro = 100 cents

Southern Germany is a largely hilly region, rising to high mountains in the Bavarian Alps on the border with Switzerland. These mountains reach a height of 9,725 ft [2,963 m] in a peak called Zugspitze. Another scenic area, which attracts many tourists, is the Black Forest (Schwarzwald). The Black Forest overlooks the River Rhine, which flows through a steep-sided rift valley in the southwest.

Europe's longest river, the Danube (called Donau in Germany), rises in the Black Forest and flows east across southern Germany to Austria, on its way to the Black Sea.

CLIMATE

Germany has a mild climate. The northwest is influenced by the warm waters of the North Sea, but the Baltic coastlands in the northeast are cooler. To the south, the climate becomes

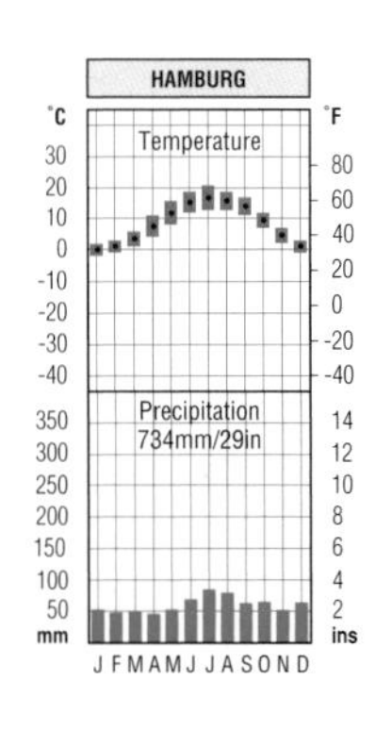

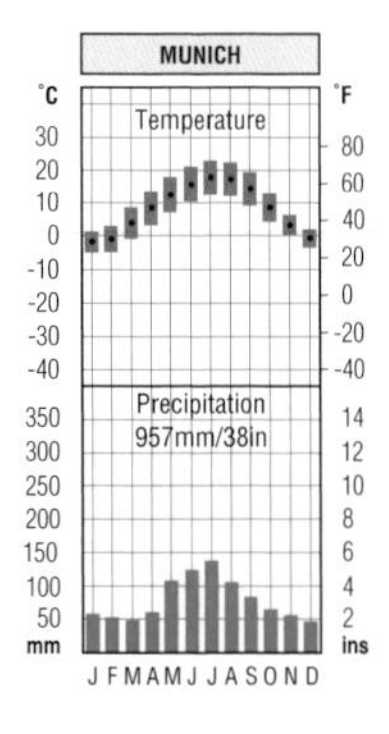

more continental, especially in the highlands. The precipitation is greatest on the hills and mountains, many of which are snow-capped in winter.

VEGETATION

The North German plain contains large areas of heathland, with such plants as grasses, heather, mosses, and lichens. The most common trees in the forests of central and southern Germany are pine, beech (on the higher mountain slopes), and oak. Industrial pollution is the cause of acid rain, which has damaged many of the trees.

HISTORY

Just over 2,000 years ago, various tribes began to move into the valleys of the Rhine and Danube, near the northern boundaries of the Roman empire. The Romans called this area *Germania*, after one of the tribes. In the 5th century AD, Germanic tribes attacked Rome. Before long, the western Roman empire began to break up into tribal kingdoms. The largest of these was the kingdom of a Germanic people called the Franks.

In 486, Clovis, a Frankish ruler, conquered Gaul (France) and ruled an area including Gaul and western Germany. Another Frankish ruler, Charlemagne, expanded the territory, uniting many tribes into a *reich* (empire), but the empire broke up in the 9th century. Much of Germany later became part of the Holy Roman Empire.

In 1618, religious and political differences were responsible for the start of the Thirty Years' War, which ravaged much of Germany. But a powerful state combining Brandenburg and Prussia emerged from this conflict. Between 1740 and 1768, Frederick II ("the Great") made Prussia a great power. German nationalism increased in the 19th century and, in 1871, a united German empire was founded.

Germany and its allies were defeated in World War I (1914–18) and the country became a republic. Adolf Hitler came to power in 1933 and ruled as a dictator. His order to invade Poland led to the start of World War II (1939–45), which ended in defeat and Germany in ruins.

In 1945, Germany was divided into four military zones. In 1949, the American, British, and French zones were amalgamated to form the

DID YOU KNOW

- that Germany was the first country to build freeways; they are called *Autobahns*
- that Germany ranks second in the world in the production of barley, beer, and lignite; it comes third in the production of cars, rye, and sugar beet
- that the first modern symphony orchestras were both founded in Germany in 1743 – one at Mannheim and the other at Leipzig
- that the Neanderthal people, who lived between 100,000 and 35,000 years ago, were named after Neander Gorge, near Düsseldorf, where their fossils were found in 1856
- that Germany is second only to the United States in the value of its foreign trade

GERMANY

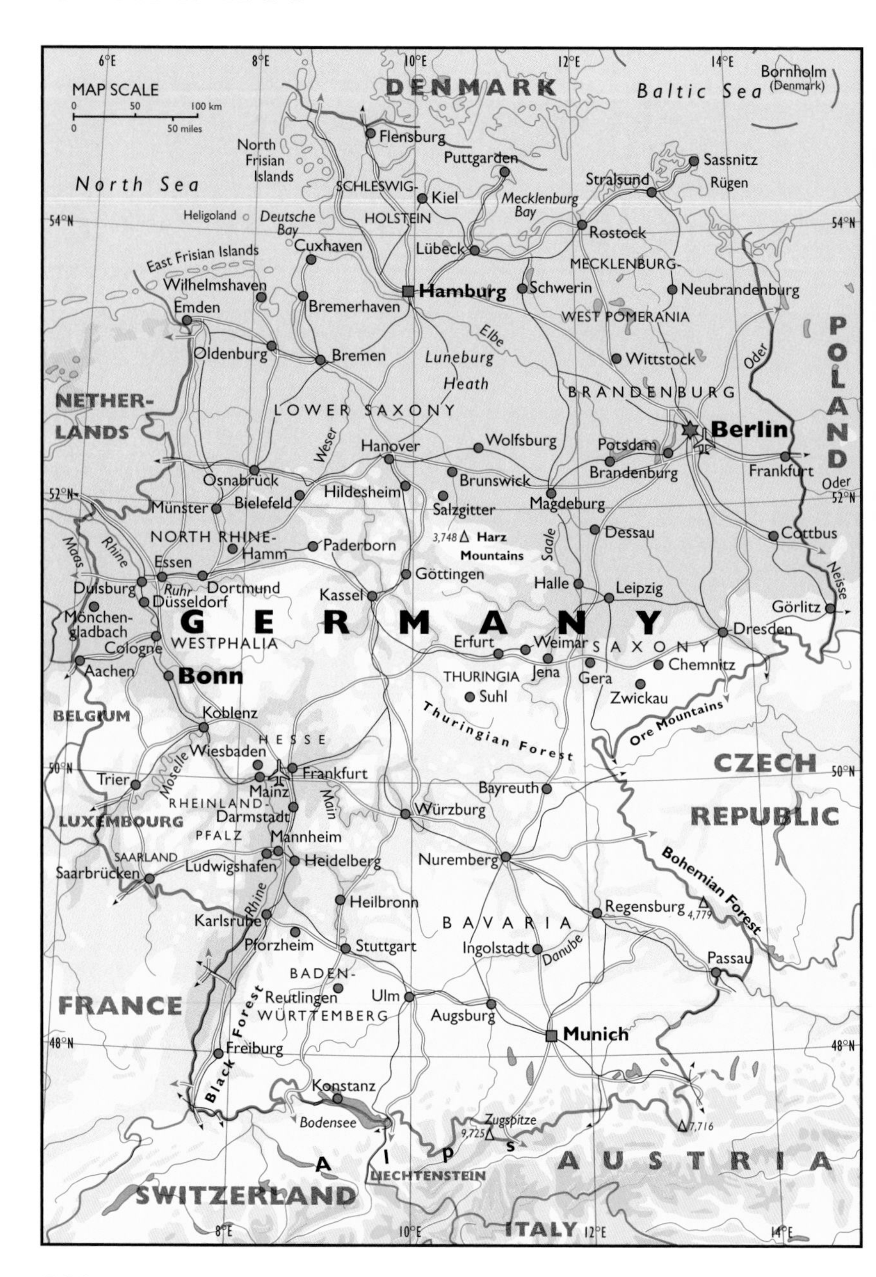

MAP SCALE
0 50 100 km
0 50 miles
DENMARK
Baltic Sea
Bornholm (Denmark)
North Sea
North Frisian Islands
Heligoland
Deutsche Bay
East Frisian Islands
Flensburg
Puttgarden
Kiel
SCHLESWIG-HOLSTEIN
Mecklenburg Bay
Stralsund
Sassnitz
Rügen
Rostock
Cuxhaven
Lübeck
Hamburg
Schwerin
MECKLENBURG-WEST POMERANIA
Neubrandenburg
Wilhelmshaven
Emden
Bremerhaven
Oldenburg
Bremen
Elbe
Luneburg Heath
Wittstock
Oder
POLAND
BRANDENBURG
NETHER-LANDS
LOWER SAXONY
Berlin
Weser
Hanover
Wolfsburg
Potsdam
Brandenburg
Frankfurt
Oder
Osnabrück
Brunswick
Hildesheim
Münster
Bielefeld
Salzgitter
Magdeburg
Cottbus
NORTH RHINE-WESTPHALIA
Paderborn
Hamm
3,748 Harz Mountains
Dessau
Saale
Maas
Rhine
Essen
Dortmund
Göttingen
Neisse
Duisburg
Ruhr
Düsseldorf
Kassel
Halle
Leipzig
Mönchen-gladbach
Görlitz
GERMANY
Dresden
Cologne
Erfurt
Weimar
SAXONY
Aachen
Bonn
THURINGIA
Jena
Gera
Chemnitz
Suhl
Zwickau
BELGIUM
Koblenz
Thuringian Forest
Ore Mountains
HESSE
Wiesbaden
CZECH REPUBLIC
Moselle
Frankfurt
Trier
Mainz
Main
Bayreuth
RHEINLAND-PFALZ
Darmstadt
LUXEMBOURG
Würzburg
Mannheim
SAARLAND
Ludwigshafen
Heidelberg
Nuremberg
Saarbrücken
Bohemian Forest
Heilbronn
Regensburg
4,779
Karlsruhe
Rhine
BAVARIA
Pforzheim
Stuttgart
Ingolstadt
Danube
Passau
BADEN-WÜRTTEMBERG
Reutlingen
Ulm
FRANCE
Black Forest
Augsburg
Munich
Freiburg
Konstanz
Bodensee
Zugspitze
9,725
7,716
Alps
AUSTRIA
LIECHTENSTEIN
SWITZERLAND
ITALY
6°E
8°E
10°E
12°E
14°E
54°N
52°N
50°N
48°N

Traditional costumes are carefully preserved in many parts of Germany. This stamp issued by the German Democratic Republic (DDR), or East Germany, in 1971 shows a costume used by Sorb dancers. The Sorbs are an ancient Slav minority who live east of the city of Dresden. The inscription at the bottom of the stamp is in the Sorb language.

Federal Republic of Germany (West Germany), while the Soviet zone became the German Democratic Republic (East Germany), a Communist state. Berlin, which had also been partitioned, became a divided city. West Berlin was part of West Germany, while East Berlin became the capital of East Germany. Bonn was the capital of West Germany.

Tension between East and West mounted during the Cold War, but West Germany rebuilt its economy quickly. In East Germany, the recovery was less rapid. In the late 1980s, reforms in the Soviet Union led to unrest in East Germany. Free elections were held in East Germany in 1990 and, on October 3, 1990, Germany was reunited.

POLITICS

The united Germany adopted West Germany's official name, the Federal Republic of Germany. Elections in December 1990 returned Helmut Kohl, West Germany's Chancellor (head of government) since 1982, to power. In 1998, following a period of economic setbacks, Kohl's party was defeated and a Social Democratic-Green coalition was set up under Gerhard Schröder. Schröder was re-elected in 2002.

ECONOMY

West Germany's "economic miracle" after the destruction of World War II was greatly helped by foreign aid. Today, despite the problems caused by reunification, Germany is one of the world's great economic powers.

Manufacturing is the most valuable part of Germany's economy and manufactured goods make up the bulk of the country's exports. Cars and other vehicles, cement, chemicals, computers, electrical equipment, processed food, machinery, scientific instruments, ships, steel, textiles, and tools are among the leading manufactures. Germany has some coal, lignite (brown coal), potash, and rock salt deposits. But it imports many of the raw materials needed by its industries.

Germany also imports food. Major agricultural products include fruits, grapes for wine-making, potatoes, sugar beet, and vegetables. Beef and dairy cattle are raised, together with many other livestock. In western Germany, most farms are small and privately owned. In eastern Germany, the government has begun to break up the large state-owned farms, selling plots to individual owners.

GHANA

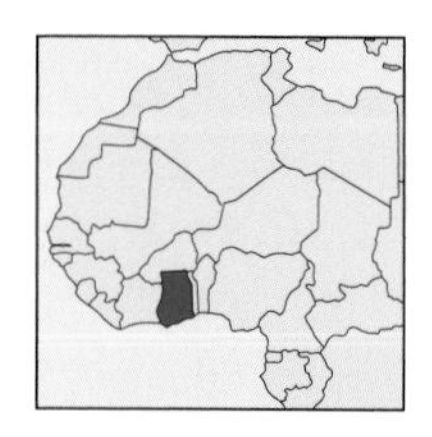

Ghana's flag has red, green, and yellow bands like the flag of Ethiopia, Africa's oldest independent nation. These colors symbolize African unity. The black star is a symbol of African freedom. Ghana's flag was adopted when the country became independent in 1957.

The Republic of Ghana faces the Gulf of Guinea in West Africa. This hot country, just north of the Equator, was formerly called the Gold Coast. Behind the thickly populated southern coastal plains, which are lined with lagoons, lies a plateau region in the southwest.

Northern Ghana is drained by the Black and White Volta rivers, which flow into Lake Volta. This lake, which has formed behind the Akosombo Dam, is one of the world's largest artificially created lakes.

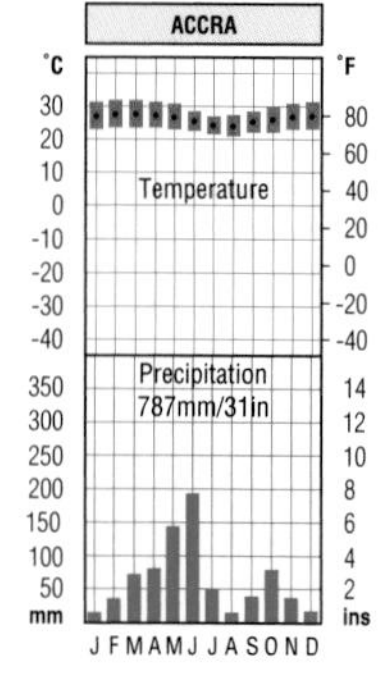

CLIMATE

Accra has a hot, tropical climate. Rain occurs all through the year, though Accra is drier than areas inland. The southwest is the wettest place. The north is warmer than the south, while the winter months (November to March) have a low average rainfall.

VEGETATION

Rain forests grow in the southwest. To the north, the forests merge into savanna (tropical grassland with some woodland). More open grasslands dominate in the far north.

AREA 92,100 sq miles [238,540 sq km]

POPULATION 19,894,000

CAPITAL (POPULATION) Accra (1,781,000)

GOVERNMENT Republic

ETHNIC GROUPS Akan 44%, Moshi-Dagomba 16%, Ewe 13%, Ga 8%. Gurma 3%, Yoruba 1%

LANGUAGES English (official), Akan, Moshi-Dagomba, Ewe, Ga

RELIGIONS Christianity 63%, traditional beliefs 21%, Islam 16%

CURRENCY Cedi = 100 pesewas

HISTORY

Portuguese explorers reached the area in 1471 and named it the Gold Coast. The area became a center of the slave trade in the 17th century. The slave trade was ended in the 1860s and, gradually, the British took control of the area.

The country became independent in 1957, when it was renamed Ghana. Ghana was a great African empire which flourished to the northwest of present-day Ghana between the AD 300s and 1000s. This name was chosen to celebrate the fact that Ghana was the first country in the Commonwealth to be ruled by black Africans.

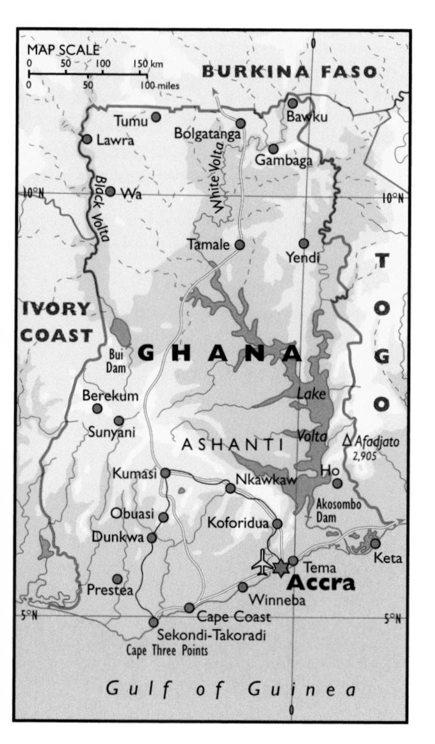

POLITICS

After independence, Ghana's government sought to develop the economy by creating large government-owned manufacturing industries. But debt and corruption, together with falls in the price of cocoa, the leading export, caused severe economic problems. This led to instability and frequent military coups. In 1981, power was invested in a Provisional National Defense Council, led by Flight-Lieutenant Jerry Rawlings.

The government steadied the economy and introduced several new policies, including the relaxation of government controls. In 1992, the government introduced a new constitution, which allowed for multiparty elections. After Rawlings retired in 2000, opposition leader John Agyekum Kufuor became president.

ECONOMY

The World Bank classifies Ghana as a "low-income" developing country. Most people are poor and farming employs 59% of the population. Food crops include cassava, groundnuts, maize, millet, plantains, rice, and yams. But cocoa is the most valuable export crop. Timber and gold are also exported. Other valuable crops are coffee, coconuts, and palm kernels.

Many small factories produce goods, such as beverages, cement, and clothing, for local consumption. The aluminum smelter at Tema, a port near Accra, is the largest factory. Electricity for southern Ghana is produced at the Akosombo Dam.

This stamp was issued in 1952, when the currency of the British colony of the Gold Coast was the pound, divided into 20 shillings. Five years later, the Gold Coast became independent as Ghana. Ghana's chief export is cocoa. The stamp shows a worker breaking cocoa pods apart, to remove the beans, which are used to make chocolate.

GREECE

Blue and white became Greece's national colors during the war of independence (1821–9). The nine horizontal stripes on the flag, which was finally adopted in 1970, represent the nine syllables of the battle cry *Eleutheria i thanatos* ("Freedom or Death").

The Hellenic Republic, as Greece is officially called, is a rugged country in southeastern Europe. Islands make up nearly a fifth of the country. Crete (Kriti) is the largest island. Mainland Greece forms the southern tip of the Balkan peninsula. Olympus, at 9,574 ft [2,917 m], is the highest peak. In the south, the Peloponnese (Pelopónnisos) is a peninsula which is attached to northern Greece only by the narrow Isthmus of Corinth. The Corinth Canal, 3 miles [4.8 km] long, cuts through this isthmus.

AREA 50,961 sq miles [131,990 sq km]
POPULATION 10,624,000
CAPITAL (POPULATION) Athens (3,097,000)
GOVERNMENT Multiparty republic
ETHNIC GROUPS Greek 98%
LANGUAGES Greek (official), English, French
RELIGIONS Greek Orthodox 98%, Islam 1%
CURRENCY Euro = 100 cents

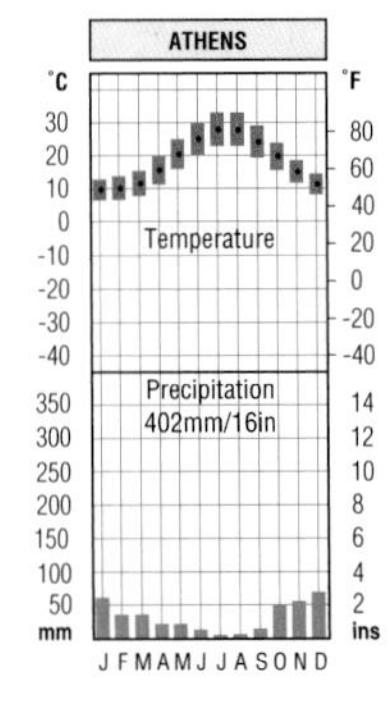

CLIMATE

Low-lying areas in Greece have mild, moist winters and hot, dry summers. The east coast has more than 2,700 hours of sunshine a year and only about half of the rainfall of the west. The mountains have a much more severe climate, with snow on the higher slopes in winter.

VEGETATION

Much of Greece's original vegetation has been destroyed. Some areas are now bare of plants and others covered by *maquis*, a thorny shrub vegetation.

HISTORY AND POLITICS

Crete was the center of the Minoan civilization, an early Greek culture, between about 3000 and 1450 BC. The Minoans were followed by the Mycenian culture which prospered on the mainland until about 1100 BC.

In about 750 BC, the Greeks began to colonize the Mediterranean, creating wealth through trade. The city state of Athens reached its peak in the 400s BC, but in 338 BC Macedonia became the dominant power. In 334–331 BC, Alexander the Great conquered southwestern Asia. Greece became a Roman province in 146 BC and, in AD 365, part of the Byzantine empire. In 1453, the Turks defeated the Byzantine empire. But between 1821 and 1829, the Greeks defeated the Turks. The country finally became an independent monarchy in 1830.

After World War II (1939–45), when Germany had occupied Greece, a civil war broke out between Communist and nationalist forces. This war ended in 1949. A military dictatorship took power in 1967. The monarchy was abolished in 1973 and democratic government was restored in 1974. Greece joined the European Community (now the EU) in 1981, and it adopted the euro in 2001. In 2002, Greece and Turkey signed agreements aimed at improving relations between them.

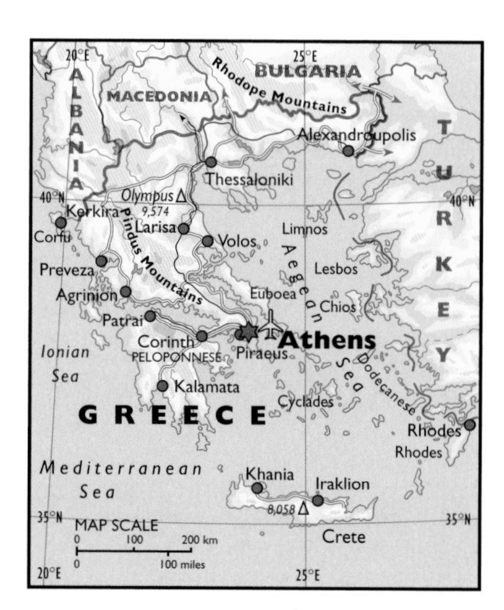

ECONOMY

Manufacturing is important. Products include processed food, cement, chemicals, metal products, textiles, and tobacco. Greece also mines lignite (brown coal), bauxite, and chromite.

Farmland covers about a third of the country, and grazing land another 40%. Major crops include barley, grapes for wine-making, dried fruits, olives, potatoes, sugar beet, and wheat. Poultry, sheep, goats, pigs, and cattle are raised. Tourist attractions include the beaches and the ruins of ancient Greece. Greece's huge shipping fleet is another source of revenue.

DID YOU KNOW

- that Greece is one of the world's leading producers of dried fruits; the word currant comes from the Greek town of Corinth
- that the first recorded Olympic Games took place in ancient Greece in 776 BC
- that the Greek cheese *feta* is made from sheep or goat's milk
- that, by ownership though not by registration, Greece has the world's largest merchant fleet
- that the *Iliad* and the *Odyssey* are the earliest known European poems; they are thought to have been written by Homer around 800 BC

The Greeks wear regional costumes during festivals. This stamp shows a woman in traditional dress from the Mégara region.

GUATEMALA

Guatemala's flag was adopted in 1871, but its origins go back to the days of the Central American Federation (1823–39), which was set up after the break from Spain in 1821. The Federation included Costa Rica, El Salvador, Guatemala, Honduras, and Nicaragua.

The Republic of Guatemala in Central America contains a thickly populated mountain region, with fertile soils. The mountains, which run in an east–west direction, contain many volcanoes, some of which are active. Volcanic eruptions and earthquakes occur from time to time in the highlands.

South of the mountains lie the thinly populated coastal lowlands facing the Pacific Ocean. Northern Guatemala is a large, thinly populated plain. The country's largest lake, Izabal, drains into the Caribbean Sea.

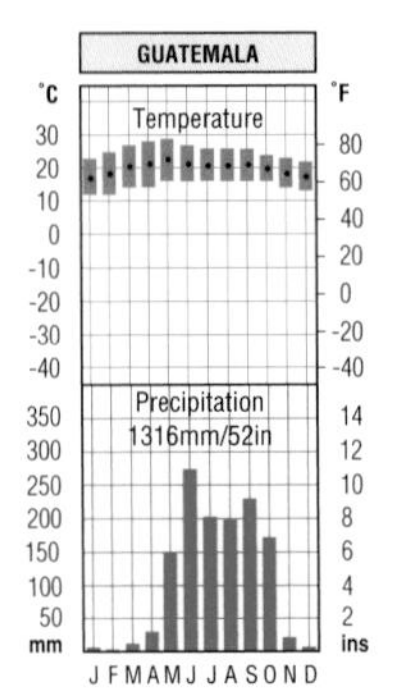

CLIMATE

Guatemala lies in the tropics. The lowlands are hot and rainy. But the central mountain region is cooler and drier. Guatemala City, at about 5,000 ft [1,500 m] above sea level, has a pleasant, warm climate, with a marked dry season between November and April.

VEGETATION

Hardwoods, including mahogany, rubber trees, palms, and chicozapote, from which the *chicle* used in chewing gum is obtained, grow in the tropical forests in the north, with mangrove swamps on the coast. Oaks and willows grow in the highlands, with firs and pines at higher levels. The Pacific plains are mostly farmed.

AREA 42,042 sq miles [108,890 sq km]

POPULATION 12,974,000

CAPITAL (POPULATION) Guatemala City (1,167,000)

GOVERNMENT Republic

ETHNIC GROUPS Ladino (mixed Hispanic and Amerindian) 55%, Amerindian 43%, other 2%

LANGUAGES Spanish (official), Amerindian languages

RELIGIONS Roman Catholic 75%, Protestant 25%

CURRENCY Guatemalan quetzal = 100 centavos; US dollar = 100 cents

HISTORY AND POLITICS

Between AD 300 and 900, Mayan Indians ruled much of Guatemala, but they abandoned their cities on the northern plains for reasons that historians do not understand. Spain conquered the area in the 1520s, but Guatemala and the other Central American states declared their independence in 1821. In 1823, Guatemala joined the Central American Federation. But it became fully independent in 1839. Since independence,

Guatemala has been plagued by instability and periodic violence.

Guatemala has a long-standing claim over Belize, but this was reduced in 1983 to the southern fifth of the country. Violence became widespread in Guatemala from the early 1960s, because of conflict between left-wing groups, including many Amerindians, and government forces. A peace accord was signed in 1996, ending a war that had lasted 36 years and claimed perhaps 200,000 lives.

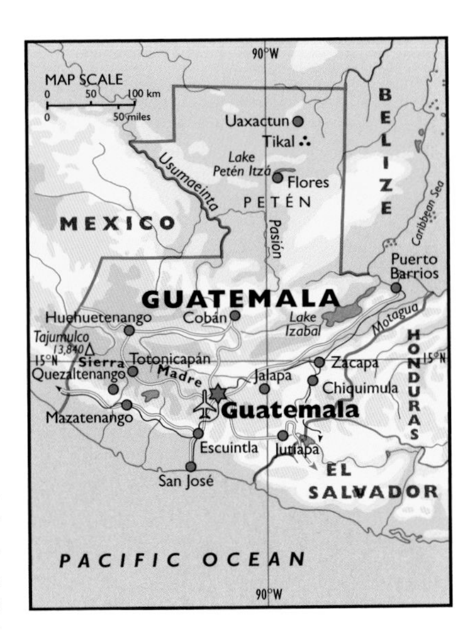

ECONOMY

The World Bank classifies Guatemala as a "lower-middle-income" developing country. Agriculture employs nearly half of the population and coffee, sugar, bananas, and beef are the leading exports. Other important crops include the spice cardamom and cotton, while maize is the chief food crop. But Guatemala still has to import food to feed the people. Forestry is important and industry is increasing. Manufactures include processed farm products, textiles, wood products, and handicrafts. Tourism is growing.

DID YOU KNOW

- that Guatemala's extinct volcano Tajumulco, at 13,840 ft [4,217 m], is the highest mountain in Central America
- that Guatemala is the world's seventh largest producer of coffee
- that the Mayan site of Tikal in northern Guatemala has the tallest temple pyramids found anywhere in the Americas
- that the town of Antigua, west of Guatemala City, is a former capital city; it was destroyed by an earthquake in 1773

The American Charles Lindbergh made the world's first solo non-stop flight across the Atlantic Ocean on May 21 and 22, 1927. His achievement was celebrated on this airmail stamp issued in 1981.

GUINEA

Guinea's flag was adopted when the country became independent from France in 1958. It uses the colors of the flag of Ethiopia, Africa's oldest nation, which symbolize African unity. The red represents work, the yellow justice, and the green solidarity.

The Republic of Guinea faces the Atlantic Ocean in West Africa. A flat, swampy plain borders the coast. Behind this plain, the land rises to a plateau region called Fouta Djalon. The Upper Niger plains, named after one of Africa's longest rivers, the Niger, which rises there, are in the northeast. The Sénégal and Gambia rivers also start in northeastern Guinea.

In the southeast of the country lie the Guinea Highlands, which reach 5,750 ft [1,752 m] at Mount Nimba on the border with Ivory Coast and Liberia.

AREA 94,927 sq miles [245,860 sq km]

POPULATION 7,614,000

CAPITAL (POPULATION) Conakry (1,508,000)

GOVERNMENT Multiparty republic

ETHNIC GROUPS Peuml 40%, Malinke 30%, Soussou 20%, other 10%

LANGUAGES French (official)

RELIGIONS Islam 85%, Christianity 8%, traditional beliefs 7%

CURRENCY Guinean franc = 100 cauris

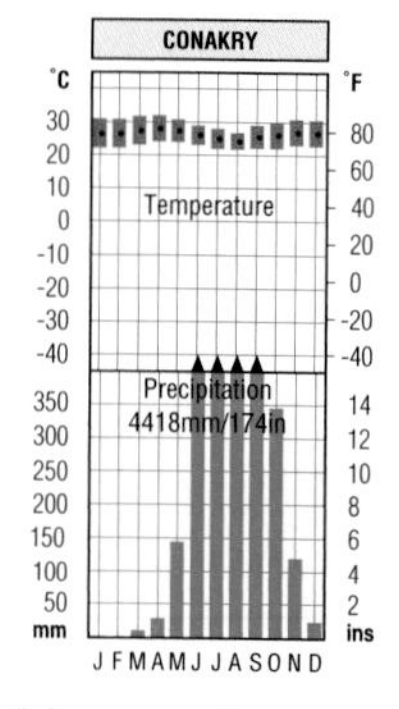

CLIMATE

Guinea has a tropical climate and Conakry, on the coast, has heavy rains between May and November. This is also the coolest period in the year. During the dry season, hot, dry harmattan winds blow southwestward from the Sahara Desert. The Fouta Djalon is cooler than the coast. The driest region is the northeast. This region and the southeastern highlands experience greater temperature variations than on the coast.

VEGETATION

Mangrove swamps grow along parts of the coast. Inland, the Fouta Djalon is largely open grassland with trees in the river valleys. Northeastern Guinea is savanna country, with such scattered trees as acacias and shea trees. Rain forests, with such trees as ebony, mahogany, and teak, grow in parts of the Guinea Highlands.

HISTORY

Between the 11th and 15th centuries, parts of Guinea belonged to the Mali and, later, the Songhai empires, which flourished in the interior of West Africa. Portuguese explorers arrived in the mid-15th century and the slave trade began soon afterward. From the 17th century, other European slave traders became active in Guinea. France became involved in the area in the mid-19th century and, in 1891, it made Guinea a French colony.

POLITICS

Guinea became independent in 1958. The first president, Sékou Touré, followed socialist policies, but most people remained poor and Touré had to introduce repressive policies to hold on to power. After his death in 1984, military leaders took over. Colonel Lansana Conté became president and his government introduced free enterprise policies. In 2000, fighting broke out as Guinea was drawn into the civil wars in neighboring Liberia and Sierra Leone.

ECONOMY

The World Bank classifies Guinea as a "low-income" developing country. Its resources include bauxite (aluminum ore), diamonds, gold, iron ore, and uranium. Bauxite and alumina are the main exports.

Agriculture, however, employs 78% of the people, many of whom produce little more than they need for their own families. Major crops include bananas, cassava, coffee, palm kernels, pineapples, rice, and sweet potatoes. Cattle and other livestock are raised in highland areas. Guinea has some manufacturing industries. Products include alumina, processed food, and textiles.

DID YOU KNOW

- that Guinea is the world's second largest producer of bauxite (aluminum ore) after Australia
- that the Malinke people of Guinea are descendants of the people who set up the ancient empire of Mali, which flourished between about 1240 and 1500
- that local history in Guinea has been kept alive by storytellers called *griots*
- that Guinea has more than 25% of the world's known reserves of bauxite

Africa's wildlife, though seriously threatened in many areas, remains one of its chief resources in attracting tourists. This stamp is one of a set of nine issued by Guinea in 1968.

GUINEA-BISSAU

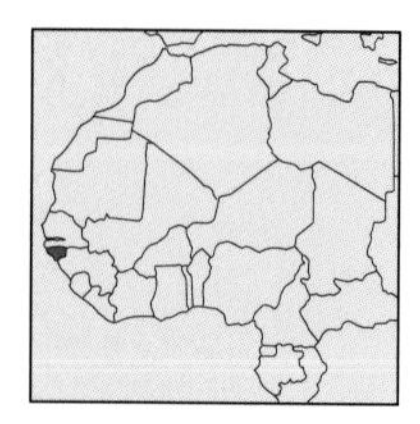

Guinea-Bissau's flag, adopted when the country became independent from Portugal in 1973, was based on the flag used by the nationalist PAIGC (African Party for the Independence of Guinea and Cape Verde). The colors, used on the flag of Ethiopia, symbolize African unity.

The Republic of Guinea-Bissau, formerly known as Portuguese Guinea, is a small country in West Africa. The land is mostly low-lying, with a broad, swampy coastal plain and many flat offshore islands, including the Bijagós Archipelago. These islands were probably once joined to the mainland but the coast has been submerged in recent geological times. This submergence has created broad river estuaries. The land rises in the east where there are low plateaux. The highest land is near the border with Guinea.

AREA 13,946 sq miles [36,120 sq km]

POPULATION 1,316,000

CAPITAL (POPULATION) Bissau (145,000)

GOVERNMENT "Interim" government

ETHNIC GROUPS Balanta 30%, Fula 20%, Manjaca 14%, Mandinga 13%, Papel 7%

LANGUAGES Portuguese (official), Crioulo

RELIGIONS Traditional beliefs 50%, Islam 45%, Christianity 5%

CURRENCY CFA franc = 100 centimes

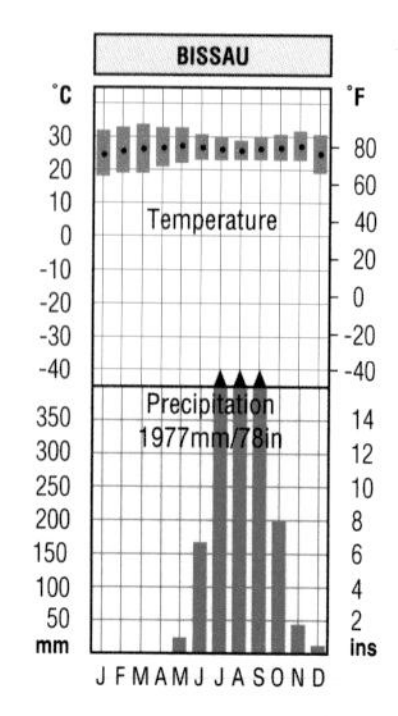

CLIMATE

The country has a tropical climate, with one dry season (December to May) and a rainy season from June to November. The rainfall on the coast averages 94 inches [2,400 mm], though the average annual rainfall inland is lower, at around 55 inches [1,400 mm].

VEGETATION

Mangrove forests grow along the coasts, while dense rain forest covers much of the coastal plain. Inland, the forests thin out and merge into savanna (tropical grassland with scattered woodland), with open grassland on the highest land. The country has a rich birdlife. Wild animals include buffaloes, crocodiles, gazelles, and leopards.

HISTORY

Historians know little of the early history of the area, which lay near the edge of the medieval West African empires of Ghana and Mali. The coast was first visited by Portuguese navigators in 1446.

Between the 17th and early 19th centuries, the Portuguese used the coast as a slave trade base. Some slaves were taken to Brazil and others to the Cape Verde Islands, where they were forced to work on Portuguese plantations.

Portugal appointed a governor to administer Guinea-Bissau and the Cape Verde Islands in 1836, but in 1879 the two territories were separated and Guinea-Bissau became a colony, then called Portuguese Guinea. But development was slow, partly because the territory did not attract settlers on the same scale as Portugal's much healthier African colonies of Angola and Mozambique.

In 1956, African nationalists in Portuguese Guinea and Cape Verde founded the African Party for the Independence of Guinea and Cape Verde (PAIGC). Because Portugal seemed determined to hang on to its overseas territories, the PAIGC began a guerrilla war in 1963. By 1968, it held two-thirds of the country. In 1972, a rebel National Assembly, elected by the people in the PAIGC-controlled area, voted to make the country independent as Guinea-Bissau. In 1974, Guinea-Bissau became independent. It was followed by Cape Verde, which became independent in 1975.

POLITICS

In 1974, Guinea-Bissau faced many problems arising from its underdeveloped economy and its lack of trained people to work in the administration. One objective of the leaders of Guinea-Bissau was to unite their country with Cape Verde. But, in 1980, army leaders overthrew Guinea-Bissau's government. The Revolutionary Council, which took over, did not favor unification with Cape Verde and instead concentrated on national policies. It followed a socialist pro-

gram, introducing some reforms. It reduced adult illiteracy from 97% to 64% between 1974 and 1990. In 1991, the PAIGC ceased to be the only party and elections were held in 1994. Civil war broke out in 1998, continuing for 11 months. In May 1999, the rebels took power, but constitutional government was soon restored and Kumba Ialá became president. Attempted coups were put down in 2000 and 2001.

ECONOMY

Guinea-Bissau is a poor country. Agriculture employs more than 70% of the people, but most farmers produce little more than they need to feed their families. Major crops include beans, coconuts, groundnuts, maize, palm kernels, and rice.

Fishing is also important, and fish, including shrimps, make up about a third of the country's exports. Groundnuts and coconuts together make up about two-fifths of exports. Livestock are raised, and some hides and skins are exported. Guinea-Bissau has few mining or manufacturing industries.

GULF STATES

KUWAIT

The State of Kuwait is a constitutional monarchy that is ruled by the Amir, a member of the Sabah family. The Sabah dynasty has ruled the area since the 18th century. A low-lying, desert nation, Kuwait came under British influence in the 19th century and, in 1899, Britain became responsible for Kuwait's defense. Kuwait became fully independent in 1961. In August 1990, Iraq invaded Kuwait, claiming that it was legally a part of Iraq. The United Nations approved the use of force if Iraqi troops were not withdrawn by January 15, 1991. When Iraq refused to withdraw, a coalition force attacked on January 17. By the end of February, the Iraqis were defeated and military operations had ceased. Despite the devastation and the setting fire to more than 500 oil wells, oil and natural gas production remains the mainstay of Kuwait's economy.

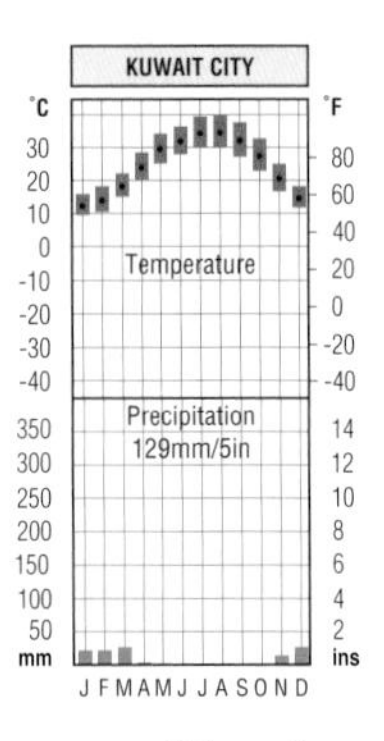

AREA 6,880 sq miles [17,820 sq km]
POPULATION 2,042,000
CAPITAL (POPULATION) Kuwait City (189,000)

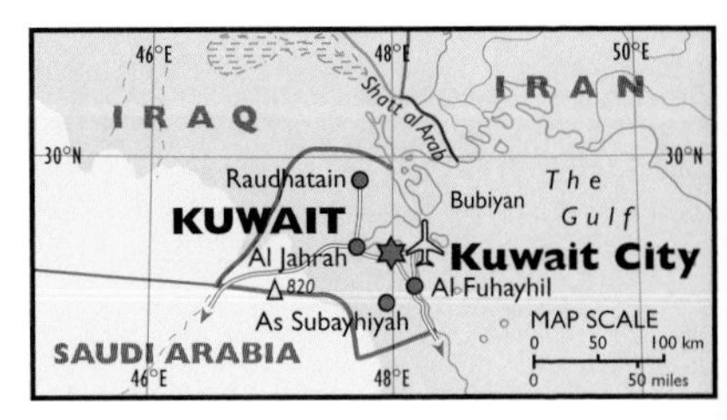

BAHRAIN

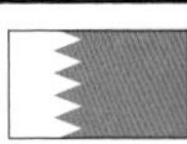

The Kingdom of Bahrain, whose flag dates from about 1932, consists of the island of Bahrain and 34 smaller islands. A causeway links this hot, mainly desert country to the Saudi Arabian mainland. A Muslim Arab country, Bahrain became a British protectorate in 1861 but became independent in 1971. Its economy depends on oil, which accounts for 80% of Bahrain's exports. Some of the money from oil sales is used to finance free social services. Bahrain has an adult literacy rate of 85%, one of the highest in the Gulf region. Free medical care is available.

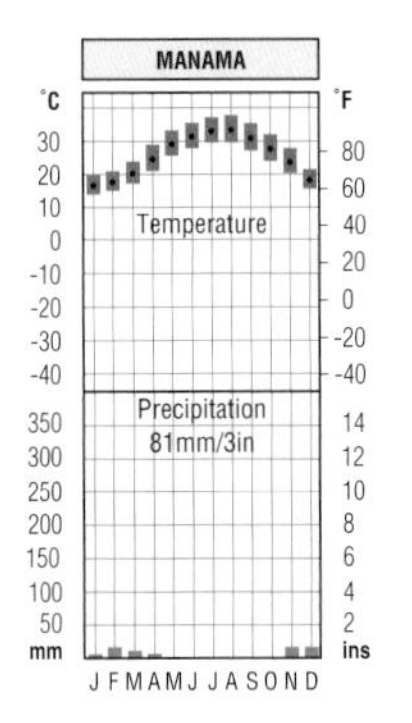

AREA 262 sq miles [678 sq km]
POPULATION 645,000
CAPITAL (POPULATION) Manama (or Al Manamah, 143,000)

QATAR

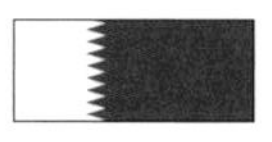

The State of Qatar is an Arab Emirate that occupies a peninsula which juts out from the Arabian peninsula into The Gulf. The low-lying land is mostly stony desert, with some barren salt flats. Much of the country's drinking water supply is distilled sea water. Qatar became a British protectorate in 1916, but in 1971 it became independent. Oil was first discovered in 1939 and oil accounts for about 90% of Qatar's exports. Money from oil sales has been used to create manufacturing industries and to develop agriculture and fishing. The oil industry also led to an expansion of Qatar's population. Many people came from other Arab countries and southern Asia to fill the new jobs.

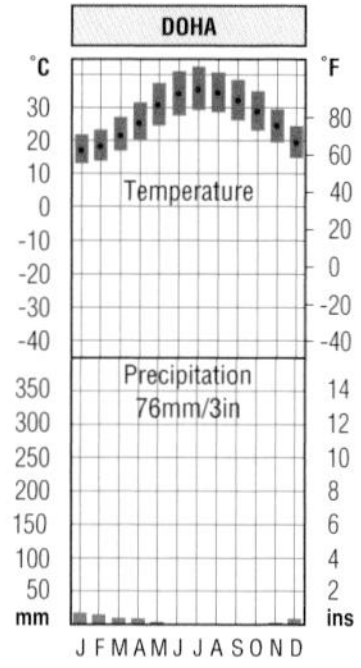

In 1995, Sheikh Hamad overthrew his father, Sheikh Khalifa bin Hamad Al-Thani, in a bloodless coup when his father was abroad. Sheikh Hamad then became Qatar's Emir.

AREA 4,247 sq miles [11,000 sq km]
POPULATION 769,000
CAPITAL (POPULATION) Doha (243,000)

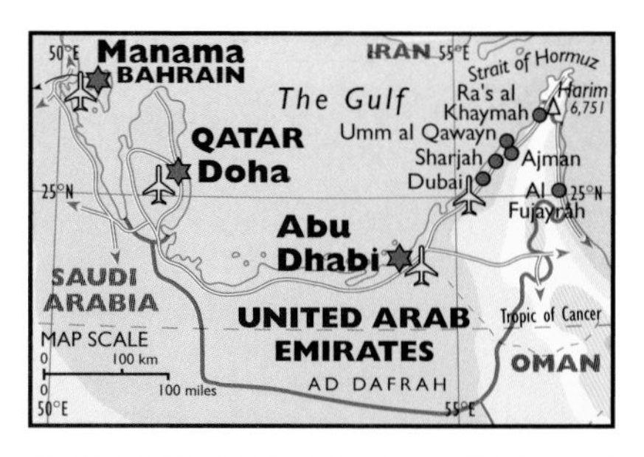

UNITED ARAB EMIRATES

The United Arab Emirates is a federation of seven Arab states on the southern end of The Gulf. They are Abu Dhabi (also called Abu Zaby), Dubai (Dubayy), Sharjah (Ash Shariqah), Ajman, Umm al Qaywayn, Ra's al Khaymah, and Al Fujayrah. Most of the land is hot low-lying desert. The country, formerly called the Trucial States, entered into treaties with Britain in the early 19th century, but it achieved full independence in 1971. The economy of this arid country depends on oil production.

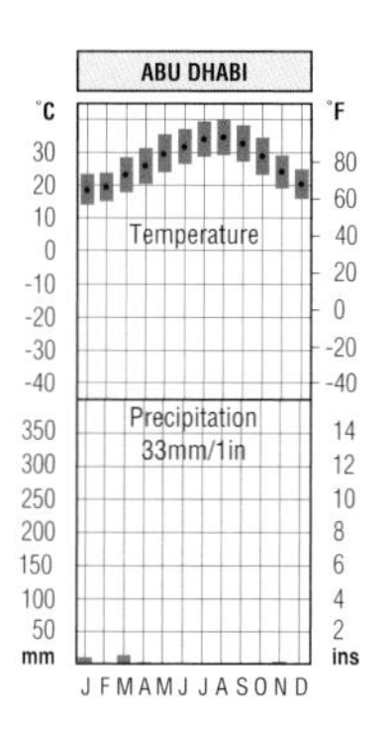

AREA 32,278 sq miles [83,600 sq km]
POPULATION 2,407,000
CAPITAL (POPULATION) Abu Dhabi (or Abu Zaby, 928,000)

GUYANA

Guyana's flag was adopted in 1966 when the country became independent from Britain. The colors symbolize the people's energy in building a new nation (red), their perseverance (black), minerals (yellow), rivers (white), and agriculture and forests (green).

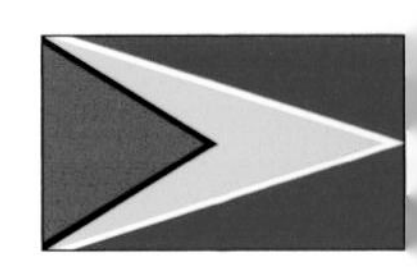

The Cooperative Republic of Guyana is a country facing the Atlantic Ocean in northeastern South America. The coastal plain is flat and much of it is below sea level. Dykes (sea walls) prevent flooding. Inland is a hilly region which rises to the Pakaraima Mountains, part of the Guiana Highlands, in the west. Other highlands are in the south.

Guyana has several spectacular waterfalls, including the King George VI Falls (1,601 ft [488 m]), the Great Falls (840 ft [256 m]), and the Kaieteur Falls (741 ft [226 m]).

AREA 83,000 sq miles [214,970 sq km]
POPULATION 697,000
CAPITAL (POPULATION) Georgetown (200,000)
GOVERNMENT Multiparty republic
ETHNIC GROUPS East Indian 49%, Black 32%, Mixed 12%, Amerindian 6%, Portuguese, Chinese
LANGUAGES English (official), Creole, Hindi, Urdu
RELIGIONS Protestant 34%, Roman Catholic 18%, Hinduism 34%, Islam 9%
CURRENCY Guyana dollar = 100 cents

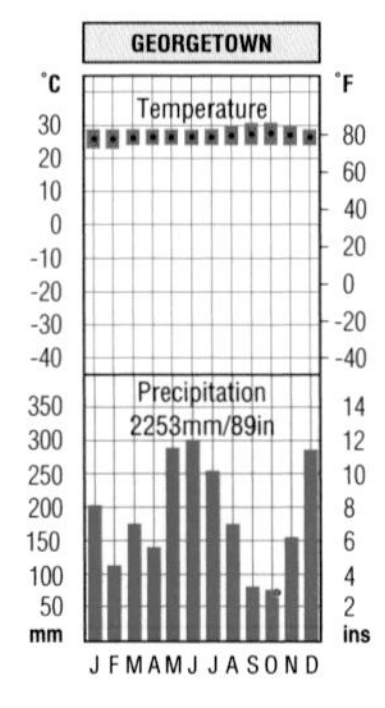

CLIMATE

Guyana has a hot and humid climate. Temperatures are lower in the highlands in the west and south. The rainfall is heavy. It occurs on more than 200 days a year. The rainfall decreases to the west and south, with a dry season between September and February.

VEGETATION

The coastal plain is mostly farmed, though there are some areas of wet savanna. But, inland, rain forests, containing valuable trees such as the greenheart, cover about 85% of the country.

HISTORY AND POLITICS

The Dutch founded a settlement in what is now Guyana in 1581. But, later on, Britain and France struggled for control of the area. Britain gained control in the early 19th century. In 1831, Britain set up the colony of British Guiana and abolished slavery in 1838, introducing Asian laborers to replace the slaves.

British Guiana became independent in 1966. A black lawyer, Forbes Burnham, became the first prime minister. Under a new constitution adopted in 1980, the powers of the president were increased. Burnham became president until his death in

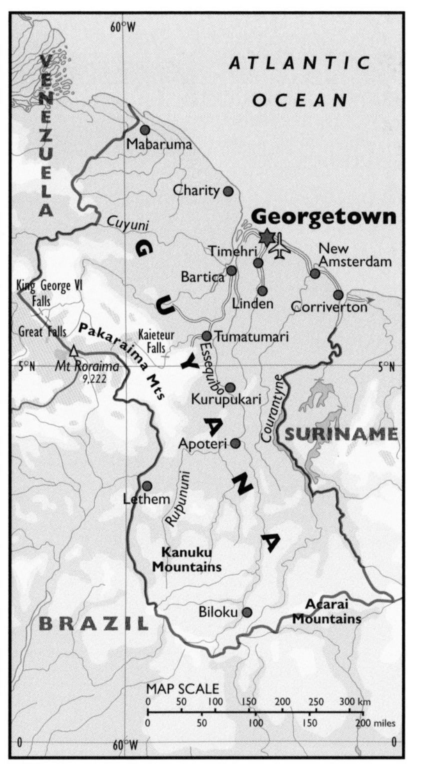

1985. In 1992, the East Indian Cheddi Jagan was elected president and, on his death in 1997, he was succeeded by his wife Janet, who retired in 1999. In 2001, Jagdeo Bharrat was elected president.

ECONOMY

Guyana is a poor, developing country. Its resources include gold, bauxite (aluminum ore), and other minerals, and its forests and fertile soils. Agriculture employs 20% of the people. Sugarcane and rice are the leading crops. Citrus fruits, cocoa, coffee, and plantains are also grown. Fishing and forestry are important. Electric power is in short supply, although the country has great hydro-electric potential.

Cricket is one of the legacies left behind in Guyana by the British colonial officials. For cricketing purposes, Guyana is considered to be part of the West Indies. This set of three stamps was issued in 1966 to commemorate a visit by an MCC (England) side to the West Indies.

HONDURAS

The flag of Honduras was officially adopted in 1949. It is based on the flag of the Central American Federation, which was set up in 1823 and included Costa Rica, El Salvador, Guatemala, Honduras, and Nicaragua. Honduras left this federation in 1838.

AREA 43,278 sq miles [112,090 sq km]

POPULATION 6,406,000

CAPITAL (POPULATION) Tegucigalpa (813,000)

GOVERNMENT Republic

ETHNIC GROUPS Mestizo 90%, Amerindian 7%, Black (including Black Carib) 2%, White 1%

LANGUAGES Spanish (official)

RELIGIONS Roman Catholic 85%, Protestant 10%

CURRENCY Honduran lempira = 100 centavos

The Republic of Honduras is a country situated in Central America. It has two coastlines. The northern coast on the Caribbean Sea extends more than 373 miles [600 km], but the Pacific coast in the southeast is only about 50 miles [80 km] long. The narrow coastal plain in the northwest is a banana-growing region, but the north-central coastlands are mainly undeveloped.

A broad plain in the northwest includes a region called Mosquitia. The interior of Honduras is mountainous. The highlands make up more than three-fifths of the country, but unlike some Central American countries, Honduras has no active volcanoes. Fertile valleys in the highlands are thickly populated. The Pacific coastlands in the southeast also contain rich farmland.

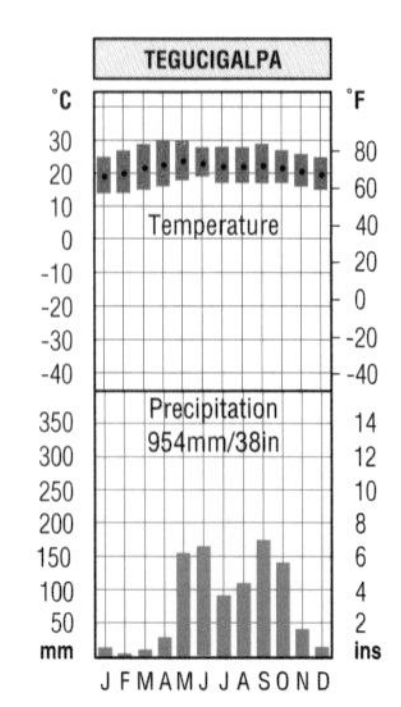

CLIMATE

Honduras has a tropical climate, but the highlands, where the capital Tegucigalpa is situated, have a cooler climate than the hot coastal plains. The months between May and November are the rainiest. The north coast is sometimes hit by fierce hurricanes that cause great damage.

VEGETATION

The northern coastal plains contain rain forest, savanna (tropical grassland with scattered pines), and mangrove swamps. Forests of evergreens, such as mahogany and rosewood, grow on the lower mountain slopes. Higher up are forests of oak and pine, with grassland at the highest levels.

HISTORY

The magnificent ruins at Copán show that western Honduras was part of the Mayan civilization until the 9th century. Christopher Columbus reached the coast in 1502 and claimed the area for Spain. Honduras declared its independence from Spain in 1821. Together with other Central American republics, it first joined Mexico, but in 1823 it became part of the Central American Federation. Honduras withdrew from the federation in 1838. However, throughout the rest of the 19th century, Honduras suffered from political instability.

In the 1890s, American companies developed plantations to grow bananas, which soon became the country's chief source of income. The companies exerted great political influence in Honduras and the country became known as a "banana republic," a name that was later applied to several other Latin American nations.

POLITICS

Instability has continued to mar the country's progress. In 1969, Honduras fought the short "Soccer War" with El Salvador. The war was sparked off by the treatment of fans during a World Cup soccer series. But the real reason was that Honduras had

forced Salvadoreans in Honduras to give up their land. In 1980, the countries signed a peace agreement.

Other problems arose when Honduras allowed US-backed "Contra" rebels from Nicaragua to operate in Honduras against Nicaragua's left-wing Sandinista government. A ceasefire between the Nicaraguan groups was signed in 1988, following which the "Contra" bases were closed down. Since 1980, Honduras has been ruled by civilian governments, though the military retain considerable influence.

> **DID YOU KNOW**
> - that the name Honduras comes from the Spanish word for "depths"; it refers to the deep waters off the northern coast
> - that Hurricane Mitch struck Honduras and Nicaragua during October 1998, causing floods and massive mudslides. The death toll was around 7,000, with perhaps twice as many people reported missing
> - that Tegucigalpa is one of the few world capitals which does not have a railroad

ECONOMY

Honduras is a developing country – one of the poorest in the Americas. It has few resources besides some silver, lead, and zinc, and agriculture dominates the economy. Bananas and coffee are the leading exports, and maize is the main food crop. Cattle are raised in the mountain valleys and on the southern Pacific plains. Shrimp and lobster fishing, and forestry are also important activities.

Honduras is the least industrialized country in Central America. Manufactures include processed food, textiles, and a wide variety of wood products.

HUNGARY

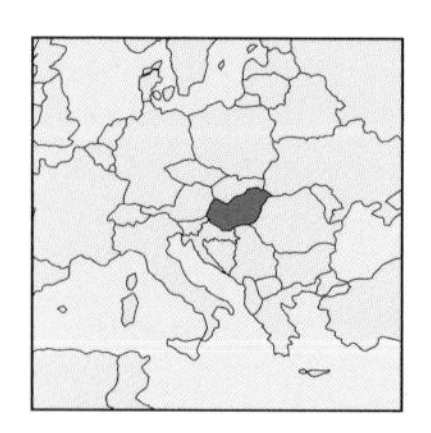

Hungary's flag was adopted in 1919. A state emblem was added in 1949 and removed in 1957. The colors of red, white, and green had been used in the Hungarian arms since the 15th century. The tricolor design became popular during the 1848 rebellion against Habsburg rule.

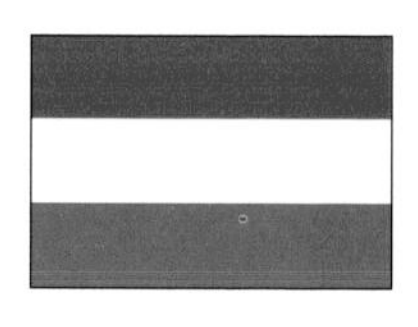

AREA 35,919 sq miles [93,030 sq km]
POPULATION 10,106,000
CAPITAL (POPULATION) Budapest (1,885,000)
GOVERNMENT Multiparty republic
ETHNIC GROUPS Magyar 90%, Gypsy, German, Croat, Romanian, Slovak
LANGUAGES Hungarian (official)
RELIGIONS Roman Catholic 64%, Protestant 23%, Orthodox 1%, Judaism 1%
CURRENCY Forint = 100 fillér

The Hungarian Republic is a landlocked country in central Europe. The land is mostly low-lying and drained by two rivers, the Danube (Duna) and its tributary, the Tisza. The main highlands, rising to a maximum height of 3,331 ft [1,015 m], are in the northeast. But most of the land east of the Danube belongs to a region called the Great Plain (Nagyalföld), which covers about half of Hungary.

West of the Danube is a hilly region, with some low mountains, called Transdanubia. This region contains the country's largest lake, Balaton. In the northwest is a small, fertile and mostly flat region called the Little Plain (Kisalföld).

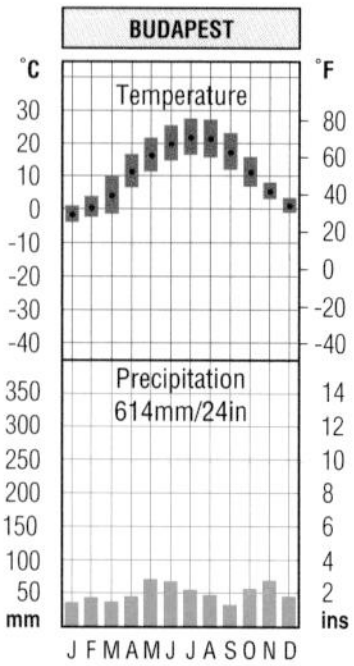

CLIMATE

Hungary lies far from the moderating influence of the sea. As a result, summers are warmer and sunnier, and the winters colder than in Western Europe. Thunderstorms make the spring a fairly wet season. In winter, snow covers the land for 30 to 40 days on average.

VEGETATION

Much of Hungary's original vegetation has been cleared to make way for farmland. The largest forests are in the scenic highlands in the northeast. Beech is common at higher levels, with oaks lower down.

HISTORY

Magyars first arrived in the area from the east in the 9th century. In the 11th century, Hungary's first king, Stephen I, made Roman Catholicism the official religion. Hungary became a powerful kingdom, but in 1526 it was defeated by Turkish forces, who later occupied much of Hungary. In the late 17th century, the Austrian Habsburgs conquered Hungary. In 1867, Austria granted Hungary equal status in a "dual monarchy," called Austria-Hungary. In 1914, a Bosnian

student killed the heir to the Austria-Hungary throne. This led to World War I, when Austria-Hungary fought alongside Germany. Defeat in 1918 led to loss of territory to neighboring countries, some of which now have Hungarian-speaking minorities.

Hungary entered World War II (1939–45) in 1941, as an ally of Germany, but the Germans occupied the country in 1944. The Soviet Union invaded Hungary in 1944 and, in 1946, the country became a republic. The Communists gradually took over the government, taking complete control in 1949.

POLITICS

From 1949, Hungary was an ally of the Soviet Union. In 1956, Soviet troops crushed an anti-Communist revolt. But in the 1980s, reforms in the Soviet Union led to the growth of anti-Communist groups in Hungary.

In 1989, Hungary adopted a new constitution making it a multiparty state for the first time in 40 years. Elections in 1990 led to a victory for the non-Communist Democratic Forum. Hungary's return to a democratic form of government led it to join the western alliance NATO in 1999. Then, in 2002, the country was invited to become a member of the European Union in May 2004.

ECONOMY

Before World War II, Hungary's economy was based mainly on agriculture. But the Communists set up many manufacturing industries. The new factories were owned by the government, as also was most of the land. However, from the late 1980s, the government has worked to increase private ownership. This change of policy caused many problems, including inflation and high rates of unemployment.

Manufacturing is the most valuable activity. The major products include aluminum made from local bauxite, chemicals, electrical and electronic goods, processed food, iron and steel, and vehicles.

Farming remains important. Major crops include grapes for wine-making, maize, potatoes, sugar beet, and wheat. Livestock include poultry, cattle, horses, and sheep.

Birdlife is abundant, especially in the northeast highlands of Hungary. This stamp, one of a set of six illustrating peafowl species which was issued in 1978, shows the peacock. The word Magyar *on the stamp means Hungarian.*

ICELAND

Iceland's flag dates from 1915. It became the official flag in 1944, when Iceland became fully independent. The flag, which uses Iceland's traditional colors, blue and white, is the same as Norway's flag, except that the blue and red colors are reversed.

The Republic of Iceland, in the North Atlantic Ocean, is a country in Europe, though it is closer to Greenland than Scotland. Iceland sits astride the Mid-Atlantic Ridge. It is slowly getting wider as the ocean is being stretched apart by continental drift. Molten lava wells up to fill the gap in the center of Iceland.

Iceland has around 200 volcanoes and eruptions are frequent. Geysers and hot springs are other common volcanic features. Ice caps and glaciers cover about an eighth of the land. The largest is Vatnajökull in the southeast. The only habitable regions are the coastal lowlands.

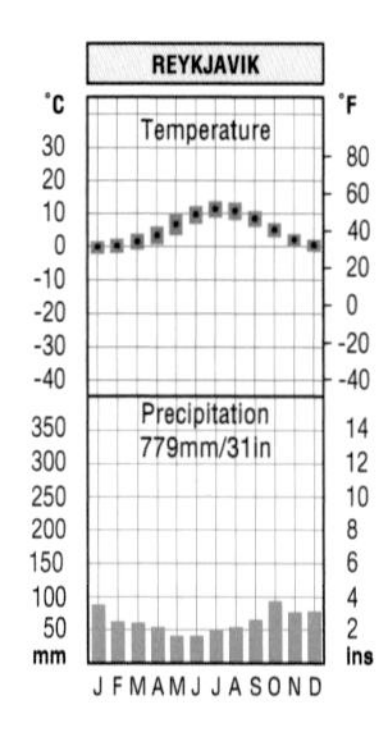

CLIMATE

Although it lies far to the north, Iceland's climate is moderated by the warm waters of the Gulf Stream. As a result, the port of Reykjavik is ice-free all the year round. The usual temperature in January, the coldest month, is 32°F [0°C], the same as in Copenhagen, Denmark.

VEGETATION

Vegetation is sparse or non-existent on three-quarters of the land. Treeless grassland or bogs cover some areas, while Iceland also has some spruce trees in sheltered areas.

AREA 39,768 sq miles [103,000 sq km]

POPULATION 278,000

CAPITAL (POPULATION) Reykjavik (103,000)

GOVERNMENT Multiparty republic

ETHNIC GROUPS Icelandic 97%, Danish 1%

LANGUAGES Icelandic (official)

RELIGIONS Evangelical Lutheran 92%, other Lutheran 3%, Roman Catholic 1%

CURRENCY Króna = 100 aurar

HISTORY AND POLITICS

Norwegian Vikings colonized Iceland in AD 874, and in 930 the settlers founded the world's oldest parliament, the Althing.

Iceland united with Norway in 1262. But when Norway united with Denmark in 1380, Iceland came under Danish rule. Iceland became a self-governing kingdom, united with Denmark, in 1918. It became a fully independent republic in 1944, following a referendum in which 97% of the people voted to break their country's ties with Denmark.

Iceland has played an important part in European affairs. It is a member of the North Atlantic Treaty Organization, which provides defense for its members in Europe and North America. When Iceland extended its territorial waters during the 1970s, it became involved in conflict with Britain. Other fishing disputes with Norway, Russia, and others continued through the 1990s.

ECONOMY

Iceland has few resources besides the fishing grounds which surround it. Fishing and fish processing are major industries which dominate Iceland's overseas trade.

Barely 1% of the land is used to grow crops, mainly root vegetables and fodder for livestock. But 23% of the country is used for grazing sheep and cattle. Iceland is self-sufficient in meat and dairy products. Vegetables and fruits are grown in greenhouses heated by water from hot springs. Manufacturing is important. Products include aluminum, cement, clothing, electrical equipment, fertilizers, and processed foods.

DID YOU KNOW

- that Iceland is sometimes called "the land of ice and fire"
- that water from hot springs is used to heat homes and offices in Reykjavik
- that Icelanders do not have family names and they are properly addressed by their first names. Their second name combines their father's first name, plus *-son* for males and *-dóttir* for females
- that the fishing industry employs 13% of Iceland's work force and accounts for about three-quarters of the country's exports
- that, in 1996, a volcanic eruption under the Vatnajökull ice cap created a huge subglacial lake, which burst causing severe flooding

Most of Iceland's people live in coastal towns, where many of them make their living either by fishing or by working in factories that process fish products. This stamp, one of a set of 13 issued in 1960 which depicted views of Iceland, shows one of the country's harbors. Iceland's dramatic scenery has been shaped partly by volcanic eruptions, spilling lava over the land, and partly by frost action and erosion by glaciers and ice caps.

INDIA

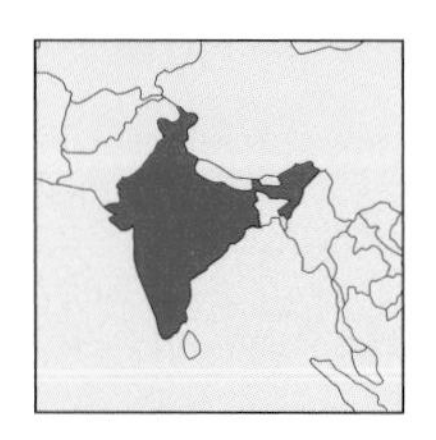

The Indian flag was adopted shortly after the country gained independence from Britain in 1947. The saffron (orange) represents renunciation, the white represents truth, and the green symbolizes mankind's relationship with nature. The central wheel represents dynamism and change.

The Republic of India is the world's seventh largest country. In population, it ranks second only to China. The north is mountainous, with mountains and foothills of the Himalayan range.

Rivers, such as the Brahmaputra and Ganges (Ganga), rise in the Himalayas and flow across the fertile northern plains. Southern India consists of a large plateau, called the Deccan. The Deccan is bordered by two mountain ranges, the Western Ghats and the Eastern Ghats.

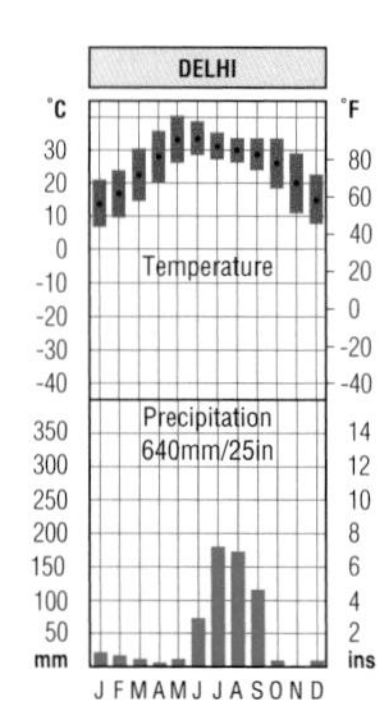

CLIMATE

India has three main seasons. The cool season runs from October to February. The hot season runs from March to June. The rainy monsoon season starts in the middle of June and continues into September. Delhi has a moderate rainfall, with about 25 inches [640 mm] a year. The southwestern coast and the northeast have far more rain. Darjeeling in the northeast has an average annual rainfall of 120 inches [3,040 mm]. But parts of the Thar Desert in the northwest have only 2 inches [50 mm] of rain per year.

AREA 1,269,338 sq miles [3,287,590 sq km]

POPULATION 1,029,991,000

CAPITAL (POPULATION) New Delhi (7,207,000)

GOVERNMENT Multiparty federal republic

ETHNIC GROUPS Indo-Aryan (Caucasoid) 72%, Dravidian (Aboriginal) 25%, other (mainly Mongoloid) 3%

LANGUAGES Hindi, English, Telugu, Bengali, Marati, Urdu, Gujarati, Malayalam, Kannada, Oriya, Punjabi, Assamese, Kashmiri, Sindhi and Sanskrit (all official)

RELIGIONS Hinduism 83%, Islam (Sunni Muslim) 11%, Christianity 2%, Sikhism 2%, Buddhism 1%

CURRENCY Rupee = 100 paisa

VEGETATION

Forests grow in the northern valleys, and bamboo forests are common in the northeast. The Thar Desert lies on the border with Pakistan, while palm trees grow on the dry Deccan plateau. The rainy southwest coast contains many tropical plants.

HISTORY

In southern India, most people are descendants of the dark-skinned Dravidians, who were among India's earliest people. In the north, most

This stamp, issued in 1976, shows Mohandas Karamchand Gandhi (1869–1948), who worked to free India of British rule through non-violent resistance. Indians call Gandhi the Mahatma, which means "Great Soul." Indians regard Gandhi as the father of their nation.

people are descendants of the lighter-skinned Indo-Aryans who arrived around 3,500 years ago.

India was the birthplace of several major religions, including Hinduism, Buddhism, and Sikhism. Islam was introduced from about AD 1000. The Muslim Mughal empire was founded in 1526. From the 17th century, Britain began to gain influence. From 1858 to 1947, India was ruled as part of the British Empire. Independence in 1947 led to the breakup of British India into India and Muslim Pakistan.

POLITICS

Although India has 15 major languages and hundreds of minor ones, together with many religions, the country has remained united as the world's largest democracy. It has faced problems with Pakistan over the disputed territory of Jammu and Kashmir, and with China. In 1998, the testing of nuclear devices in India and Pakistan led to fears of war. The situation remained tense into the 21st century, with periodic outbreaks of violence, but they did not lead to wider conflict.

ECONOMY

According to the World Bank, India is a "low-income" developing country. Many people in the crowded cities and the villages are extremely poor.

Farming employs more than 60% of the people. The main food crops are rice, wheat, millet, and sorghum, together with beans and peas. India has more cattle than any other country. These animals provide milk, but Hindus do not eat beef.

India has large mineral reserves, including coal, iron ore, and oil. Manufacturing has expanded greatly since 1947. Products include iron and steel, machinery, refined petroleum, textiles, and transport equipment.

BHUTAN

Bhutan is a kingdom between India and Tibet. The monarch is head of state and head of the government. The country has no constitution. The king appoints the Council of Ministers. The National Assembly (Tshogdu) has 150 members, two-thirds of whom are elected by the people.

The land is mountainous, being part of the high Himalayan range. Most people live in the warm, wet valleys in the south. According to the World Bank, Bhutan is one of the world's poorest developing countries. Farming employs nearly 90% of the people, whose chief food crops are barley, rice, and wheat.

AREA 18,147 sq miles [47,000 sq km]
POPULATION 2,049,000
CAPITAL (POPULATION) Thimphu (30,000)
LANGUAGES Dzongkha (official)
RELIGIONS Buddhism 75%, Hinduism
CURRENCY Ngultrum = 100 chetrum

NEPAL

Nepal is another remote kingdom between India and Tibet, but an increasing number of tourists are now visiting the country. Mount Everest, the world's highest peak at 29,035 ft [8,850 m], stands on the northern border. But most people live in the valleys south of the Himalayas. Nepal was the birthplace of the Buddha (Prince Siddhartha Gautama), but most Nepalis are Hindus. The king used to be an absolute ruler, but he introduced democratic reforms in 1990.

Nepal is a poor country, where farming employs more than 80% of the people. Exports include clothing, carpets, leather goods, and grain.

AREA 54,363 sq miles [140,800 sq km]
POPULATION 25,284,000
CAPITAL (POPULATION) Katmandu (535,000)
LANGUAGES Nepali (official)
CURRENCY Nepalese rupee = 100 paisa

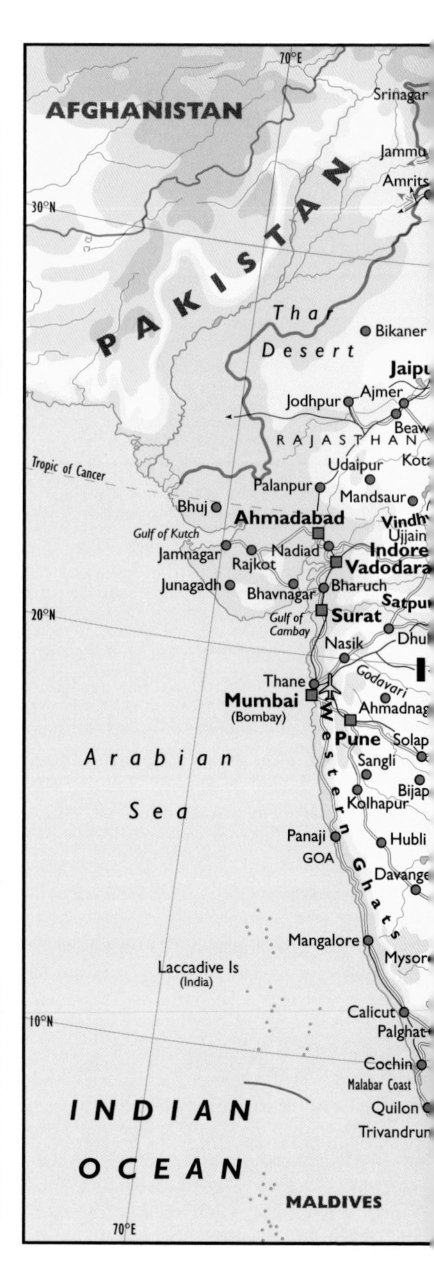

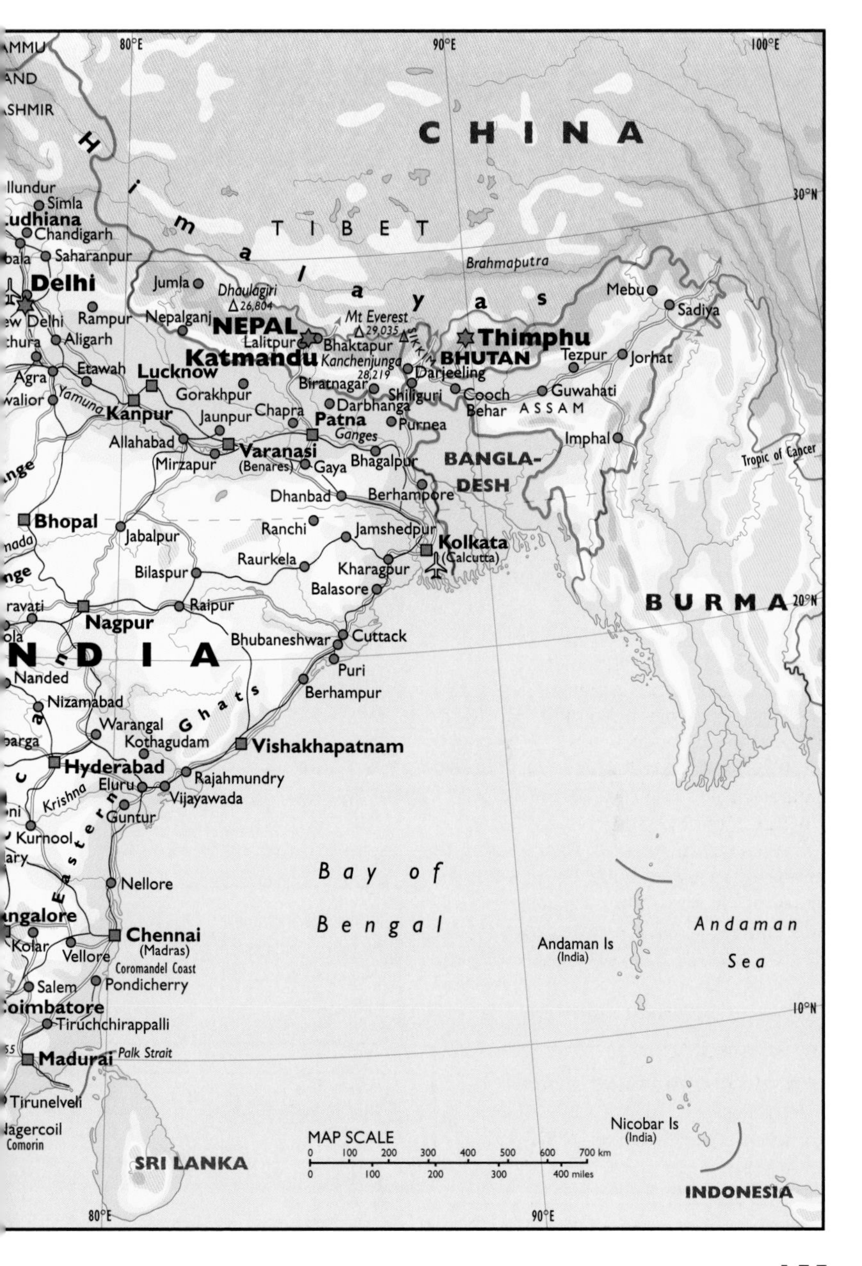

80°E
90°E
100°E
30°N
20°N
10°N
CHINA
TIBET
Himalayas
Brahmaputra
Simla
Chandigarh
Saharanpur
Delhi
Rampur
Aligarh
Agra
Etawah
Yamuna
Lucknow
Kanpur
Allahabad
Mirzapur
Jumla
Dhaulagiri
26,804
NEPAL
Nepalganj
Katmandu
Lalitpur
Bhaktapur
Mt Everest
29,035
Kanchenjunga
28,219
SIKKIM
Biratnagar
Gorakhpur
Jaunpur
Chapra
Darbhanga
Patna
Ganges
Varanasi
(Benares)
Gaya
Bhagalpur
Purnea
Shiliguri
Darjeeling
Thimphu
BHUTAN
Cooch
Behar
ASSAM
Tezpur
Jorhat
Guwahati
Mebu
Sadiya
Imphal
Tropic of Cancer
BANGLA-
DESH
Berhampore
Dhanbad
Ranchi
Jamshedpur
Kolkata
(Calcutta)
Raurkela
Kharagpur
Balasore
Bhopal
Jabalpur
Bilaspur
Raipur
Nagpur
BURMA
INDIA
Bhubaneshwar
Cuttack
Puri
Berhampur
Nanded
Nizamabad
Warangal
Kothagudam
Eastern Ghats
Vishakhapatnam
Hyderabad
Rajahmundry
Eluru
Vijayawada
Krishna
Guntur
Kurnool
Nellore
Chennai
(Madras)
Coromandel Coast
Kolar
Vellore
Salem
Pondicherry
Tiruchchirappalli
Madurai
Palk Strait
Tirunelveli
SRI LANKA
Bay of Bengal
Andaman Is
(India)
Andaman Sea
Nicobar Is
(India)
INDONESIA
MAP SCALE
0 100 200 300 400 500 600 700 km
0 100 200 300 400 miles

INDIAN OCEAN

COMOROS

The Federal Islamic Republic of the Comoros consists of three large, mountainous islands and some small coral islands. The period May to October is a cool, dry season. The other months are hot and rainy. France gained control of the islands by 1866. Independence was achieved in 1974, but one island, Mayotte, remained French. In 1997, the islands of Anjouan and Mohéli attempted to secede, but were offered instead a greater degree of self-rule. This poor country exports vanilla, cloves, and perfume oils.

AREA 861 sq miles [2,230 sq km]
POPULATION 596,000
CAPITAL (POPULATION) Moroni (22,000)

MALDIVES

The Republic of the Maldives is an archipelago of about 1,200 low-lying coral islands to the southwest of India. The climate is hot and humid. The country came under British control in 1887, but independence came in 1965. Tourism and fishing are the main industries.

AREA 115 sq miles [298 sq km]
POPULATION 311,000
CAPITAL (POPULATION) Malé (55,000)

MAURITIUS

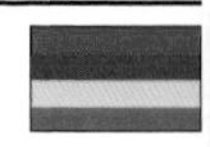

The Republic of Mauritius consists of the main island of Mauritius, Rodrigues, and various islets. The hot summer months (November to April) are also the wettest. The country was British from 1810 until it became an independent constitutional monarchy in 1968. Mauritius became a republic in 1992. Sugar production is in decline. However, tourism and textile manufacturing are growing in importance.

AREA 718 sq miles [1,860 sq km]
POPULATION 1,190,000
CAPITAL (POPULATION) Port Louis (144,000)

RÉUNION

Réunion, a French overseas department, is a mountainous island, which has been ruled by France since 1640. Sugar and sugar products account for about 80% of the exports.

AREA 969 sq miles [2,510 sq km]
POPULATION 733,000
CAPITAL (POPULATION) St-Denis (123,000)

SEYCHELLES

The Republic of Seychelles consists of four large islands and many islets. Mahé is the largest island. The

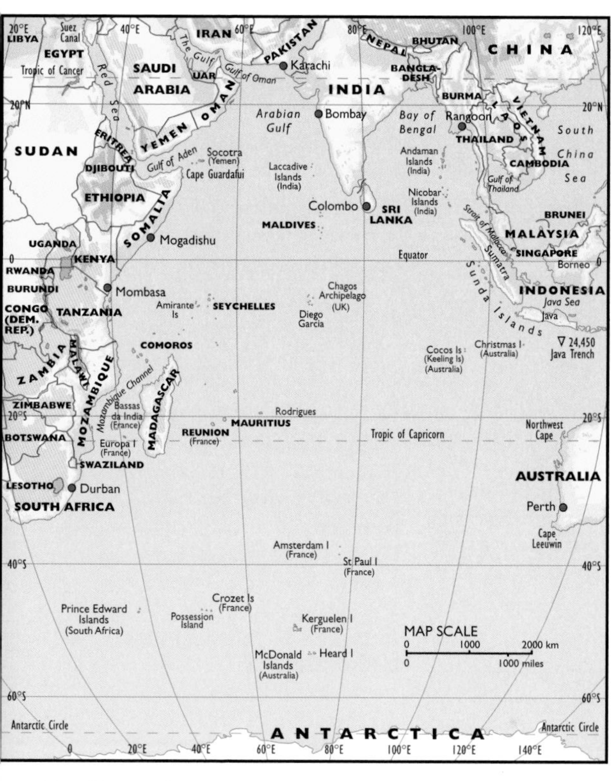

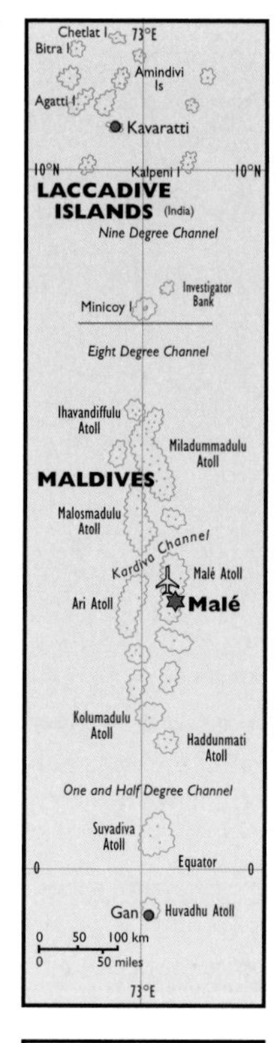

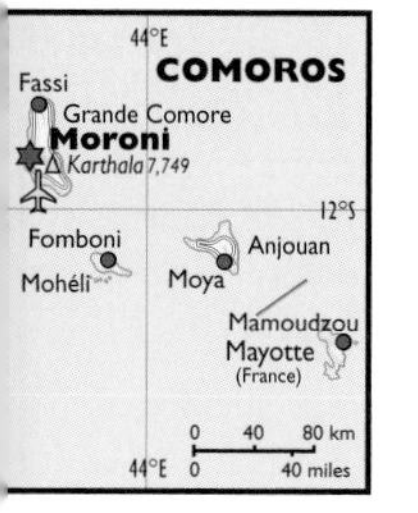

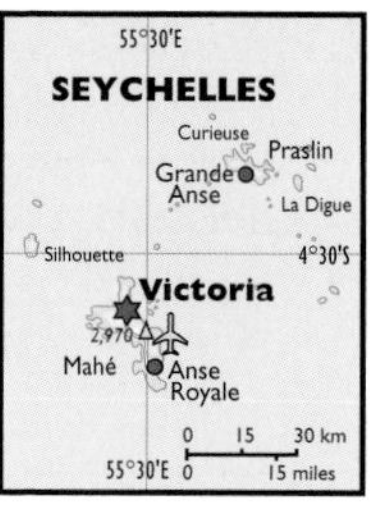

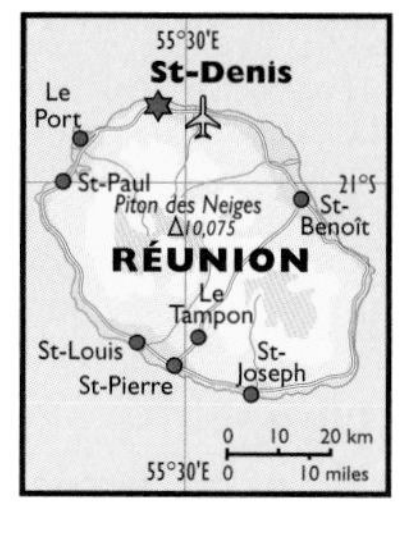

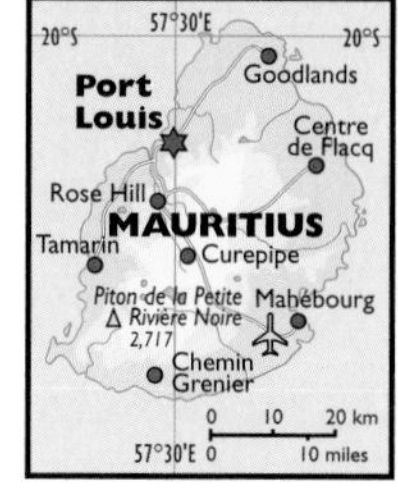

tropical climate is ideal for tourism, which is a major industry, together with fishing and farming. Major products include cinnamon, coconuts, and copra. Britain ruled the islands from 1814 until 1976.

AREA 176 sq miles [455 sq km]
POPULATION 80,000
CAPITAL (POPULATION) Victoria (30,000)

INDONESIA

This flag was adopted in 1945, when Indonesia proclaimed itself independent from the Netherlands. The colors, which date back to the Middle Ages, were adopted in the 1920s by political groups in their struggle against Dutch rule.

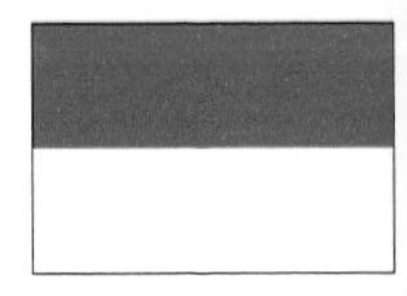

AREA 729,613 sq miles [1,889,700 sq km]
POPULATION 227,701,000
CAPITAL (POPULATION) Jakarta (11,500,000)
GOVERNMENT Multiparty republic
ETHNIC GROUPS Javanese 45%, Sundanese 14%, Madurese 7%, Coastal Malays 7%, more than 300 others
LANGUAGES Bahasa Indonesian (official)
RELIGIONS Islam 88%, Roman Catholic 3%, Hinduism 2%, Buddhism 1%
CURRENCY Indonesian rupiah = 100 sen

The Republic of Indonesia is an island nation in Southeast Asia. In all, Indonesia contains about 13,600 islands, less than 6,000 of which are inhabited. Three-quarters of the country is made up of five main areas: the islands of Sumatra, Java, and Celebes (Sulawesi), together with Kalimantan (southern Borneo) and Irian Jaya (western New Guinea). The islands are mountainous and Indonesia contains more active volcanoes than any other country, but the bigger islands also contain large coastal lowlands.

CLIMATE AND VEGETATION

Indonesia lies on the Equator and conditions are hot and humid throughout the year. The rainfall is generally heavy and only Java and the Sunda Islands have a relatively dry season.

Mangrove swamps line the coast while rain forests cover the lowlands. However, many forested areas have been cleared for farming.

HISTORY AND POLITICS

Indonesia is the world's most populous Muslim nation, though Islam was introduced as recently as the 15th century. The Dutch became active in the area in the early 17th century and Indonesia became a Dutch colony in 1799. After a long struggle, the Netherlands recognized Indonesia's independence in 1949. Since then, its progress has been marred by instability, though it has gradually expanded its economy. Around the turn of the century, conflict developed as various groups sought to secede from the country.

ECONOMY

Indonesia's resources include oil, natural gas, tin and other minerals, fertile soils, and its forests. Oil and gas are major exports. Timber, textiles, rubber, coffee, and tea are also exported. The chief food crop is rice. Manufacturing is increasing.

EAST TIMOR

The Republic of East Timor became independent on May 20, 2002. The land is mainly rugged. The climate is hot, but the rainfall is moderate. East Timor became Portuguese in the 19th century. But, when Portugal withdrew in 1975, Indonesia seized the area. Resistance steadily increased and the people voted for independence in 1999, which was achieved after great loss of life. East Timor is heavily dependent on foreign aid, but oil and natural gas deposits lie offshore.

AREA 5,731 sq miles [14,870 sq km]
POPULATION 737,000
CAPITAL (POPULATION) Dili (154,600)
ETHNIC GROUPS Timorese
LANGUAGES Tetum and Portuguese (both official), Indonesian and English (working languages)
CURRENCY US dollar = 100 cents

BRUNEI

The Islamic Sultanate of Brunei is a small country on the coast of northern Borneo. The climate is tropical and rain forests cover large areas. Formerly a British protectorate, Brunei became independent in 1984. It is now a prosperous country because of its oil and gas production. Revenue from oil and gas exports has made the Sultan possibly the world's richest man. The people pay no income tax. They enjoy many free services and have a high standard of living.

AREA 2,228 sq miles [5,770 sq km]
POPULATION 344,000
CAPITAL (POPULATION) Bandar Seri Begawan (55,000)
ETHNIC GROUPS Malay 69%, Chinese 18%, Indian
LANGUAGES Malay and English (both official), Chinese
CURRENCY Brunei dollar = 100 cents

IRAN

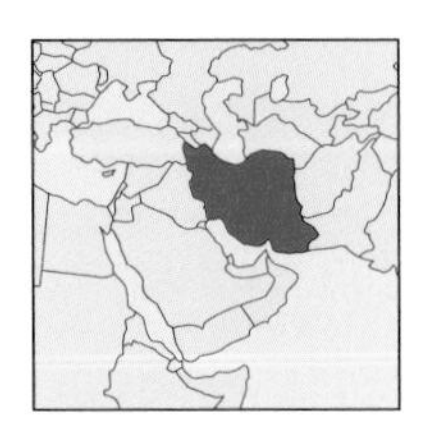

Iran's flag was adopted in 1980 by the country's Islamic government. The white strip contains the national emblem, which is the word for *Allah* (God) in formal Arabic script. The words *Allah Akbar* (God is Great) is repeated 11 times on both the green and red stripes.

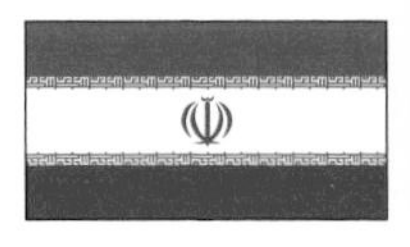

The Islamic Republic of Iran is a large country in southwestern Asia. The barren central plateau covers about half of the country. It includes the Dasht-e-Kavir (Great Salt Desert) and the Dasht-e-Lut (Great Sand Desert). The Elburz (or Alborz) Mountains north of the plateau contain Iran's highest point, Damavand. The narrow Caspian Sea lowlands lie between the mountains and the Caspian Sea. West of the plateau are the Zagros Mountains, beyond which the land descends to plains bordering The Gulf.

AREA 636,293 sq miles [1,648,000 sq km]
POPULATION 66,129,000
CAPITAL (POPULATION) Tehran (6,759,000)
GOVERNMENT Islamic republic
ETHNIC GROUPS Persian 51%, Azeri 24%, Gilaki/Mazandarani 8%, Kurd 7%, Arab, Lur, Baluchi, Turkmen
LANGUAGES Persian 58%, Turkic 26%, Kurdish
RELIGIONS Islam 99%
CURRENCY Rial = 100 dinars

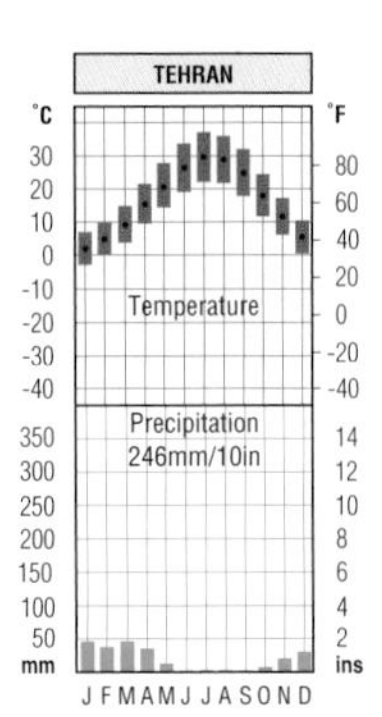

CLIMATE

Much of Iran has a severe, dry climate, with hot summers and cold winters. In Tehran, rain falls on only about 30 days in the year and the annual temperature range is more than 45°F [25°C]. The climate in the lowlands, however, is generally milder.

VEGETATION

Forests of such trees as beech and oak cover about a tenth of Iran, but semidesert and desert cover most of the country.

HISTORY AND POLITICS

Iran was called Persia until 1935. Ancient Persia was a powerful empire which flourished between 550 and 330 BC, when it fell to Alexander the Great. Islam was introduced in AD 641.

Britain and Russia competed for influence in the area in the 19th century, and in the early 20th century the British began to develop the country's oil resources. In 1925, the Pahlavi family took power. Reza Khan became *shah* (king) and worked to modernize the country. The Pahlavi dynasty was ended in 1979 when a religious leader, Ayatollah Ruhollah Khomeini, made Iran an Islamic republic. In 1980–8, Iran and Iraq fought a war over disputed borders. Khomeini died in 1989, but his fundamentalist views

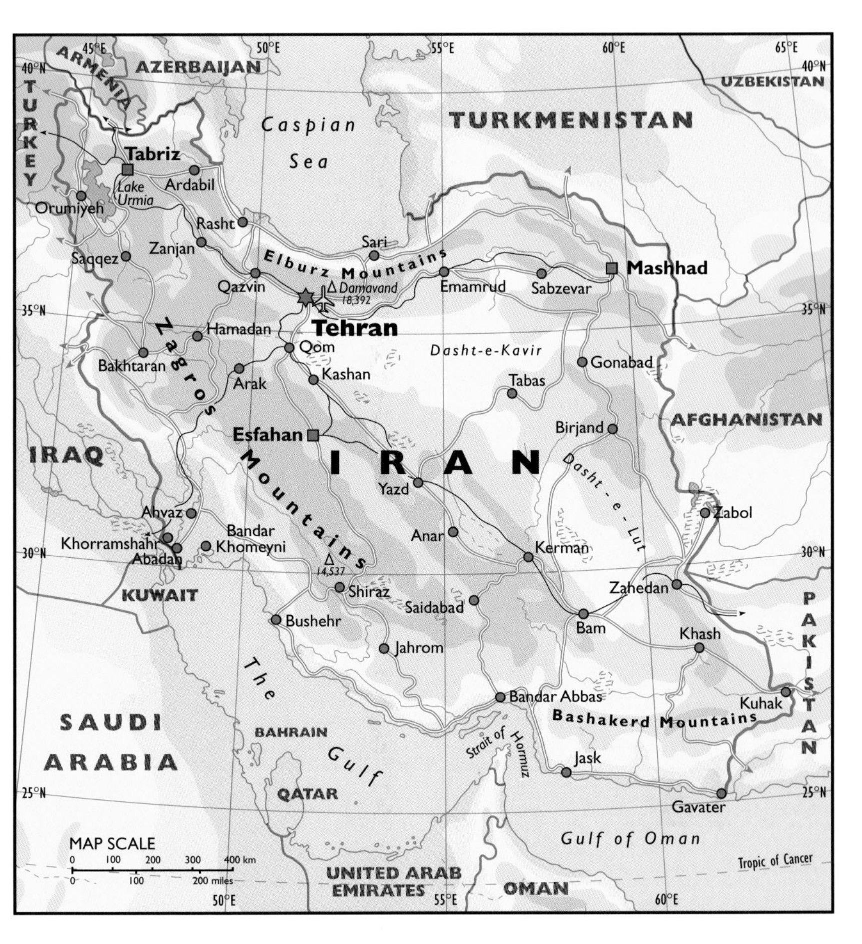

continued to be influential on Muslims around the world. In 1997, a liberal, Mohammed Khatami, was elected president. But many Muslim clerics opposed his policies and made it difficult for him to introduce reforms. Khatami's problems continued even after his re-election in 2001.

ECONOMY

Iran has large oil reserves, and oil and natural gas dominate its exports. Revenue from oil sales have been used to develop manufacturing industries.

Farms cover only about a tenth of the land, but agriculture remains an important part of the economy. The main crops are wheat and barley. Livestock and fishing in the Caspian Sea and The Gulf are other leading activities, though Iran has to import much of its food.

IRAQ

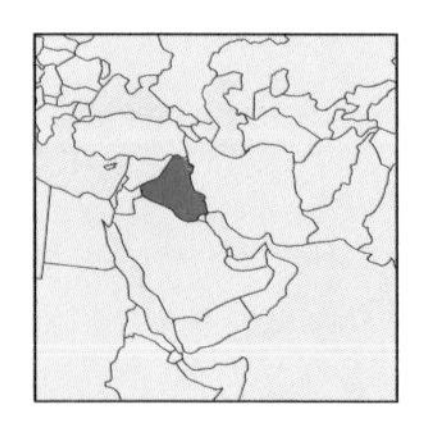

Iraq's flag was adopted in 1963, when the country was planning to federate with Egypt and Syria. It uses the four Pan-Arab colors. The three green stars symbolize the three countries. Iraq retained these stars even though the union failed to come into being.

The Republic of Iraq is a southwest Asian country at the head of The Gulf. Rolling deserts cover western and southwestern Iraq, with mountains in the northeast. The northern plains, across which flow the rivers Euphrates (Nahr al Furat) and Tigris (Nahr Dijlah), are dry. But the southern plains, including Mesopotamia, and the delta of the Shatt al Arab, the river formed south of Al Qurnah by the combined Euphrates and Tigris, contain irrigated farmland, together with swamps and marshes.

AREA 169,235 sq miles [438,320 sq km]

POPULATION 23,332,000

CAPITAL (POPULATION) Baghdad (3,841,000)

GOVERNMENT Republic

ETHNIC GROUPS Arab 77%, Kurdish 19%, Turkmen, Persian, Assyrian

LANGUAGES Arabic (official), Kurdish (official in Kurdish areas)

RELIGIONS Islam 96%, Christianity 4%

CURRENCY Iraqi dinar = 20 dirhams = 1,000 fils

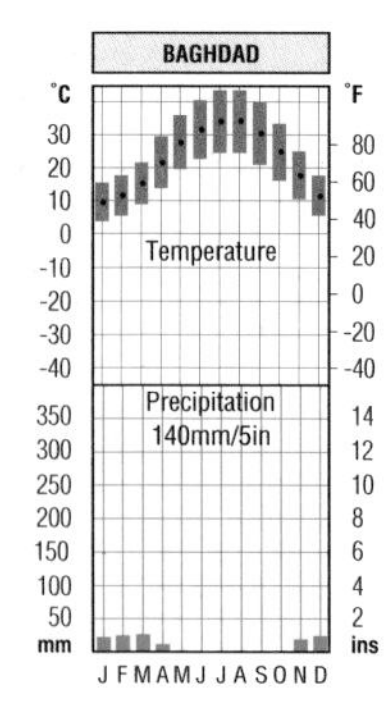

CLIMATE

The climate of Iraq varies from temperate in the north to subtropical in the south and east. Baghdad, in central Iraq, has cool winters, with occasional frosts, and hot summers. The rainfall is generally low. Most rain falls between the months of November and April.

VEGETATION

Forests grow on only 3% of the land, with dry grassland and low shrubs in the north, and thorny plants in the western and southern deserts. Grasses, sedges, and reeds grow in the southern marshes.

HISTORY

Mesopotamia was the home of several great civilizations, including Sumer, Babylon, and Assyria. It later became part of the Persian empire. Islam was introduced in AD 637 and Baghdad became the brilliant capital of the powerful Arab empire. But Mesopotamia declined after the Mongols invaded it in 1258. From 1534, Mesopotamia became part of the Turkish Ottoman empire. Britain invaded the area in 1916. In 1921, Britain renamed the country Iraq and set up an Arab monarchy. Iraq finally became independent in 1932.

By the 1950s, oil dominated Iraq's

economy. In 1952, Iraq agreed to take 50% of the profits of the foreign oil companies. This revenue enabled the government to pay for welfare services and development projects. But many Iraqis felt that they should benefit more from their oil.

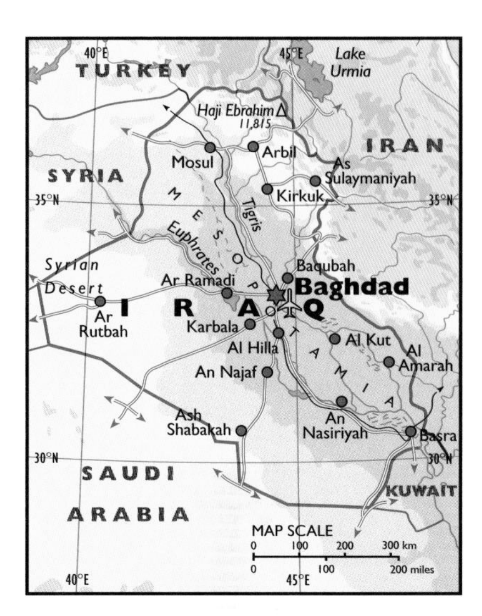

POLITICS

Since 1958, when army officers killed the king and made Iraq a republic, the country has undergone turbulent times. In the 1960s, the Kurds, who live in northern Iraq and also in Iran, Turkey, Syria, and Armenia, asked for self-rule. The government rejected their demands and war broke out. A peace treaty was signed in 1975, but conflict has continued.

In 1979, Saddam Hussein became Iraq's president. Under his leadership, Iraq invaded Iran in 1980, starting an eight-year war. During this war, Iraqi Kurds supported Iran and the Iraqi government attacked Kurdish villages with poison gas.

DID YOU KNOW

- that Mesopotamia in Iraq means "land between rivers"
- that the Garden of Eden in the Bible is said to have been near the meeting point of the Tigris and Euphrates rivers
- that the world's oldest civilization developed at Sumer in Mesopotamia in about 3500 BC
- that Iraq was the second largest oil producer before the wars in the 1980s and 1990s
- that the Marsh Arabs live in raft houses in the lagoons and reed beds of southern Iraq

In 1990, Iraqi troops occupied Kuwait but an international force drove them out in 1991. Since 1991, Iraqi troops have attacked Shiite Marsh Arabs and Kurds. In 1998, Iraq failed to permit UN officials, charged with disposing of Iraq's deadliest weapons, access to all suspect sites. In 2003, a Western force led by the US and the UK invaded Iraq with the aim of overthrowing Saddam's regime.

ECONOMY

Civil war, war damage, UN sanctions, economic mismanagement, and the invasion in 2003 created a state of economic chaos. Oil is the chief resource. Farmland, including pasture, covers about a fifth of the country. Products include barley, cotton, dates, fruits, livestock, wheat, and wool, but Iraq has to import food. Oil refining and the manufacturing of petrochemical and consumer goods are also important.

IRELAND

Ireland's flag was adopted in 1922 after the country had become independent from Britain, though nationalists had used it as early as 1848. Green represents Ireland's Roman Catholics, orange the Protestants, and the white a desire for peace between the two.

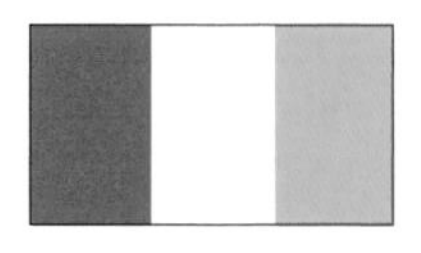

The Republic of Ireland occupies five-sixths of the island of Ireland. The country consists of a large lowland region surrounded by a broken rim of low mountains. The lowlands include peat bogs, where the peat (formed of partly decayed plants) is dug up and used as fuel. The uplands include the Mountains of Kerry where Carrauntoohill, Ireland's highest peak at 3,417 ft [1,041 m], is situated. The River Shannon is the longest in the British Isles. It flows through three large lakes, loughs Allen, Ree, and Derg.

AREA 27,135 sq miles [70,280 sq km]
POPULATION 3,841,000
CAPITAL (POPULATION) Dublin (1,024,000)
GOVERNMENT Multiparty republic
ETHNIC GROUPS Irish 94%
LANGUAGES Irish and English (both official)
RELIGIONS Roman Catholic 93%, Protestant 3%
CURRENCY Euro = 100 cents

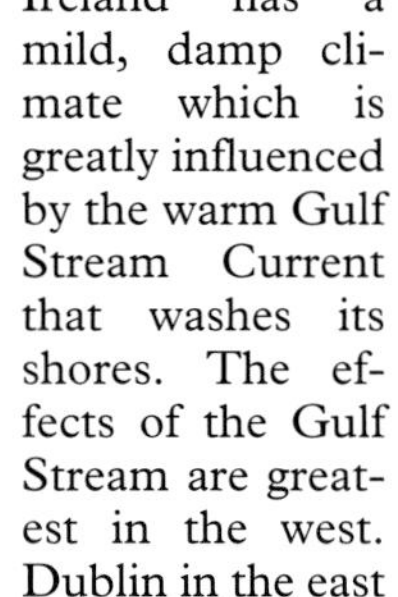

CLIMATE

Ireland has a mild, damp climate which is greatly influenced by the warm Gulf Stream Current that washes its shores. The effects of the Gulf Stream are greatest in the west. Dublin in the east is cooler than places on the west coast. Rain occurs throughout the year.

VEGETATION

Forests cover only about 5% of Ireland. More than 70% of the land is pasture and another 14% is under crops.

HISTORY

Most Irish people are descendants of waves of immigrants who settled on the island over a long period. Celts settled in Ireland from about 400 BC. They were followed later by Vikings, Normans, and the English.

Vikings raided Ireland from the 790s and in the 8th century they established settlements. But Norse domination was ended in 1014 when they were defeated by Ireland's king, Brian Boru. The Normans arrived in 1169 and, gradually, Ireland came under English influence. Much of Ireland's history after that time was concerned with the struggle against British rule and, from the 1530s, the preservation of Roman Catholicism.

In 1801, the Act of Union created the United Kingdom of Great Britain and Ireland. But Irish discontent intensified in the 1840s when a potato blight caused a famine in which a million people died and nearly a million emigrated. Britain was blamed for not having done enough to help.

In 1916, an uprising in Dublin was crushed, but between 1919 and 1922 civil war occurred. In 1922, the Irish Free State was created as a Dominion in the British Commonwealth. But Northern Ireland remained part of the UK.

POLITICS

Ireland became a republic in 1949. Since then, Irish governments have sought to develop the economy, and it was for this reason that Ireland joined the European Community in 1973. Irish reunification remains a major issue, supported by all of the country's political parties. But Ireland opposes the terrorist tactics of the IRA (Irish Republican Army). In 1998, it helped to negotiate an agreement aimed at ending conflict in Northern Ireland,

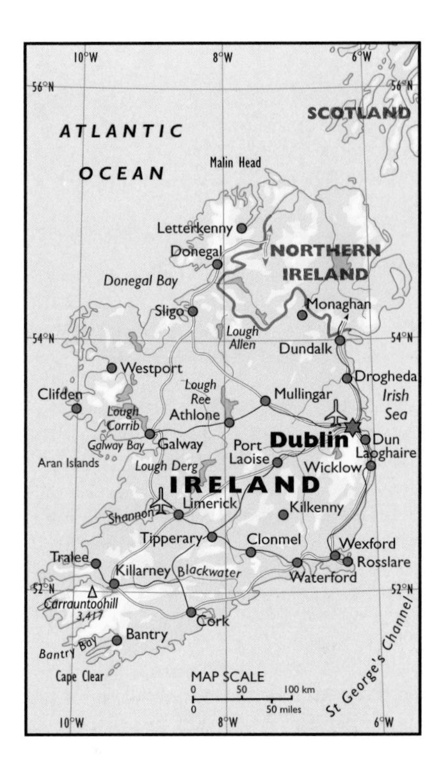

and it removed claims on Northern Ireland from its constitution.

DID YOU KNOW

- that Ireland has no snakes; according to legend, Saint Patrick, Ireland's patron saint, drove them out; Saint Patrick introduced Christianity to Ireland in 432
- that the title of Ireland's prime minister is the *taoiseach*, which is pronounced "tee-shuk"
- that Ireland's name in Gaelic (Irish) is Eire; another poetic name is Erin
- that Ireland's national symbol, the three-leaved shamrock, was used by Saint Patrick to explain the idea of the Holy Trinity

ECONOMY

Major farm products in Ireland include barley, cattle and dairy products, pigs, potatoes, poultry, sheep, sugar beet, and wheat, while fishing provides another valuable source of food. Farming is now a prosperous activity, aided by grants from the European Union, but manufacturing is now the most valuable economic sector. Many factories use farm products to make beverages and processed food. Others produce chemicals and pharmaceuticals, electronic equipment, machinery, paper, and textiles.

ISRAEL

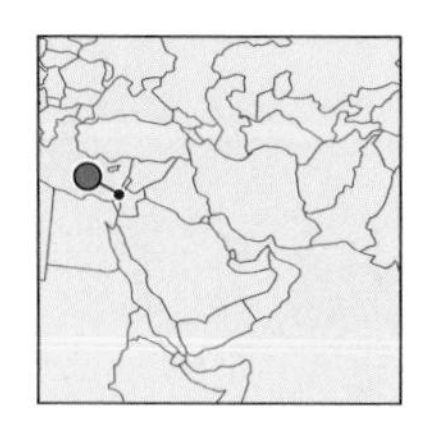

Israel's flag was adopted when the Jewish state declared itself independent in 1948. The blue and white stripes are based on the *tallit*, a Hebrew prayer shawl. The ancient, six-pointed Star of David is in the center. The flag was designed in America in 1891.

The State of Israel is a small country in the eastern Mediterranean. It includes a fertile coastal plain, where Israel's main industrial cities, Haifa (Hefa) and Tel Aviv-Jaffa, are situated. Inland lie the Judaeo-Galilean highlands, which run from northern Israel to the northern tip of the Negev Desert in the south. To the east lies part of the Great Rift Valley which runs through East Africa as far south as Mozambique. In Israel, the Rift Valley contains the River Jordan, the Sea of Galilee, and the Dead Sea.

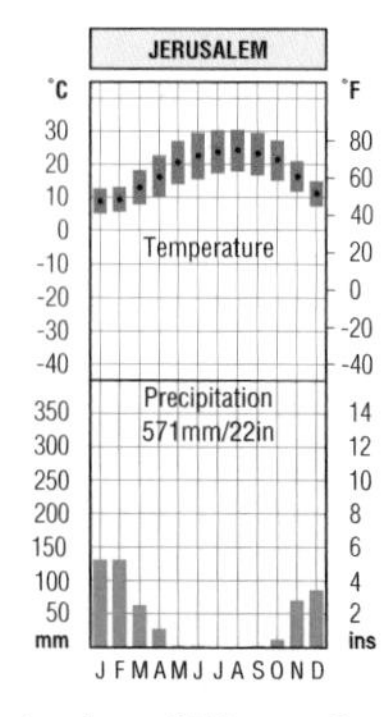

CLIMATE

Israel has hot, dry, sunny summers. Winters are mild and moist on the coast, but the total rainfall decreases from west to east and also from north to south, where the Dead Sea region has only 2.5 inches [70 mm] a year. Jerusalem has lower temperatures than coastal areas.

VEGETATION

Most of Israel's forest has been cut down. Farms covers about a fifth of Israel and pasture another two-fifths. Israel also has areas of *maquis* (shrub vegetation) and desert scrub.

AREA 7,960 sq miles [20,600 sq km]
POPULATION 5,938,000
CAPITAL (POPULATION) Jerusalem (591,000)
GOVERNMENT Multiparty republic
ETHNIC GROUPS Jewish 82%, Arab and others 18%
LANGUAGES Hebrew and Arabic (both official)
RELIGIONS Judaism 80%, Islam (mostly Sunni) 14%, Christianity 2%, Druse and others 2%
CURRENCY New Israeli sheqel = 100 agorat

HISTORY

Israel is part of a region called Palestine. Some Jews have always lived in the area, though most modern Israelis are descendants of immigrants who began to settle there from the 1880s. Britain ruled Palestine from 1917. Large numbers of Jews escaping Nazi persecution arrived in the 1930s. This provoked a local Arab uprising against British rule.

In 1947, the UN agreed to partition Palestine into an Arab and a Jewish state. The Arabs rejected the plan and fighting broke out. The State of Israel came into being in May 1948, but fighting continued into 1949. Other Arab-Israeli wars in 1956, 1967, and 1973 led to land gains for Israel.

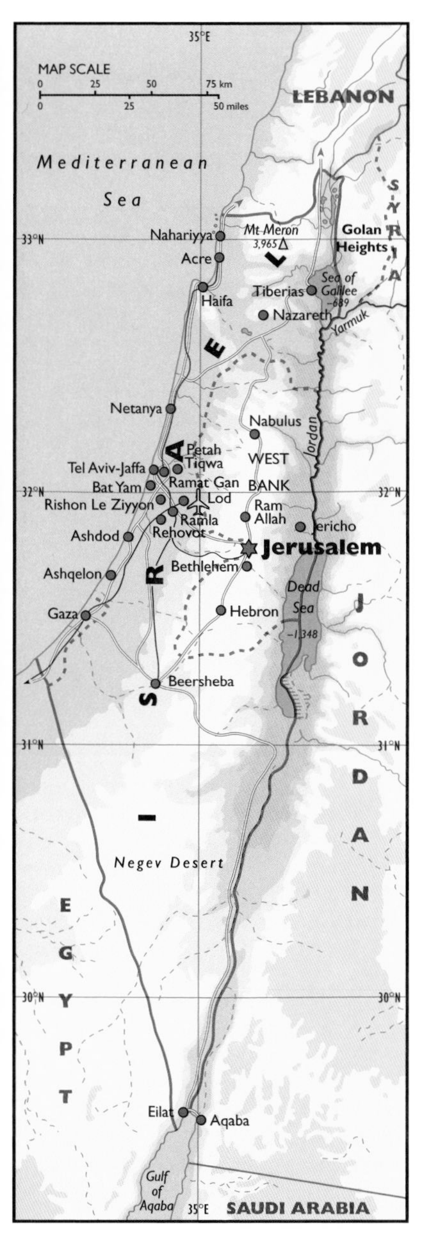

POLITICS

In 1978, Israel signed a treaty with Egypt which led to the return of the occupied Sinai peninsula to Egypt in 1979. But conflict continued between Israel and the PLO (Palestine Liberation Organization). In 1993, the PLO and Israel agreed to establish Palestinian self-rule in two areas: the occupied Gaza Strip, and in the town of Jericho in the occupied West Bank. The agreement was extended in 1995 to include more than 30% of the West Bank. Israel's prime minister, Yitzhak Rabin, was assassinated in 1995. His successor, Simon Peres, was defeated in elections in 1996. The peace process stalled. Following the victory of Ariel Sharon in 2001, conflict between government troops and Palestinians intensified in the occupied territories.

ECONOMY

Israel is a prosperous country. The most valuable activity is manufacturing, and the country's products include chemicals, electronic equipment, fertilizers, jewelry, military equipment, processed food, and scientific instruments. Manufactures are the main exports. Farming and tourism are also important.

This stamp, issued in 1968, shows some of Israel's exports.

ITALY

The Italian flag is based on the military standard carried by the French Republican National Guard when Napoleon invaded Italy in 1796, causing great changes in Italy's map. It was finally adopted as the national flag after Italy was unified in 1861.

The Republic of Italy is famous for its history and traditions, its art and culture, and its beautiful scenery. Northern Italy is bordered in the north by the high Alps, with their many climbing and skiing resorts. The Alps overlook the northern plains which are drained by the River Po. This is Italy's most fertile and densely populated region.

The Apennines, a long mountain range, forms the backbone of southern Italy. Bordering the Apennines are hilly areas and coastal plains.

Southern Italy contains a string of volcanoes, stretching from Vesuvius, near Naples (Nápoli), through the Lipari Islands, to Mount Etna on Sicily. Sicily is the largest island in the Mediterranean. Sardinia is also part of Italy.

AREA 116,320 sq miles [301,270 sq km]
POPULATION 57,680,000
CAPITAL (POPULATION) Rome (2,654,000)
GOVERNMENT Multiparty republic
ETHNIC GROUPS Italian 94%, German, French, Slovene, Albanian, Greek
LANGUAGES Italian 94% (official), German, French, Slovene
RELIGIONS Roman Catholic 83%, Protestant, Judaism, Islam
CURRENCY Euro = 100 cents

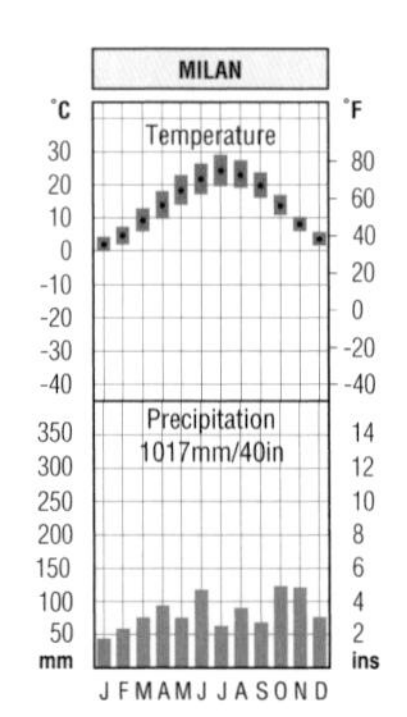

CLIMATE

Milan (Milano), in the north, has cold, often snowy winters. But the summer months are warm and sunny. Rain is plentiful all year. In the summer, it often falls during brief but powerful thunderstorms. Southern Italy has mild, moist winters and warm, dry summers.

VEGETATION

Forests of beech, chestnut, and oak grow on the lower slopes of the Alps. The northern plains are a patchwork of fields and grasslands. In southern Italy, the destruction of forests and overgrazing by goats and sheep have reduced some areas to scrubland.

HISTORY

Magnificent ruins throughout Italy testify to the glories of the ancient Roman Empire, which was founded, according to legend, in 753 BC. It reached its peak in the AD 100s. It finally collapsed in the 400s, although the Eastern Roman Empire, also called the Byzantine Empire, survived for another 1,000 years.

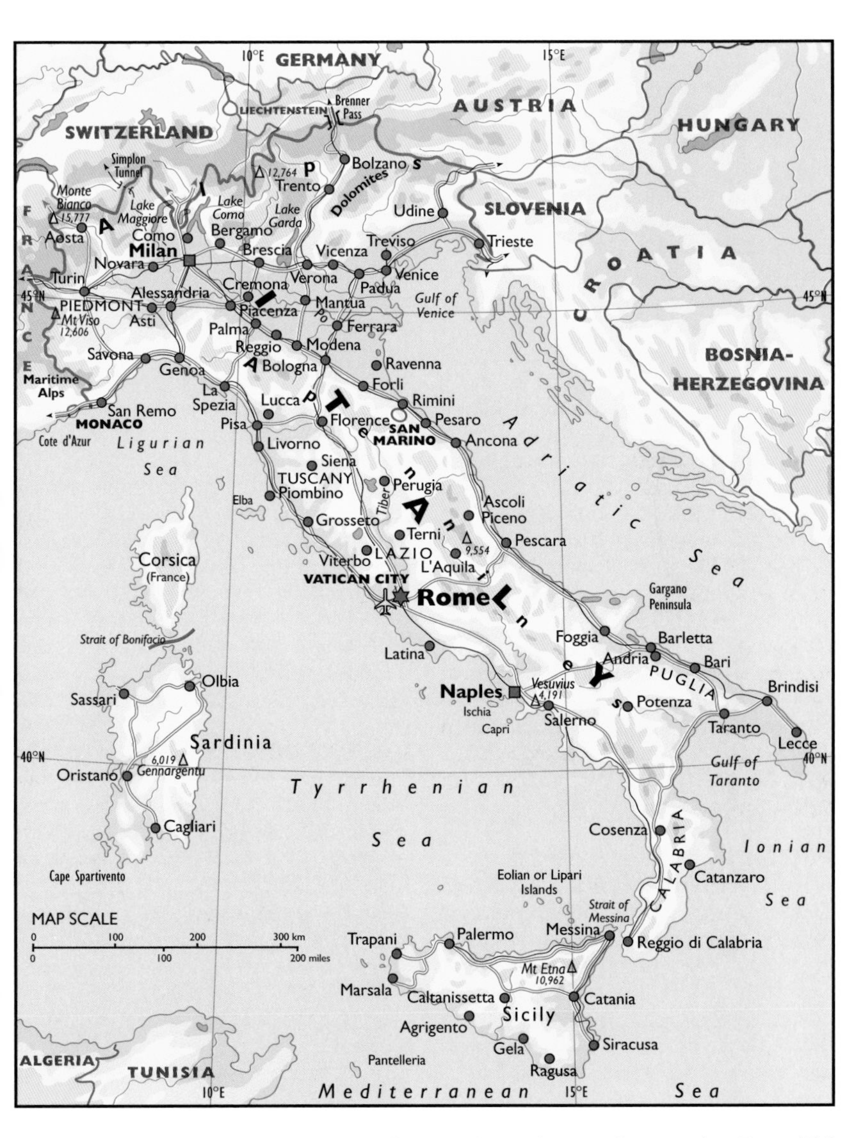

In the Middle Ages, Italy was split into many tiny states. But they made a great contribution to the revival of art and learning, called the Renaissance, in the 14th to 16th centuries. Beautiful cities, such as Florence (Firenze) and Venice (Venézia), testify to the artistic achievements of this period.

ITALY

Italy finally became a united kingdom in 1861, although the Papal Territories (a large area ruled by the Roman Catholic Church) was not added until 1870. The Pope and his successors disputed the takeover of the Papal Territories. The dispute was finally resolved in 1929, when the Vatican City was set up in Rome as a fully independent state.

Italy fought in World War I (1914–18) alongside the Allies – Britain, France, and Russia. In 1922, the dictator Benito Mussolini, leader of the Fascist party, took power. Under Mussolini, Italy conquered Ethiopia. During World War II (1939–45), Italy at first fought on Germany's side against the Allies. But in late 1943, Italy declared war on Germany.

Italy abolished the monarchy in 1946 and became a republic. The country played an important part in European affairs. It was a founder member of the defense alliance called the North Atlantic Treaty Organization (NATO) in 1949, and also of the European Union in 1958.

POLITICS

After the setting up of the European Union, Italy's economy developed quickly. But the country faced many problems. For example, much of the economic development was in the north. This forced many people to leave the poor south to find jobs in the north or abroad. Social problems, corruption at high levels of society, and a succession of weak coalition governments all contributed to instability. In 1994, new political groupings emerged. A right-wing coalition was formed, but its leader, Silvio Berlusconi, was forced to resign in December. Elections in 1996 were won by the left-wing Olive Tree alliance. But, in 2001, a right-wing alliance led by Berlusconi won a majority in parliament.

ECONOMY

Only 50 years ago, Italy was a mainly agricultural society. But today it is a leading industrial power. The country lacks mineral resources, and most raw materials used in industry, including oil, are imported. Leading manufactures include textiles and clothing, processed foods, machinery, cars, and chemicals. The main industrial region is in the northwest, in the area bounded by Milan (Milano), Turin (Torino), and Genoa (Génova).

Farmland covers around 42% of the land, pasture 17%, and forest and woodland 22%. Major crops include citrus fruits, grapes which are used to make wine, olive oil, sugar beet, and vegetables. Livestock farming is important, though meat is imported.

The Pontifical Academy of Sciences is the subject of this stamp issued by Vatican City in 1984.

VATICAN CITY

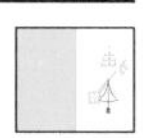

Vatican City is the world's smallest independent nation. It covers only 0.17 sq miles [0.44 sq km] and has a population of around 890. The Vatican City, which lies in north-western Rome, is the base for the Holy See, the governing body of the Roman Catholic Church, which is headed by the Pope. Vatican City has its own newspaper and radio station. It also has its own police and railroad station, and it issues its own stamps and coins.

Vatican City includes St Peter's Square in front of the huge St Peter's Basilica. The state's treasures include the frescos (wall paintings) in the Sistine Chapel by the great artist Michelangelo. Vatican City also has museums full of great paintings, statues, and archaeological treasures. The Vatican Library contains a major collection of early manuscripts and many valuable books.

The state's flag, which shows the triple tiara of the Popes above the keys of heaven given to St Peter, was adopted in 1929.

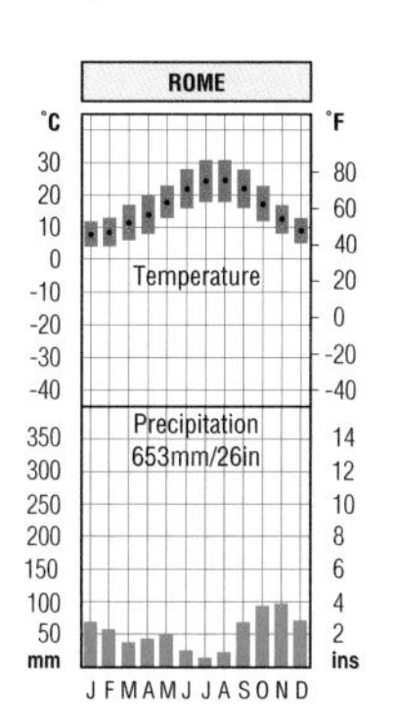

CLIMATE

Vatican City, like southern Italy, has warm and dry summers. The winters are usually mild and moist. But cold spells may sometimes occur between the months of November and March.

SAN MARINO

San Marino is a small independent republic. It is completely surrounded by Italy. Its official name is "The Most Serene Republic of San Marino." The white band in the flag symbolizes snowy mountains, and the blue band represents the sky. These colors are derived from the state coat of arms.

According to tradition, San Marino was founded in the AD 300s. The country has been independent since 885 and a republic since the 14th century. This makes it the world's oldest and smallest republic. The country has a 60-member legislature which is called the Great and General Council.

San Marino lies in the eastern Apennine Mountains. Summers are warm, but the winter months can be fairly cold. Tourism is the chief industry and it provides the country's main source of income. Farmers produce barley, fruit (notably grapes), animal hides, wheat, and vegetables. Ceramics, leather goods, textiles, and tiles are manufactured.

AREA 24 sq miles [61 sq km]
POPULATION 27,000
CAPITAL (POPULATION) San Marino (2,395)
GOVERNMENT Republic
ETHNIC GROUPS San Marinese, Italian
LANGUAGES Italian (official)
RELIGIONS Roman Catholic
CURRENCY Euro = 100 cents

IVORY COAST

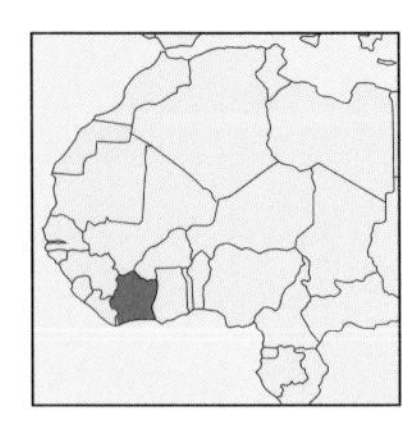

This flag was adopted in 1960 when the country became independent from France. It combines elements from the French tricolor and the Pan-African colors. The orange represents the northern savanna, the white peace and unity, and the green the forests in the south.

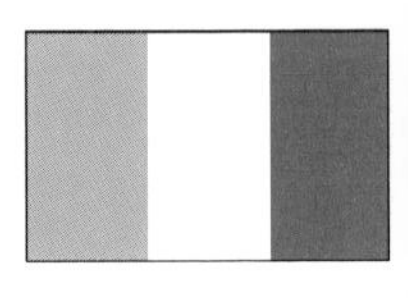

The Republic of the Ivory Coast, in West Africa, is officially known as Côte d'Ivoire. The southeast coast is bordered by sand bars that enclose lagoons, on one of which the former capital and chief port of Abidjan is situated. But the southwestern coast is lined by rocky cliffs. Behind the coast is a coastal plain, but the land rises inland to high plains. The highest land is an extension of the Guinea Highlands in the northwest, along the borders with Liberia and Guinea. The rivers run generally north–south.

AREA 124,502 sq miles [322,460 sq km]
POPULATION 16,393,000
CAPITAL (POPULATION) Yamoussoukro (120,000)
GOVERNMENT Multiparty republic
ETHNIC GROUPS Akan 42%, Voltaic 18%, Northern Mande 16%, Kru 11%, Southern Mande 10%
LANGUAGES French (official), Akan, Voltaic
RELIGIONS Christianity 34%, Islam 27%, traditional beliefs 17%
CURRENCY CFA franc = 100 centimes

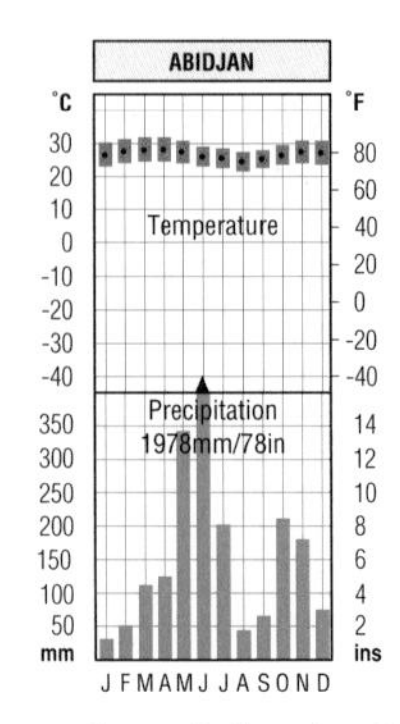

CLIMATE

Ivory Coast has a hot and humid tropical climate, with high temperatures throughout the year. The south of the country has two distinct rainy seasons: between May and July, and from October to November. Inland, the rainfall decreases. Northern Ivory Coast has a dry season and only one rainy season.

VEGETATION

Rain forests, containing such valuable trees as mahogany and African teak, once covered the southern lowlands, but much of the land has been cleared for farming. To the north, the forests merge into savanna (tropical grassland with scattered trees). Forests cover much of the Guinea Highlands.

HISTORY AND POLITICS

European contact with the region dates back to the late 15th century. Trade in ivory and slaves soon became important. French trading posts were founded in the late 17th century and Ivory Coast became a French colony in 1893. From 1895, Ivory Coast was governed as part of French West Africa, a massive union which also included what are now Benin, Burkina Faso, Guinea, Mali, Mauritania, Niger, and Senegal. In 1946, Ivory Coast became a territory in the French Union.

Ivory Coast became fully independent in 1960 and its first president, Félix Houphouët-Boigny, became the longest serving head of state in Africa with an uninterrupted period in office which ended with his death in 1993. Houphouët-Boigny was a paternalistic, pro-Western leader, who made his country a one-party state. In 1983, the National Assembly agreed to move the capital from Abidjan to Yamoussoukro, the president's birthplace, which officially became the capital in 1990. In 1993, Henri Konan Bédié became president and he was re-elected in 1995. A coup led by General Robert Guei occurred in 1999, but a civilian government under veteran Laurent Gbagbo was restored in 2000. In 2002, civil war broke out after an army rebellion.

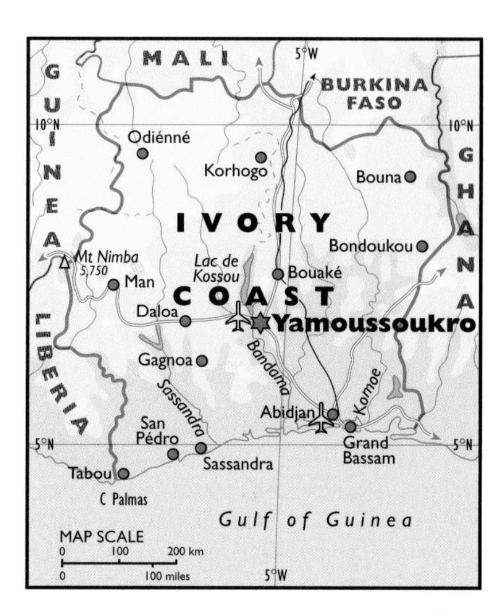

ECONOMY

Agriculture employs about two-thirds of the people, and farm products, notably cocoa beans, coffee, cotton, and cotton cloth, make up nearly half of the value of the total exports. Other important export items include bananas, palm oil, pineapples, and tropical woods. Food crops include cassava, rice, vegetables, and yams. Manufacturing has grown in importance since independence and products include fertilizers, processed food, refined oil, textiles, and timber.

DID YOU KNOW

- that Ivory Coast is the world's leading producer of cocoa beans; it ranks sixth in coffee production
- that Ivory Coast got its name from the trade in elephants' tusks between the 15th and 19th centuries; slaves were the area's other major export
- that the new Roman Catholic church at Yamoussoukro, the new capital of Ivory Coast, is the largest in the world

More than 6% of the land in Ivory Coast is now protected in order to ensure the survival of many animal species. This stamp, showing an antelope called a duiker, is one of a set of three entitled "Animals in danger of extinction." This set was issued in 1979.

JAPAN

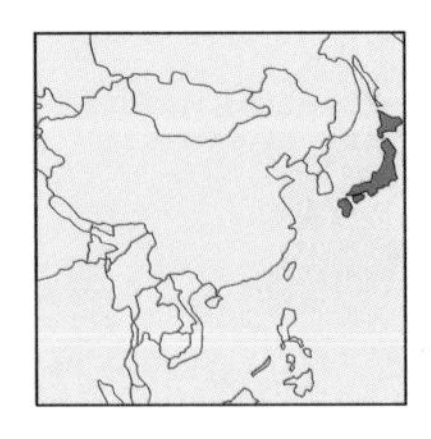

Japan's flag was officially adopted in 1870, though Japanese emperors had used this simple design for many centuries. The flag shows a red sun on a white background. The geographical position of Japan is also expressed in its name *Nippon* or *Nihon*, meaning "source of the Sun."

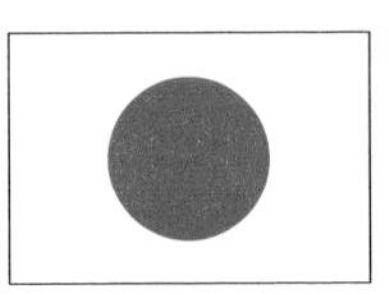

Japan is an island nation in northeastern Asia. It is a constitutional monarchy, with an emperor as its head of state.

Japan contains four large islands. In order of size, they are Honshu, Hokkaido, Kyushu, and Shikoku. These islands make up more than 98% of the country. But Japan also has thousands of small islands, including the Ryukyu island chain which extends south of Kyushu toward the island of Taiwan.

The four main islands are mainly mountainous, while many of the small islands are the tips of volcanoes rising from the sea bed. Japan has more than 150 volcanoes, about 60 of which are active. Volcanic eruptions, earthquakes and *tsunamis* (destructive sea waves triggered by underwater eruptions and earthquakes) often occur in Japan. For example, an earthquake in 1995 killed over 5,000 people in Kobe. This is because the islands lie on a unstable part of the Earth where the land is constantly moving.

Around the coast are small, fertile plains mainly covered by fertile alluvium deposited by the short rivers that rise in the mountains. Most of Japan's people live on these plains. One densely populated zone stretches from the Kanto plain, where Tokyo is situated, along the narrow plains on the southern coasts of Honshu, to northern Kyushu.

AREA 145,869 sq miles [377,800 sq km]

POPULATION 126,772,000

CAPITAL (POPULATION) Tokyo (17,950,000)

GOVERNMENT Constitutional monarchy

ETHNIC GROUPS Japanese 99%, Chinese, Korean, Ainu

LANGUAGES Japanese (official)

RELIGIONS Shintoism and Buddhism 84% (most Japanese consider themselves to be both Shinto and Buddhist)

CURRENCY Yen = 100 sen

CLIMATE

The climate of Japan varies greatly from north to south. Hokkaido in the north has cold, snowy winters. At Sapporo, temperatures below 4°F

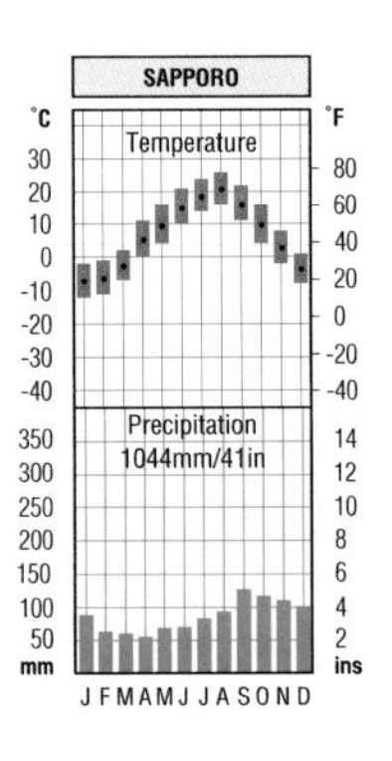

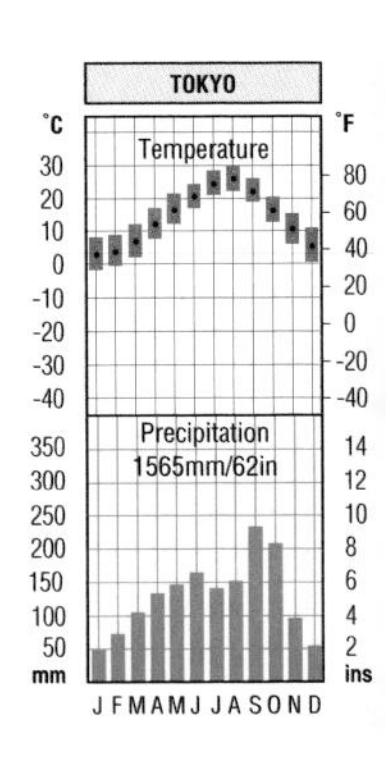

[–20°C] have been recorded between December and March. But summers are warm, with temperatures sometimes exceeding 86°F [30°C]. Rain falls throughout the year, though Hokkaido is one of the driest parts of Japan.

Tokyo has higher rainfall and temperatures, though frosts may occur as late as April when northwesterly winds are blowing. The southern islands of Shikoku and Kyushu have warm temperate climates. Summers are long and hot. Winters are mild.

VEGETATION

Forests and woodland cover about two-thirds of the land. In the north, the forests are mostly made up of such trees as fir and spruce, which can withstand the cold winters. Central Japan has mixed forests, with beech, maple, and oak blanketing the scenic mountain slopes. Deciduous trees, which shed their leaves in the fall, dominate in the south. The cherry tree, known for its spring blossoms, is found throughout the country.

HISTORY

Among the earliest people in Japan were the ancestors of the Ainu, whose origins are disputed by scientists. Today, this group numbers only about 15,000, most of whom live on Hokkaido.

However, the ancestors of most Japanese came from mainland Asia, probably through Korea, though some may also have come from islands to the south. Early areas to be settled were northern Kyushu and the coasts of Setonaiki (the Inland Sea).

According to legend, Japan's monarchy dates back to 660 BC. From the early 12th century, power passed increasingly to warrior leaders called *shoguns*. The *shoguns* claimed to be the protectors of the emperors and ruled in their names. Under the Tokugawa *shogunate*, which began in 1603, Japan enjoyed a long period of stability and prosperity. The Tokugawa family was finally overthrown in 1867 and, in 1868, the traditional monarchy was restored.

The name of the country inscribed on this stamp, which was issued in 1987, is Nippon, *meaning Japan. The kimono is the traditional garment worn by both women and men in Japan. It is now worn on special occasions. For everyday use most city dwellers wear modern clothes.*

European contact began when Portuguese sailors reached Japan in 1543. In the 1630s, Japan began a period of isolation from the rest of the world. However, in 1853 and 1854, Commodore Matthew C. Perry forced the Tokugawa *shogunate* to open its ports to Western trade.

In the late 19th century, Japan began a program of modernization. Under its new imperial leaders, it began to look for lands to conquer. In 1894–5, it fought a war with China

JAPAN

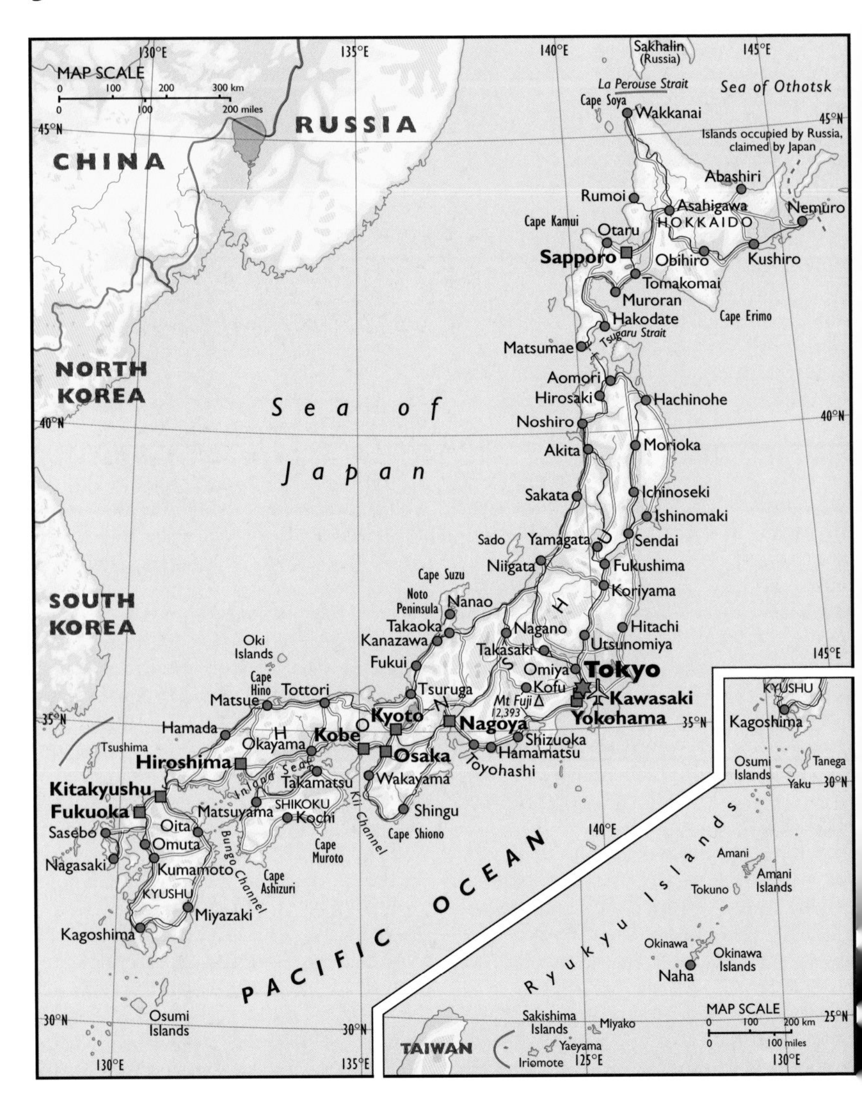

and, in 1904–5, it defeated Russia. Soon its overseas empire included Korea and Taiwan. In 1930, Japan invaded Manchuria (northeast China) and, in 1937, it began a war against China. In 1941, Japan launched an attack on the US base at Pearl Harbor in Hawaii. This drew both Japan and the United States into World War II.

Japan surrendered in 1945 when

the Americans dropped atomic bombs on two cities, Hiroshima and Nagasaki. The United States occupied Japan until 1952. During this period, Japan adopted a democratic constitution. The emperor, who had been regarded as a god, became a constitutional monarch. Power was vested in the prime minister and cabinet, who are chosen from the Diet (elected parliament).

DID YOU KNOW

- that only the United States has a higher economic output than Japan
- that the world's most destructive earthquake ever recorded struck Tokyo and Yokohama in 1923; about 575,000 dwellings were destroyed
- that Japan claims the southern Kuril Islands, which lie to the northeast of Hokkaido, from Russia; the Japanese call these islands the Northern Territories
- that the undersea Saikan Tunnel, linking Honshu and Hokkaido, is the world's longest commercial rail tunnel; it is 33.5 miles [53.9 km] long
- that Mount Fuji (Fuji-San), Japan's famous snow-capped volcano, last erupted in 1707

POLITICS

From the 1960s, Japan experienced many changes as the country rapidly built up new industries. By the early 1990s, Japan had become the world's second richest economic power after the US. But economic success has brought problems. For example, the rapid growth of industrial cities has led to housing shortages and pollution. Another problem is that the proportion of people over 65 years old is steadily increasing.

Japan, whose economic success depends on exporting its industrial goods, faces problems with its trading partners, who have tried to persuade Japan to lift trade barriers that restrict imports from other countries. The growth of the economy started to slow down in the 1990s and, in the early 2000s, the government was working to restore the economy.

ECONOMY

Japan has the world's second highest gross domestic product (GDP) after the United States. [The GDP is the total value of all goods and services produced in a country in one year.] The most important sector of the economy is industry. Yet Japan has to import most of the raw materials and fuels it needs for its industries. Its success is based on its use of the latest technology, its skilled and hard-working labor force, its vigorous export policies, and its comparatively small government spending on defense. Manufactures dominate its exports, which include machinery, electrical and electronic equipment, vehicles and transport equipment, iron and steel, chemicals, textiles, and ships.

Japan is one of the world's top fishing nations and fish is an important source of protein. Because the land is so rugged, only 15% of the country can be farmed. Yet Japan produces about 70% of the food it needs. Rice is the chief crop, taking up about half of the total farmland. Other major products include fruits, sugar beet, tea, and vegetables. Livestock farming has increased since the 1950s.

JORDAN

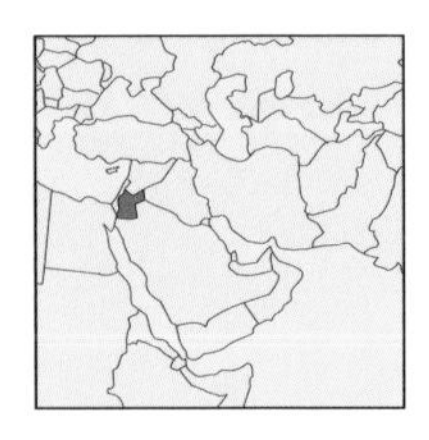

The green, white, and black on this flag are the colors of the three tribes who led the Arab Revolt against the Turks in 1917. Red is the color of the Hussein dynasty. The star was added in 1928. Its seven points represent the first seven verses of the sacred book, the Koran.

The Hashemite Kingdom of Jordan is an Arab country in southwestern Asia. The Great Rift Valley in the west contains the River Jordan and the Dead Sea, which Jordan shares with Israel. The Great Rift Valley is part of a huge gash in the Earth's crust, which runs south through East Africa to Mozambique. East of the Rift Valley is the Transjordan plateau, where most Jordanians live. To the east and south lie vast areas of desert. Jordan has a short coastline on an arm of the Red Sea, the Gulf of Aqaba.

AREA 34,444 sq miles [89,210 sq km]
POPULATION 5,153,000
CAPITAL (POPULATION) Amman (1,752,000)
GOVERNMENT Constitutional monarchy
ETHNIC GROUPS Arab 99%, of which Palestinians make up roughly half
LANGUAGES Arabic (official)
RELIGIONS Islam 93%, Christianity 5%
CURRENCY Jordan dinar = 1,000 fils

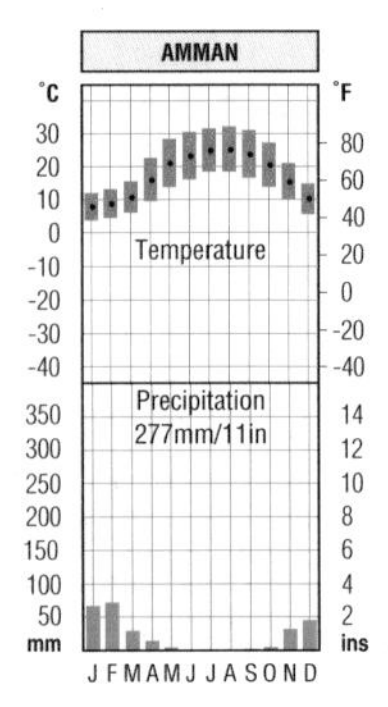

CLIMATE

Amman has a much lower rainfall and longer dry season than the Mediterranean lands to the west. The Transjordan plateau, on which Amman stands, is a transition zone between the Mediterranean climate zone to the west and the true desert climate to the east.

VEGETATION

Parts of the western plateau have a scrub vegetation, much like that in Mediterranean lands. Jordan also has areas of dry grassland, though few plants grow in the eastern deserts.

HISTORY

Jordan's early history is closely linked with that of Israel. It was first settled by Semitic peoples about 4,000 years ago, and was later conquered by Egyptian, Assyrian, Chaldean, Persian, and Roman forces. The area fell to the Muslim Arabs in AD 636 and the Arab culture they introduced survives to this day.

Most of Jordan came under the Ottoman empire from 1517 until World War I (1914–18), when Arab and British forces defeated the Turks. After the war, the League of Nations appointed Britain to rule the area.

In 1921, Britain created a territory called Transjordan east of the River Jordan. In 1923, Transjordan became self-governing, but Britain retained control of its defenses, finances, and foreign affairs. This territory became fully independent as Jordan in 1946.

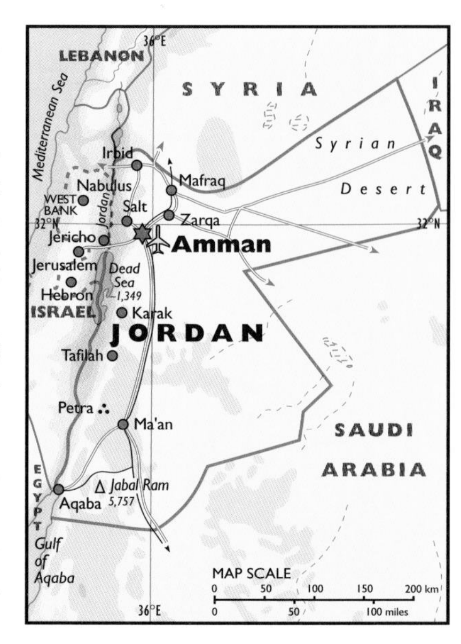

POLITICS

Jordan has suffered from instability arising from the Arab-Israeli conflict since the creation of the State of Israel in 1948. After the first Arab-Israeli War in 1948–9, Jordan acquired East Jerusalem and a fertile area called the West Bank. In 1967, Israel occupied this area. In Jordan, the presence of Palestinian refugees led to civil war in 1970–1.

In 1974, Arab leaders declared that the PLO (Palestine Liberation Organization) was the sole representative of the Palestinian people. In 1988, King Hussein of Jordan renounced his country's claims to the West Bank and passed responsibility for it to the PLO. Multiparty elections were held in 1993 and, in 1994, Jordan and Israel signed a peace treaty, ending a state of war that had been going on for 40 years. Hussein, who had commanded great respect for his role in Middle Eastern affairs, died in 1999. His eldest son became King Abdullah II.

DID YOU KNOW

- that the shoreline of the Dead Sea, at 1,349 ft [411 m] below mean sea level, is the world's lowest point on land
- that the name Hashemite in Jordan's official name denotes someone who is a descendant of the Prophet Muhammad, founder of Islam
- that the so-called rose-red city of Petra in Jordan contains ancient buildings carved out of rose-colored stone cliffs
- that the Dead Sea is the world's saltiest body of water; it is about nine times saltier than the water in the oceans

ECONOMY

Jordan lacks natural resources, apart from phosphates and potash, and the country's economy depends substantially on aid. The World Bank classifies Jordan as a "lower-middle-income" developing country. Because of the dry climate, under 6% of the land is farmed or used as pasture. Major crops include barley, citrus fruits, grapes, olives, vegetables, and wheat.

Jordan has an oil refinery and manufactures include cement, pharmaceuticals, processed food, fertilizers, and textiles. Main exports are phosphates, fertilizers, fruits, and vegetables.

KAZAKHSTAN

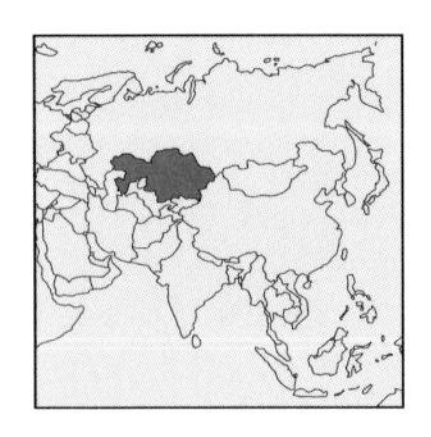

Kazakhstan's flag was adopted on June 4, 1992, about six months after it had become independent. The blue represents cloudless skies, while the golden sun and the soaring eagle represent love of freedom. A vertical strip of gold ornamentation is on the left.

Kazakhstan is a large country in west-central Asia. In the west, the Caspian Sea lowlands include the Karagiye depression, which reaches 433 ft [132 m] below sea level. The lowlands extend eastward through the Aral Sea area. The north contains high plains, but the highest land is along the eastern and southern borders. These areas include parts of the Altai and Tian Shan mountain ranges.

Eastern Kazakhstan contains several freshwater lakes, the largest of which is Lake Balkhash. The water in the rivers has been used for irrigation, causing ecological problems. For example, the Aral Sea, deprived of water, shrank from 25,830 sq miles [66,900 sq km] in 1960 to 12,989 sq miles [33,642 sq km] in 1993. Areas which once provided fish have dried up and are now barren desert.

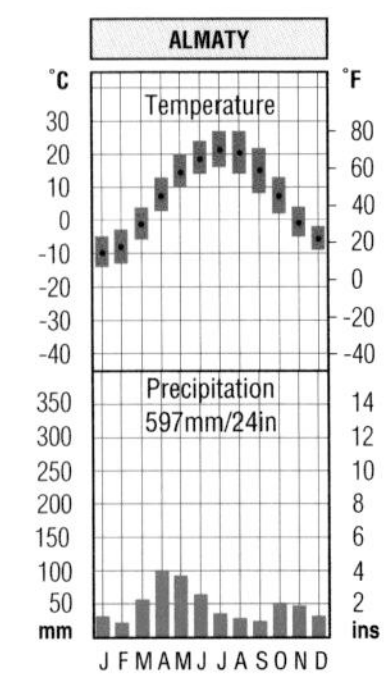

CLIMATE

The climate of Kazakhstan reflects its position in the heart of a huge land mass, far from the moderating influence of the oceans. Winters are cold and snow covers the land for about 100 days, on average, at Almaty. Summers are warm. The rainfall is generally low.

AREA 1,049,150 sq miles [2,717,300 sq km]

POPULATION 16,731,000

CAPITAL (POPULATION) Astana (280,000)

GOVERNMENT Multiparty republic

ETHNIC GROUPS Kazakh 53%, Russian 30%, Ukrainian 4%, German 2%, Uzbek 2%

LANGUAGES Kazakh (official); Russian, the former official language, is widely spoken

RELIGIONS Islam 47%, Russian Orthodox 44%

CURRENCY Tenge = 100 tiyn

VEGETATION

Kazakhstan has very little woodland. Steppe (grassland) covers much of the north, while the south is desert or semidesert. However, large dry areas between the Aral Sea and Lake Balkhash are irrigated farmland.

HISTORY

Little is known of the early history of Kazakhstan, except that it was the home of nomadic peoples, including Turks from the west and Mongols from the east. It is from these two groups that most Kazakhs are descended.

From the late 15th century, the Kazakhs built up a large nomadic empire ruled by *khans*. But Kazakh power declined in the 17th century, and in the early 18th century Russia became influential in the area. In 1731, the Kazakhs in the west accepted Russian rule to gain protection from attack from neighboring peoples. By the mid-1740s, Russia ruled most of the region and, in the early 19th century, Russia abolished the *khanates*. They also encouraged Russians and Ukrainians to settle in Kazakhstan.

After the Russian Revolution of 1917, many Kazakhs wanted to make their country independent. But the Communists prevailed and in 1936 Kazakhstan became a republic of the Soviet Union, called the Kazakh Soviet Socialist Republic. During World War II and also after the war, the Soviet government moved many people from the west into Kazakhstan. From the 1950s, people were encouraged to work on a "Virgin Lands" project, which involved bringing large

areas of grassland under cultivation.

POLITICS

Reforms in the Soviet Union in the 1980s led to the breakup of the country in December 1991. Kazakhstan kept contacts with Russia and most of the other republics in the former Soviet Union by joining the Commonwealth of Independent States (CIS). Kazakhstan's first elected president, Nursultan Nazarbayev, a former Communist party leader, introduced a multiparty system. However, in the early 2000s, he was criticised for his repressive measures against political opponents.

DID YOU KNOW

- that Kazakhstan is the world's ninth largest country
- that Bayqongyr, northeast of the Aral Sea, was the rocket-launching site for the Soviet Union's space program; in 1994 Russia agreed to rent it from Kazakhstan for 20 years
- that before a Kazakh marriage, the groom's family negotiates a bride price, or *kalym*, with the bride's family; it often amounts to the salary of one person over several years

ECONOMY

The World Bank classifies Kazakhstan as a "lower-middle-income" developing country. Livestock farming, especially sheep and cattle, remains an important activity, while major crops include barley, cotton, rice, and wheat.

Kazakhstan is rich in mineral resources. It has coal and oil reserves, and such metals as bauxite (aluminum ore), copper, lead, tungsten, and zinc. Manufactures include chemicals, processed food, machinery, and textiles. Exports include oil, metals, chemicals, grain, wool, and meat.

KENYA

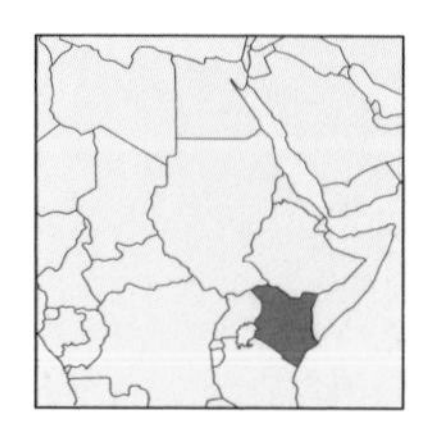

Kenya's flag dates from 1963, when the country became independent. It is based on the flag of KANU (Kenya African National Union), the political party which led the nationalist struggle. The Masai warrior's shield and crossed spears represent the defense of freedom.

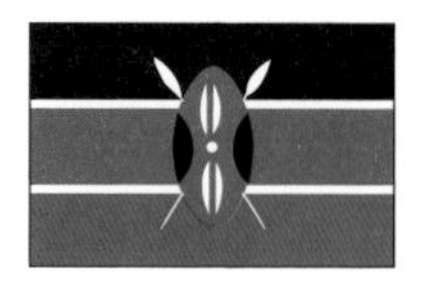

The Republic of Kenya is a country in East Africa which straddles the Equator. It is slightly larger in area than France. Behind the narrow coastal plain on the Indian Ocean, the land rises to high plains and highlands, broken by volcanic mountains, including Mount Kenya, the country's highest peak at 17,063 ft [5,199 m].

Crossing the country is an arm of the Great African Rift Valley. On the floor of this steep-sided valley are several lakes, including Magadi, Naivasha, Nakuru, Baringo, and, on the northern frontier, Lake Turkana (Lake Rudolf).

AREA 224,081 sq miles [580,370 sq km]
POPULATION 30,766,000
CAPITAL (POPULATION) Nairobi (2,000,000)
GOVERNMENT Multiparty republic
ETHNIC GROUPS Kikuyu 21%, Luhya 14%, Luo 13%, Kalenjin 12%, Kamba 11%
LANGUAGES Kiswahili and English (both official)
RELIGIONS Protestant 45%, Roman Catholic 33%, traditional beliefs 10%, Islam 10%
CURRENCY Kenya shilling = 100 cents

CLIMATE

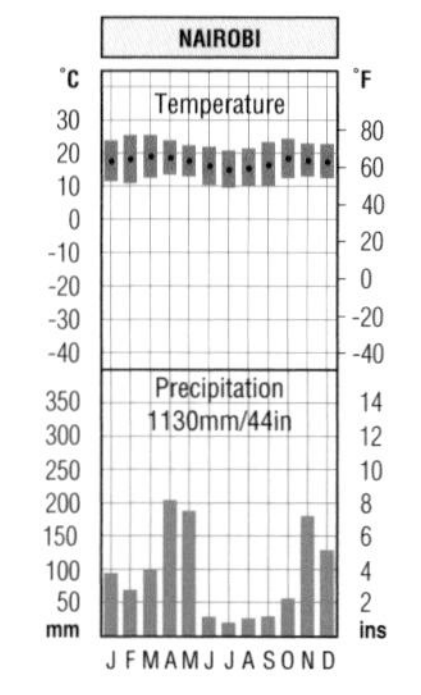

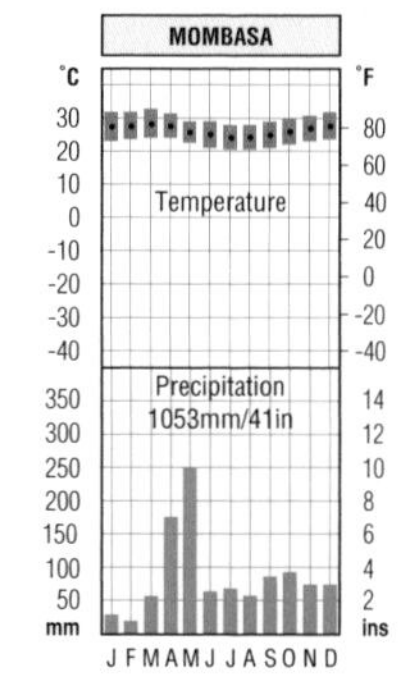

Mombasa on the coast is hot and humid. But inland, the climate is moderated by the height of the land. As a result, Nairobi, in the thickly populated southwestern highlands, has summer temperatures which are 18°F [10°C] lower than Mombasa. Nights can be cool, but temperatures do not fall below freezing. Nairobi's main rainy season is from April to May, with "little rains" in November and December. However, only about 15% of the country has a reliable rainfall of 31 inches [800 mm].

VEGETATION

The coast is lined by lagoons, mangrove swamps, and palm trees. The inland plains are covered by grasses and low shrubs, but much of the north is semidesert. Forests and grasslands are found in the south-

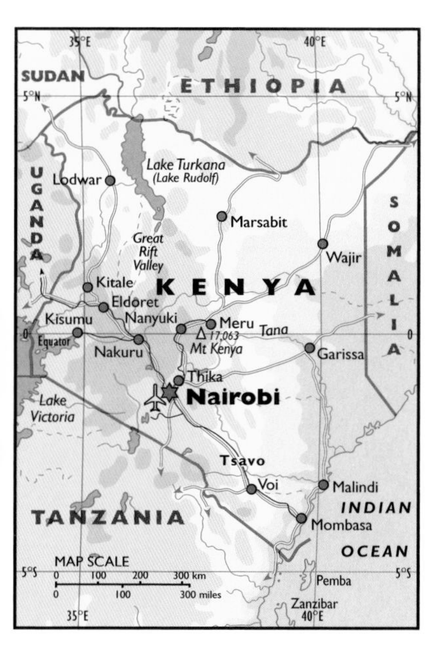

western highlands. Kenya's national parks and wildlife reserves now attract many tourists.

HISTORY AND POLITICS

The Kenyan coast has been a trading center for more than 2,000 years. In the 8th century, the Arabs founded settlements there. The Portuguese explorer Vasco da Gama reached the coast in 1498, and Portuguese traders took over the area in the 16th century. But the Arabs regained control in the late 17th century. Britain took over the coast in 1895 and soon extended its influence inland. In the 1950s, a secret movement, called Mau Mau, launched an armed struggle against British rule. Although Mau Mau was defeated, Kenya became independent in 1963.

Many Kenyans wanted Kenya to have a strong central government and Kenya was a one-party state for much of the time since 1963. In 2002, the victory of the opposition leader Mwai Kibaki, who became president, was hailed as a triumph for democracy.

An important development occurred in 1999 when Kenya, Tanzania, and Uganda set up the East African Community to increase cooperation between the countries.

ECONOMY

According to the United Nations, Kenya is a "low-income" developing country. Agriculture employs about 80% of the people, but many Kenyans are subsistence farmers, growing little more than they need to support their families. The chief food crop is maize. Bananas, beans, cassava, and sweet potatoes are also grown. The main cash crops and leading exports are coffee and tea. Manufactures include chemicals, leather and footwear, processed food, petroleum products, and textiles.

The Citrus Swallowtail is one of a set of stamps showing butterflies issued in 1989–90. It is a reminder of the fascinating wildlife found in Kenya.

KOREA, NORTH

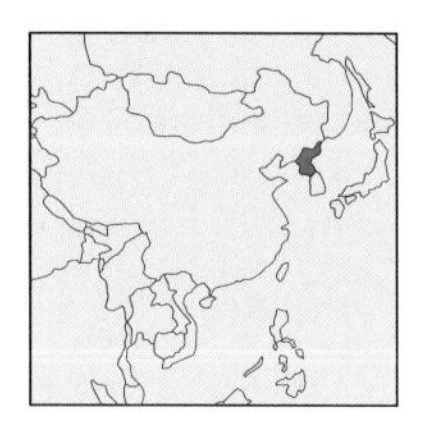

The flag of the Democratic People's Republic of Korea (North Korea) has been flown since Korea was split into two states in 1948. The colors are traditional ones in Korea. The design, with the red star, indicates that North Korea is a Communist country.

AREA 46,540 sq miles [120,540 sq km]
POPULATION 21,968,000
CAPITAL (POPULATION) Pyongyang (2,741,000)
GOVERNMENT Single-party people's republic
ETHNIC GROUPS Korean 99%
LANGUAGES Korean (official)
RELIGIONS Buddhism and Confucianism; some Christian and Chondogyo
CURRENCY North Korean won = 100 chon

The Democratic People's Republic of Korea occupies the northern part of the Korean peninsula which extends south from northeastern China. Mountains form the heart of the country, with the highest peak, Paektu-san, reaching 9,006 ft [2,744 m] on the northern border. East of the mountains lie the eastern coastal plains, which are densely populated, as also are the coastal plains in the west, which contain the capital, Pyongyang. Another small highland region is in the southeast, bordering South Korea.

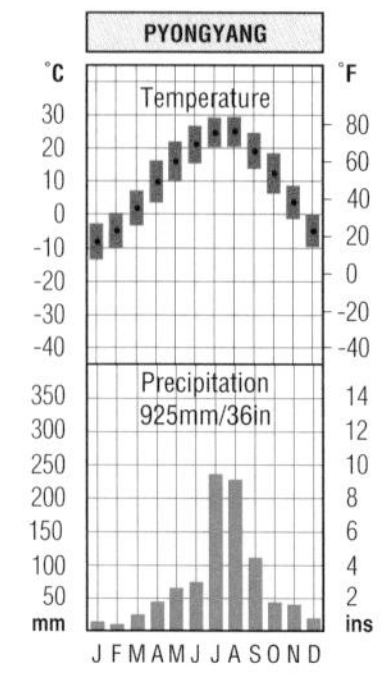

CLIMATE

North Korea has a fairly severe climate, with bitterly cold winters when winds blow from across central Asia, bringing snow. Rivers freeze over and sea-ice may block harbors on the coast. In the summer months, moist winds blow from the oceans, bringing rain.

VEGETATION

The coastal plains are mostly farmed, but some patches of chestnut, elm, and oak woodland survive on the hilltops. The mountains contain forests of such trees as cedar, fir, pine, and spruce.

HISTORY

North Korea's early history is described on pages 186–7 [*see Korea, South*]. North Korea was created in 1945, when the peninsula, a Japanese colony since 1910, was divided into two parts. Soviet forces occupied the north, with US forces in the south. Soviet occupation led to a Communist government being established in 1948 under the leadership of Kim Il Sung.

The Korean War began in June 1950 when North Korean troops invaded the south. North Korea, aided by China and the Soviet Union, fought with South Korea, which was supported by the United States and

other members of the United Nations. The war ended in July 1953. An armistice agreement was signed, but there was no permanent peace treaty. The war caused great destruction and loss of life – around 1.6 million Communist troops were killed, wounded, or reported missing.

POLITICS

Between 1948 and his death in 1994, Kim Il Sung was a virtual dictator, ruling along similar lines to Joseph Stalin in the Soviet Union. In pursuit of its aim of reunifying Korea, North Korea adopted a hostile policy toward South Korea. At times, the situation grew so tense that it became a matter of international concern.

The ending of the Cold War in the late 1980s eased the situation, and both North and South Korea joined the United Nations in 1991. The two countries made several agreements, including one in which they agreed not to use force against each other.

In 1993, North Korea began a new international crisis by announcing that it was withdrawing from the Nuclear Non-Proliferation Treaty. This led to suspicions that North Korea, which had signed the Treaty in 1985, was developing its own nuclear weapons. When Kim Il Sung died in 1994, he was succeeded by his son, Kim Jong Il. In early 2003, Korea's renunciation of the Nuclear Non-Proliferation Treaty and its threat of resuming its nuclear weapons program caused alarm.

ECONOMY

North Korea has considerable resources, including coal, copper, iron ore, lead, tin, tungsten, and zinc. Under Communism, North Korea has concentrated on developing heavy, state-owned industries. Manufactures include chemicals, iron and steel, machinery, processed food, and textiles. Agriculture employs about a third of the people of North Korea and rice is the leading crop.

Many Koreans regard education as the key to the country's future. This North Korean stamp, showing children studying chemistry, is one of a set issued in 1965.

KOREA, SOUTH

South Korea's flag, adopted in 1950, is white, the traditional symbol for peace. The central *yin-yang* symbol signifies the opposing forces of nature. The four black symbols stand for the four seasons, the points of the compass, and the Sun, Moon, Earth, and Heaven.

The Republic of Korea, as South Korea is officially known, occupies the southern part of the Korean peninsula. Mountains cover much of the country. The western coastlands, where the capital and main industrial city, Seoul, is situated, is a region of low hills and plains dotted with farms. The southern coast is another major farming region. Many islands are found along the west and south coasts. The largest is Cheju-do, which contains South Korea's highest peak, which rises to 6,400 ft [1,950 m].

AREA 38,232 sq miles [99,020 sq km]

POPULATION 47,904,000

CAPITAL (POPULATION) Seoul (or Soul, 10,231,000)

GOVERNMENT Multiparty republic

ETHNIC GROUPS Korean 99%

LANGUAGES Korean (official)

RELIGIONS Christianity 49%, Buddhism 47%, Confucianism 3%

CURRENCY South Korean won = 100 chon

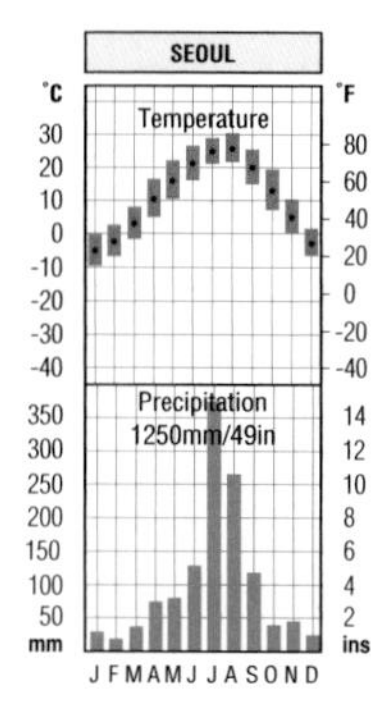

CLIMATE

Like North Korea, South Korea is chilled in winter by cold, dry winds blowing from central Asia. Snow often covers the mountains in the east of the country. The summers are hot and wet, especially in July and August, when it rains, on average, every other day.

VEGETATION

Forests cover two-thirds of the land, though much of the original forest has been destroyed and used as fuel. Most areas contain broadleaf and coniferous trees. The plains in the south contain subtropical species.

HISTORY

In the last 2,000 years, China has had a great influence on the people of Korea. The Chinese conquered the north in 108 BC and ruled until they were thrown out in AD 313. Mongol armies attacked Korea in the 13th century, but in 1388, a general, Yi Songgye, founded a dynasty of rulers which lasted until 1910.

From the 17th century, Korea prevented foreigners from entering the country. Korea was often called the "Hermit Kingdom" until 1876, when Japan forced it to open some of its ports. Soon, the United States, Russia, and some European countries were trading with Korea. In 1910, Korea became a Japanese colony.

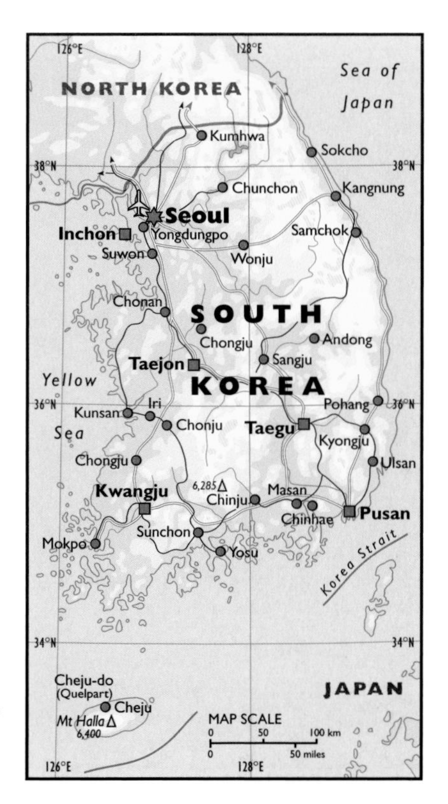

After Japan's defeat in World War II (1939–45), North Korea was occupied by troops from the Soviet Union, while South Korea was occupied by United States forces. Attempts to reunify Korea failed and, in 1948, a National Assembly was elected in South Korea. This Assembly created the Republic of Korea, while North Korea became a Communist state. North Korean troops invaded the South in June 1950, sparking off the Korean War (1950–3). [*See Korea, North, on pages 184–5.*]

POLITICS

In the 1950s, South Korea had a weak economy, which had been further damaged by the destruction caused by the Korean War. From the 1960s to the 1980s, South Korean governments worked to industrialize the economy. The governments were dominated by military leaders, who often used authoritarian methods, imprisoning opponents and restricting freedom of speech. In 1987, a new constitution was approved enabling presidential elections to be held every five years. This led in 1992 to the election of Kim Young Sam, South Korea's first president for 30 years with no ties to the military. In 1993, he began an anti-corruption campaign.

In 1991, both South and North Korea became members of the United Nations. In the early 2000s, the government of South Korea pursued a "sunshine policy" toward North Korea in an attempt to improve relations between the two countries.

ECONOMY

The World Bank classifies South Korea as an "upper-middle-income" developing country. It is also one of the world's fastest growing industrial economies. The country's resources include coal and tungsten, and its main manufactures are processed food and textiles. Since partition, heavy industries have been built up, making chemicals, fertilizers, iron and steel, machinery, and ships. Most recently, South Korea has developed the production of such things as computers, cars, and television sets.

Farming remains important. Rice is the chief crop, together with fruits, grains, and vegetables. Fishing is another major industry.

KYRGYZSTAN

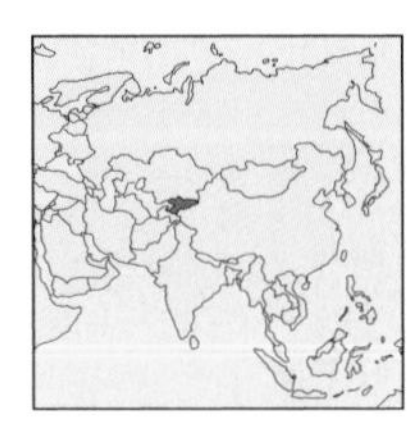

Kyrgyzstan's flag was adopted in March 1992. The flag depicts a bird's-eye view of a *yurt* (circular tent) within a radiant sun. The *yurt* recalls the traditional nomadic way of life. The 40 rays of the sun stand for the 40 traditional tribes.

The Republic of Kyrgyzstan, or Kirghizia as it is also known, is a landlocked country between China, Tajikistan, Uzbekistan, and Kazakhstan. The country is mountainous, with spectacular scenery. The highest mountain, Pik Pobedy in the Tian Shan range, reaches 24,406 ft [7,439 m] above sea level in the east. Less than a sixth of the country is below 2,950 ft [900 m]. The largest of the country's many lakes is Ozero (Lake) Issyk Kul in the northeast. Its shoreline is 5,279 ft [1,609 m] above sea level.

AREA 76,640 sq miles [198,500 sq km]
POPULATION 4,753,000
CAPITAL (POPULATION) Bishkek (589,000)
GOVERNMENT Multiparty republic
ETHNIC GROUPS Kyrgyz 52%, Russian 18%, Uzbek 13%, Ukrainian 3%, German, Tatar
LANGUAGES Kyrgyz and Russian (both official), Uzbek
RELIGIONS Islam
CURRENCY Som = 100 tyiyn

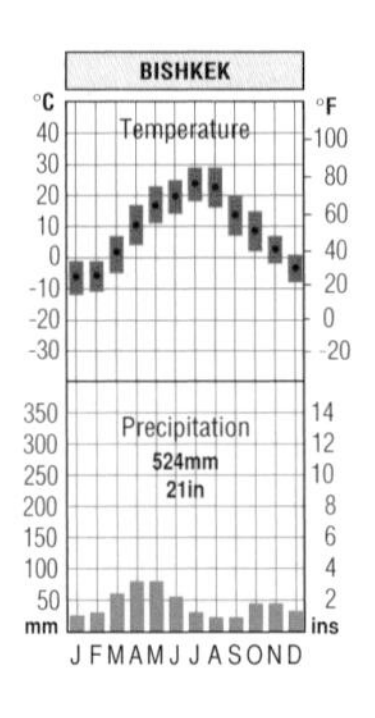

CLIMATE

The lowlands of Kyrgyzstan have warm summers and cold winters. But the altitude influences the climate in the mountains, where the January temperatures plummet to –18°F [–28°C]. Because of its remoteness from the sea, much of Kyrgyzstan has a low annual rainfall.

VEGETATION

Mountain pasture covers much of Kyrgyzstan. Woodland covers only a small area, mainly in the lower valleys. The trees are mainly conifers, including white spruce. Less than a tenth of the land is used for crops.

HISTORY AND POLITICS

The area that is now Kyrgyzstan was populated in ancient times by nomadic herders. Mongol armies conquered the region in the early 13th century. They set up areas called *khanates*, ruled by chieftains, or *khans*. Islam was introduced in the 17th century.

China gained control of the area in the mid-18th century, but, in 1876, Kyrgyzstan became a province of Russia. Russia encouraged settlement by Russians, Ukrainians, and others, a policy resented by the local people whose grazing land was reduced in area. In 1916, Russia put down a rebellion. Many local people fled to China.

In 1917, Communist leaders seized control of Russia and, in 1922, when the Soviet Union was formed, Kyrgyzstan became an autonomous *oblast* (self-governing region). In 1936, it became a Soviet Socialist Republic in the Soviet Union. Under Communism, nomads were forced to live on government-run farms. The government tried to suppress some local customs, including religious worship. But Communist rule also led to improvements in such areas as health and education.

In 1991, Kyrgyzstan became an independent country following the breakup of the Soviet Union. The Communist party was dissolved, but the country maintained ties with Russia through an organization called the Commonwealth of Independent States. Under a new constitution, adopted in 1994, parliamentary elections were held in 1995 and 2000.

ECONOMY

In the early 1990s, when Kyrgyzstan was working to reform its economy, the World Bank classified it as a "lower-middle-income" developing country. Agriculture, especially livestock rearing, is the chief activity. Sheep are the main domestic animals, but cattle, goats, pigs, and yaks are also raised. The chief products include cotton, eggs, fruits, grain, tobacco, vegetables, and wool. But food must be imported. Industries are mainly concentrated around the capital Bishkek. Manufactures include machinery, processed food, metals, and textiles. Exports include wool, chemicals, cotton, and metals.

DID YOU KNOW

- that many people in Kyrgyzstan claim they are descendants of the Mongols who invaded the area in the 13th century
- that *kumiss*, or fermented mare's milk, is a popular food in Kyrgyzstan
- that Bishkek was named Frunze under Communist rule
- that the name of the spectacular mountain range in southeastern Kyrgyzstan, Tian Shan, means "heavenly mountains"

An outline map of Kyrgyzstan, together with the national symbol which appears on the flag. On the map is the capital Bishkek, whose name is written in the Cyrillic alphabet.

LAOS

Since 1975, Laos has flown the flag of the Pathet Lao, the Communist movement which won control of the country after a long struggle. The blue stands for the River Mekong, the white disk for the Moon, and the red for the unity and purpose of the people.

The Lao People's Democratic Republic is a landlocked country in Southeast Asia. Mountains and plateaux cover much of the country. The highest point is Mount Bia, which reaches 9,245 ft [2,817 m] in central Laos.

Most people live on the plains bordering the River Mekong and its tributaries. This river, one of Asia's longest, forms much of the country's northwestern and southwestern borders. A range of mountains, called the Annam Cordillera, runs along the eastern border with Vietnam.

AREA 91,428 sq miles [236,800 sq km]
POPULATION 5,636,000
CAPITAL (POPULATION) Vientiane (532,000)
GOVERNMENT Single-party republic
ETHNIC GROUPS Lao Loum 68%, Lao Theung 22%, Lao Soung 9%
LANGUAGES Lao (official), Khmer, Tai, Miao
RELIGIONS Buddhism 58%, traditional beliefs 34%, Christianity 2%, Islam 1%
CURRENCY Kip = 100 at

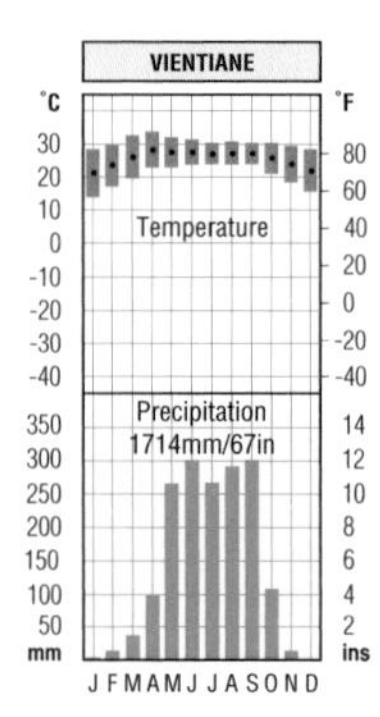

CLIMATE

Laos has a tropical monsoon climate. Winters are dry and sunny, with winds blowing in from the northeast. The temperatures rise until April, when the wind directions are reversed and moist southwesterly winds reach Laos, heralding the start of the wet monsoon season.

VEGETATION

Forests cover about three-fifths of the land. Common plants in the north include laurels, oaks, and pines. The southern forests contain such trees as bamboo, ebony, rosewood, sandalwood, and teak. Coarse grasses and scrub grow in areas where the forests have been cut down.

HISTORY

From the 9th century AD, Lao and Tai people set up a number of small states ruled by princes. But the area that is now Laos was united in 1353, in a kingdom called Lan Xang ("land of a million elephants"). Around 1700, the region was divided into three separate kingdoms.

France made Laos a protectorate in the late 19th century and ruled it as part of French Indochina, a region which also included Cambodia and Vietnam. Laos became a member of the French Union in 1948 and an independent kingdom in 1954.

POLITICS

After independence, Laos suffered from instability caused by a long power struggle between royalist government forces and a pro-Communist group called the Pathet Lao. A civil war broke out in 1960 and continued off and on into the 1970s. The Pathet Lao finally took control of Laos in 1975 and the king abdicated. Laos was greatly influenced by the Communist government in Vietnam, which had used Laos as a supply line during the Vietnam War (1957–75).

From the late 1980s, along with other Communist countries, Laos began a program of economic reforms, including the encouragement of private enterprise. But there were few indications that the Communist government was prepared to introduce political reforms or some measure of democracy.

ECONOMY

Laos is one of the world's poorest countries. Agriculture employs about 72% of the people. Rice is the main crop, while timber and coffee are both exported. But the most valuable source of foreign revenue is electricity, which is produced at hydroelectric stations on the Mekong River and exported to Thailand. Problems arose in 2001, when Thailand reduced its purchases of electricity. Laos also produces opium. In the 1990s, Laos was thought to be the world's third biggest source of this illegal drug.

Elephants are important beasts of burden in southern Asia. This Laotian stamp, showing an elephant with a howdah *on its back, is one of a set of seven issued in 1958, when Laos was a monarchy (as the name* Royaume du Laos *indicates), to celebrate this magnificent animal.*

LATVIA

The burgundy and white Latvian flag, which dates back to at least 1280, was revived after Latvia achieved its independence in 1991. According to one legend, the flag was first made from a white sheet which had been stained with the blood of a Latvian hero.

The Republic of Latvia is one of three states on the southeastern corner of the Baltic Sea which were ruled as parts of the Soviet Union between 1940 and 1991. Latvia consists mainly of flat plains separated by low hills, composed of moraine (ice-worn rocks). The moraine was dumped there by ice sheets during the Ice Age. The country's highest point is only 1,021 ft [311 m] above sea level. Small lakes and peat bogs are common. The country's main river, the Daugava, is also known as the Western Dvina.

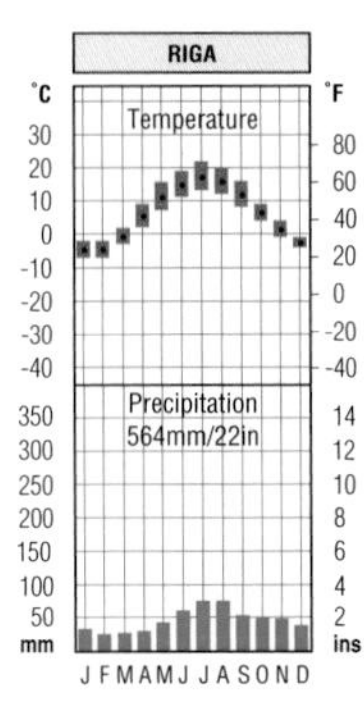

CLIMATE

Riga has warm summers, but the winter months (from December to March) are subzero. In the winter, the sea often freezes over. The rainfall is moderate and it occurs throughout the year, with light snow in winter. But the snow seldom lasts long before it thaws.

VEGETATION

Forests cover about two-fifths of the country. Pine and spruce forests are widespread, but mixed forests with alder, aspen, and birch also occur.

AREA 24,938 sq miles [64,589 sq km]

POPULATION 2,385,000

CAPITAL (POPULATION) Riga (840,000)

GOVERNMENT Multiparty republic

ETHNIC GROUPS Latvian 56%, Russian 30%, Belarussian 4%, Ukrainian 3%, Polish 2%, Lithuanian, Jewish

LANGUAGES Latvian (official), Russian

RELIGIONS Lutheran, Russian Orthodox, Roman Catholic

CURRENCY Lats = 10 santimi

About 27% of the land is under crops, while 13% is used for grazing livestock.

HISTORY

The ancestors of most modern Latvians settled in the area about 2,000 years ago. Between the 9th and 11th centuries, the region was attacked by Vikings from the west and Russians from the east. In the 13th century, German invaders took over, naming the country Livland, or Livonia in Latin.

From 1561, the area was partitioned between various groups, including Poles, Lithuanians, and Swedes. In 1710, the Russian Peter the Great took Riga, and by the end of the 18th century Latvia was under Russian rule. Nationalist movements developed in the 19th century and, just after the end of World War I (1914–18), Latvia declared itself independent. In 1939, Germany and the Soviet Union made a secret agreement to divide up parts of Eastern Europe, and in 1940 Soviet troops invaded Latvia, which became part of the Soviet Union. German forces seized Latvia in 1941, but Soviet troops returned in 1944.

POLITICS

Under Soviet rule, many Russian immigrants settled in Latvia and many Latvians feared that Russians would become the dominant ethnic group. In the late 1980s, when reforms were being introduced in the Soviet Union, Latvia's government ended absolute Communist rule and made Latvian the official language. In 1990, it declared the country to be independent, an act which was finally recognized by the Soviet Union in September 1991.

Latvia held its first free elections to its parliament (the *Saeima*) in 1993. Voting was limited only to citizens of Latvia on June 17, 1940, and their descendants. This meant that about 34% of Latvian residents were unable to vote. In 1994, Latvia adopted a law restricting the naturalization of non-Latvians, but restrictions were eased in 1998. In 2002, Latvia was invited to become a member of the European Union in May 2004.

DID YOU KNOW

- that the Latvian language is related to Sanskrit, an ancient Indian language
- that Latvians are also called Letts
- that Latvia's official name is *Latvijas Republika* (Republic of Latvia)
- that Riga was a major port and cosmopolitan city in the 19th century, when it was often called the "Paris of the Baltic"

ECONOMY

The World Bank classifies Latvia as a "lower-middle-income" country and, in the 1990s, it faced many problems in transforming its government-run economy into a free-market one. The country lacks natural resources apart from its land and forests, and it has to import many raw materials required by manufacturing.

Its industries cover a wide range, with products including electronic goods, farm machinery, fertilizers, processed food, plastics, radios, washing machines, and vehicles. But Latvia produces only about a tenth of the electricity it needs. The rest has to be imported from Belarus, Russia, and Ukraine. Farm products include barley, dairy products, beef, oats, potatoes, and rye.

LEBANON

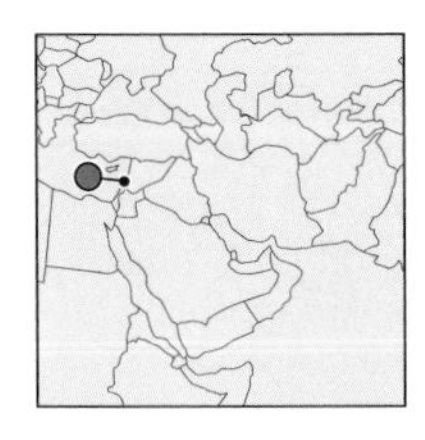

Lebanon's flag was adopted in 1943. It uses the colors of Lebanese nationalists in World War I (1914–18). The cedar tree on the white stripe has been a Lebanese symbol since Biblical times. Because of deforestation, only a few of Lebanon's giant cedars survive.

The Republic of Lebanon is a country on the eastern shores of the Mediterranean Sea. The country's narrow coastal plain contains the main cities, including Beirut (Bayrut) and Tripoli (Tarabulus). Behind the coastal plain are the rugged Lebanon Mountains (Jabal Lubnan), which rise to 10,135 ft [3,088 m]. Another range, the Anti-Lebanon Mountains (Al Jabal Ash Sharqi), form the eastern border with Syria. Between the two ranges is the Bekaa (Beqaa) Valley, a fertile farming region.

AREA 4,015 sq miles [10,400 sq km]
POPULATION 3,628,000
CAPITAL (POPULATION) Beirut (or Bayrut, 1,500,000)
GOVERNMENT Multiparty republic
ETHNIC GROUPS Lebanese 80%, Palestinian 12%, Armenian 5%, Syrian, Kurdish
LANGUAGES Arabic (official)
RELIGIONS Islam 70%, Christianity 30%
CURRENCY Lebanese pound = 100 piastres

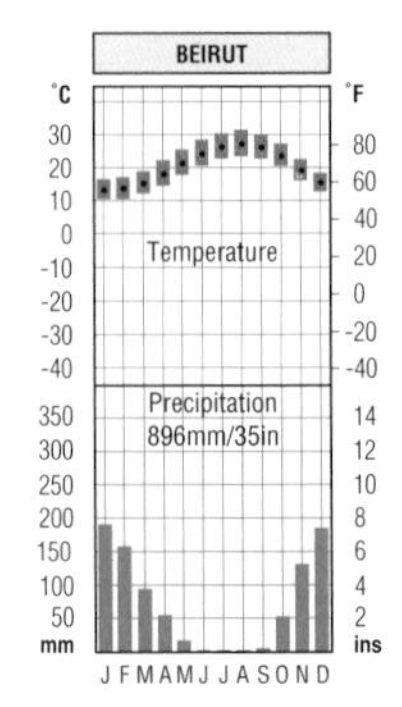

CLIMATE

The Lebanese coast has the hot, dry summers and mild, wet winters that are typical of many Mediterranean lands. Inland, onshore winds bring heavy rain to the western slopes of the mountains in the winter months, with snow at the higher altitudes.

VEGETATION

Lebanon was famous in ancient times for its cedar forests in the mountains, but these have been largely cut down. Forest and woodland now cover about 8% of the land.

HISTORY

The Phoenicians, who lived in the area from about 3000 BC, were traders and explorers who operated throughout the Mediterranean region. From about 800 BC, the area came under various foreign rulers. Christianity was introduced in AD 325. But in the early 7th century, the region was taken by Arabs who converted many people to Islam, though Christianity survived in the mountains. From about 1100 to 1300, the area was a battlefield between Christian crusaders and Muslim armies. In 1516, Lebanon became part of the Turkish Ottoman empire.

Turkish rule continued until World War I (1914–18). France ruled the country from 1923, but Lebanon became fully independent in 1946. After independence, the Muslims and Christians agreed to share power, and Lebanon made rapid economic progress.

POLITICS

From the late 1950s, development was slowed down by periodic conflict between Sunni and Shia Muslims, Druses, and Christians. The situation was further complicated by the presence of Palestinian refugees who used bases in Lebanon to attack Israel.

In 1975, civil war broke out as private armies representing the many factions struggled for power. This led to intervention by Israel in the south and Syria in the north. UN peacekeeping forces arrived in 1978, but bombings, assassinations, and kidnappings became almost everyday events in the 1980s. From 1991, Lebanon enjoyed an uneasy peace. However, Israel continued to occupy an area in the south and in 1993, 1996–7, and again in early 2000, Israel launched attacks on pro-Iranian Hezbollah guerrillas in Lebanon.

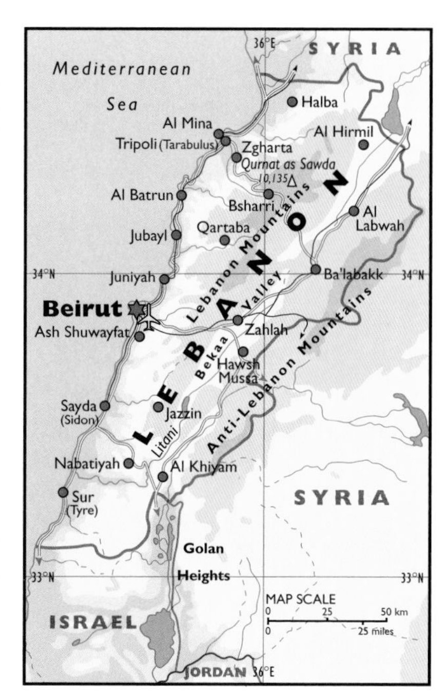

> **DID YOU KNOW**
> - that Lebanon has a higher percentage of Christians than any other Arab country
> - that in winter people can swim in the warm Mediterranean Sea and then travel a short distance inland to ski in the mountains
> - that Lebanon was the homeland of the Phoenicians who set up trading colonies throughout the Mediterranean region, reaching their peak between the 12th and 9th centuries BC
> - that the Druses of Lebanon follow a secret religion related to Islam, but they are not regarded as Muslims

ECONOMY

Lebanon's civil war almost destroyed the valuable trade and financial services which had been Lebanon's chief source of income, together with tourism. The manufacturing industry, formerly another major activity, was also badly hit and many factories were damaged.

Manufactures include chemicals, electrical goods, processed food, and textiles. Farm products include citrus and other fruits, potatoes and other vegetables, and sugar beet.

LESOTHO

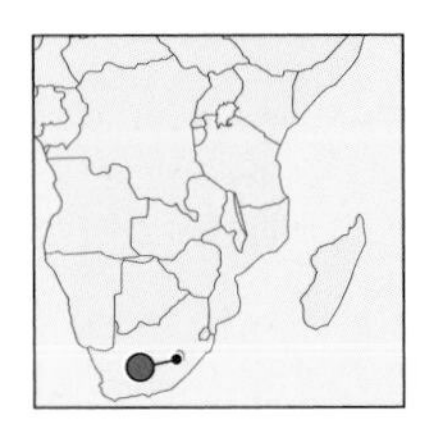

Lesotho's flag, adopted in 1987, replaced one used since 1966 when the country became independent from Britain. The white, blue, and green represent peace, rain, and prosperity – the words of the national motto. The emblem is part of the country's coat of arms.

The Kingdom of Lesotho is a landlocked country, completely enclosed by South Africa. The land is mountainous, rising to 11,428 ft [3,482 m] on the northeastern border. The Drakensberg range covers most of the country, giving Lesotho some spectacular scenery. But most people live in the western lowlands, where the capital Maseru is situated, or in the southern valley of the River Orange. This river rises in northeastern Lesotho and then flows through South Africa to the Atlantic Ocean.

AREA 11,718 sq miles [30,350 sq km]

POPULATION 2,177,000

CAPITAL (POPULATION) Maseru (130,000)

GOVERNMENT Constitutional monarchy

ETHNIC GROUPS Sotho 99%

LANGUAGES Sesotho and English (both official)

RELIGIONS Christianity 80%, traditional beliefs 20%

CURRENCY Loti = 100 lisente

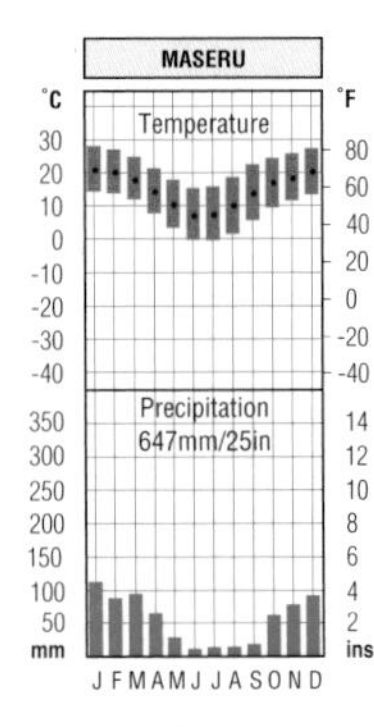

CLIMATE

The climate of Lesotho is greatly affected by the altitude, because most of the country lies above 4,921 ft [1,500 m]. Maseru has warm summers, but the temperatures fall below freezing in the winter. The mountains are colder. The rainfall varies, averaging around 28 inches [700 mm].

VEGETATION

Mountain grassland covers much of Lesotho. Trees and shrubs, such as wild olives and willows, grow only in sheltered valleys in remote areas.

HISTORY AND POLITICS

From the late 18th century, when tribal wars swept southern Africa, people sought safety in the remote mountains of what is now Lesotho. They were united into the Basotho nation in the 1820s by a chief who became their king, Moshoeshoe I. (Basotho is the name for the people of Lesotho and Sesotho is the name for their Bantu language.)

In 1868, Britain took over the area as a protectorate and, in 1871, it was placed under the British Cape Colony in South Africa. But in 1884, after the British tried to disarm the Basotho, the country, then called Basutoland, was reconstituted as a British protectorate. Under British

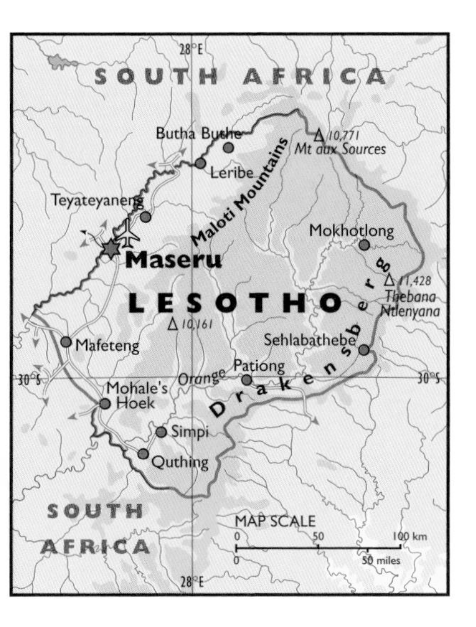

rule, whites were not allowed to own land.

The country became independent in 1966 as the Kingdom of Lesotho, with Moshoeshoe II, great-grandson of Moshoeshoe I, as its king. Since independence, Lesotho has suffered instability. The military seized power in 1986 and stripped Moshoeshoe II of his powers in 1990, installing his son, Letsie III, as the monarch. Multiparty elections in 1993 were won by the Basotho Congress Party led by Ntsu Mokhele. Moshoeshoe II was restored to the throne in 1995, but after his death in 1996, Letsie III again became king. In 1998, an army revolt, following an election in which the ruling party won 79 out of the 80 seats, caused much damage, despite the intervention of a South African force, which withdrew in 1999.

ECONOMY

The World Bank classifies Lesotho as a "low-income" developing country. Lesotho lacks natural resources apart from water. Agriculture, mainly at subsistence level, light manufacturing, and money sent home by Basotho working abroad, mainly in the mines of South Africa, are the main sources of income. Major farm crops include beans, cattle, hides and skins, maize, mohair, wool, and wheat. Manufactures include processed food, handicrafts, and textiles. Tourism is developing.

These stamps issued in 1989 show the Maloti Mountains, which form part of the Drakensberg range, in northern Lesotho.

LIBERIA

Liberia was founded in the early 19th century as an American colony for freed slaves who wanted to return to Africa. Its flag was adopted in 1847, when Liberia became independent. The 11 red and white stripes represent the 11 men who signed Liberia's Declaration of Independence.

The Republic of Liberia is a country in West Africa. Behind the coastline, 311 miles [500 km] long, lies a narrow coastal plain. Beyond, the land rises to a plateau region, with the highest land along the border with Guinea.

A number of rivers flow across the country generally from the northeast to the southwest. The most important rivers are the Cavally (or Cavalla), which forms the border with Ivory Coast, and the St Paul, which reaches the Atlantic Ocean near the capital, Monrovia.

MONROVIA

Temperature

Precipitation 4413mm/174in

J F M A M J J A S O N D

CLIMATE

Liberia has a tropical climate with high temperatures and high humidity all through the year. The rainfall is abundant all year round, but there is a particularly wet period from June to November. The rainfall generally increases from east to west.

VEGETATION

Mangrove swamps line parts of the coast. Inland, forests cover nearly two-fifths of the land. Trees include red ironwood and various kinds of mahogany. Liberia also has areas of savanna (grassland with scattered trees). Only 5% of the land is used for crops or pasture.

AREA 43,000 sq miles [111,370 sq km]

POPULATION 3,226,000

CAPITAL (POPULATION) Monrovia (962,000)

GOVERNMENT Multiparty republic

ETHNIC GROUPS Kpelle 19%, Bassa 14%, Grebo 9%, Gio 8%, Kru 7%, Mano 7%

LANGUAGES English (official), Mande, Mel, Kwa

RELIGIONS Christianity 40%, Islam 20%, traditional beliefs and others 40%

CURRENCY Liberian dollar = 100 cents

HISTORY

In the late 18th century, some white Americans in the United States wanted to help freed black slaves to return to Africa. In 1816, they set up the American Colonization Society, which bought land in what is now Liberia.

In 1822, the Society landed former slaves at a settlement on the coast which they named Monrovia. In 1847, Liberia became a fully independent republic with a constitution much like that of the United States. For many years, the Americo-Liberians controlled the country's government. US influence remained strong and the American Firestone Company, which ran Liberia's rubber plantations, was especially influential. Foreign companies were also involved in exploiting Liberia's mineral resources, including its huge iron-ore deposits.

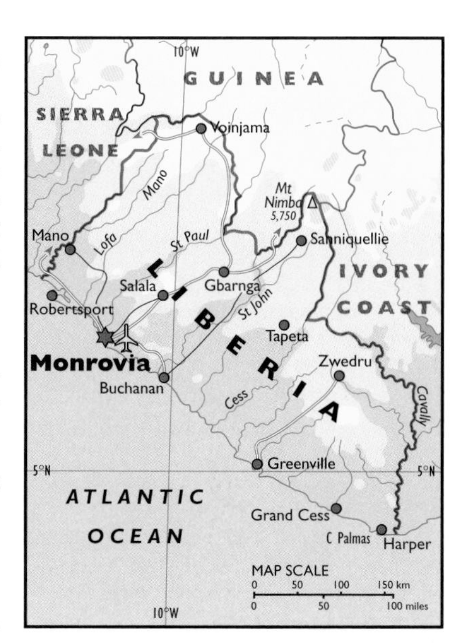

POLITICS

In 1980, a military group composed of people from the local population killed the Americo-Liberian president, William R. Tolbert. An army sergeant, Samuel K. Doe, was made president of Liberia. Elections held in 1985 resulted in victory for Doe.

From 1989, the country was plunged into civil war between various ethnic groups. Doe was assassinated in 1990, but his successor, Amos Sawyer, continued to struggle with rebel groups. Peacekeeping forces from other West African countries intervened and a ceasefire was agreed in 1995, when a council of state, composed of former warlords, was set up. Around 200,000 people were estimated to have been killed in the civil war. In 1997, one of the warlords, Charles Taylor, was elected president. But fighting continued and, in 2002, Taylor declared a state of emergency.

DID YOU KNOW

- that Liberia is Africa's oldest independent republic
- that Liberia got its name from a Latin word *liber*, meaning "free"
- that Monrovia was named after James Monroe, the US president when the first settlement for freed slaves was founded in 1822
- that Americo-Liberians, the descendants of freed slaves who settled in the country in the 19th century, now make up about 5% of the population

ECONOMY

Liberia's civil war devastated its economy. Three out of four people depend on agriculture, though many grow little more than they need to feed their families. Chief food crops include cassava, rice, and sugarcane, while rubber, cocoa, and coffee are grown for export. Timber is also exported. The most valuable export is iron ore.

Liberia's low taxes enable it to obtain revenue from its "flag of convenience," which is used by about one-sixth of the world's commercial shipping. However, only a few of the ships are actually owned by Liberians.

LIBYA

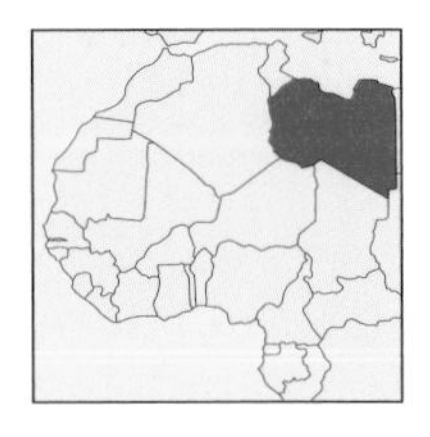

Libya's flag was adopted in 1977. It replaced the flag of the Federation of Arab Republics which Libya left in that year. Libya's flag is the simplest of all world flags. It represents the country's quest for a green revolution in agriculture.

The Socialist People's Libyan Arab Jamahiriya, as Libya is officially called, is a large country in North Africa. Most people live on the Mediterranean coastal plains in the northeast and northwest. The Sahara, the world's largest desert which occupies 95% of Libya, reaches the Mediterranean coast along the Gulf of Sidra (Khalij Surt). The Sahara is mostly uninhabited except around scattered oases. The land rises toward the south, reaching 7,503 ft [2,286 m] at Bette Peak on the border with Chad.

AREA 679,358 sq miles [1,759,540 sq km]
POPULATION 5,241,000
CAPITAL (POPULATION) Tripoli (or Tarabulus, 960,000)
GOVERNMENT Single-party socialist state
ETHNIC GROUPS Libyan Arab and Berber 97%
LANGUAGES Arabic (official), Berber
RELIGIONS Islam (Sunni)
CURRENCY Libyan dinar = 1,000 dirhams

CLIMATE

The coastal plains in the northeast and northwest of the country have Mediterranean climates, with hot, dry summers and mild winters, with some rain in the winter months. Inland, the average yearly rainfall drops to 4 inches [100 mm] or less. Daytime temperatures are high, but the nights are cool.

VEGETATION

Shrubs and grasses grow on the northern coasts, with some trees in wetter areas. Few plants grow in the desert, except at oases where date palms provide protection from the hot sun.

HISTORY

The earliest known inhabitants of Libya were the Berbers. Between the 7th century BC and the 5th century AD, the region came under the rule of Greeks, Carthaginians, Romans and Vandals. The Romans left magnificent ruins, but the Arabs who invaded the country in AD 642 imposed their religion, Islam, and their Arab culture survives to this day. From 1551, Libya was part of the Turkish Ottoman empire. Barbary corsairs (pirates) used bases on the Libyan coast and, in the early 1880s, the United States fought a war against the corsairs.

Italy took over Libya in 1911, but lost it during World War II

(1939–45). Britain and France then jointly ruled Libya until 1951, when the country became an independent kingdom.

POLITICS

In 1969, a military group headed by Colonel Muammar Gaddafi deposed the king and set up a military government. Under Gaddafi, the government took control of the economy and used money from oil exports to finance welfare services and development projects. Gaddafi attracted international criticism for his support for terrorist movements around the world and Libya became isolated from the mid-1980s. In 1999, Gaddafi sought to restore good relations with other countries by surrendering for trial two Libyans suspected of planting a bomb on a PanAm plane, which exploded over the Scottish town of Lockerbie in 1988. He also paid compensation to the family of a British policewoman killed in 1984 by shots fired from within the Libyan People's Bureau in London. His actions were intended to help to end Libya's pariah status. In 2002, Libya withdrew from the Arab League, but it became deeply involved in the creation of the African Union, which took over from the Organization of African Unity.

DID YOU KNOW

- that the world's highest shade temperature, 136°F [58°C], was recorded at Al Aziziyah, south of Tripoli in 1922
- that the "Great Man-made River" is a project to bring water pumped from wells in southern Libya through a system of pipelines to the north; Phase I of this project was completed in 1991
- that Tripoli was once a base for pirates called the Barbary corsairs
- that the word *Jamahiriya* in Libya's official name means "state of the masses"

ECONOMY

The discovery of oil and natural gas in 1959 led to the transformation of Libya's economy, making the country one of Africa's richest. But Libya remains a developing country, overdependent on oil, which accounts for nearly all of its export revenues.

Agriculture is important, although Libya has to import food. Crops include barley, citrus fruits, dates, olives, potatoes, and wheat. Cattle, sheep, and poultry are raised. Libya has oil refineries and petrochemical plants. Other manufactures include cement, processed food, and steel.

LITHUANIA

This flag was created in 1918 when Lithuania became an independent republic. After the Soviet Union annexed Lithuania in 1940, the flag was suppressed. It was revived in 1988 and again became the national flag when Lithuania became fully independent in 1991.

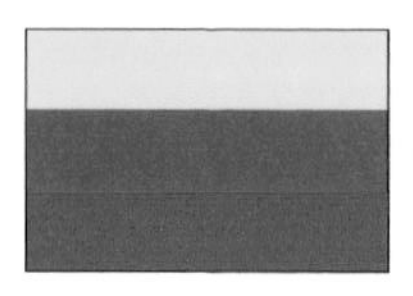

The Republic of Lithuania is the southernmost of the three Baltic states which were ruled as part of the Soviet Union between 1940 and 1991. Much of the land is flat or gently rolling, with the highest land in the southeast.

Most of the land is covered by moraine (ice-worn rocks) which was deposited there by ice sheets during the Ice Age. Hollows in the moraine contain about 3,000 lakes. The longest of the many rivers is the Neman, which rises in Belarus and flows through Lithuania to the Baltic Sea.

AREA 25,200 sq miles [65,200 sq km]
POPULATION 3,611,000
CAPITAL (POPULATION) Vilnius (580,000)
GOVERNMENT Multiparty republic
ETHNIC GROUPS Lithuanian 80%, Russian 9%, Polish 7%, Belarussian 2%
LANGUAGES Lithuanian (official), Russian, Polish
RELIGIONS Mainly Roman Catholic
CURRENCY Litas = 100 centai

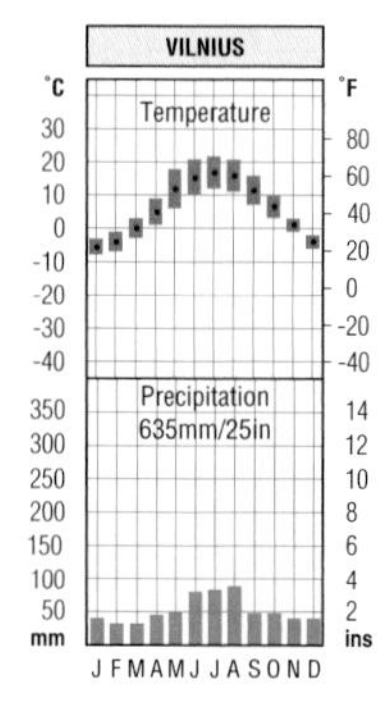

CLIMATE

Winters are cold. January's temperatures average 27°F [–3°C] in the west and 21°F [–6°C] in the east. Summers are warm, with average temperatures in July of 63°F [17°C]. The average rainfall in the west is about 25 inches [630 mm]. Inland areas are drier.

VEGETATION

Farmland covers nearly three-quarters of Lithuania and forests only 16%. Pine is predominant in the west and south, with oak and birch in the center, and spruce in the east.

HISTORY

The Lithuanian people were united into a single nation in the 12th century and, in the 14th century, they pushed their eastern borders almost as far as Moscow. In 1386, Lithuania joined a union with Poland and this became a single state in 1569. The union ended in the 17th century and, in 1795, Lithuania came under Russian rule. The Russians sought unsuccessfully to suppress Lithuanian culture.

After World War I (1914–18), Lithuania declared itself independent, and in 1920 it signed a peace treaty with the Russians, though Poland held Vilnius until 1939. In 1940, the Soviet Union occupied all of

Lithuania. The Germans invaded in 1941, but Soviet forces returned in 1944, integrating Lithuania into the Soviet Union.

POLITICS

In 1988, when the Soviet Union was introducing reforms, the Lithuanians demanded independence. In 1990, pro-independence groups won the national elections and, in 1991, the Soviet Union was forced to recognize Lithuania's independence.

After independence, Lithuania faced many problems as it sought to change its Communist-run economy into a private enterprise one. In the late 1990s, Lithuania sought closer ties with the West. Its progress toward transforming itself into a free enterprise democracy was rewarded in 2002, when it was invited, together with the other two Baltic states, Estonia and Latvia, to become a member of the European Union in May 2004.

DID YOU KNOW

- that before World War II, Jews made up about 8% of Lithuania's population; nearly all of them were killed by the Nazis during the war
- that a Russian enclave, sandwiched between Lithuania and Poland, contains the naval base of Kaliningrad, Russia's westernmost port
- that under Soviet rule, Lithuania's flag was mainly red and contained the Communist hammer and sickle symbol; religious worship and the practice of local customs were discouraged

ECONOMY

The World Bank classifies Lithuania as a "lower-middle-income" developing country. Lithuania lacks natural resources, but manufacturing, based on imported materials, is the most valuable activity and it supplies the country's main exports. Products include chemicals, electronic goods, processed food, and machine tools. Dairy and meat farming are important, as also is fishing.

The word Lietuva *on the stamp means Lithuania. This stamp was issued in 1923, five years after the country had declared itself independent. It is one of a set of 13 showing views.*

LUXEMBOURG

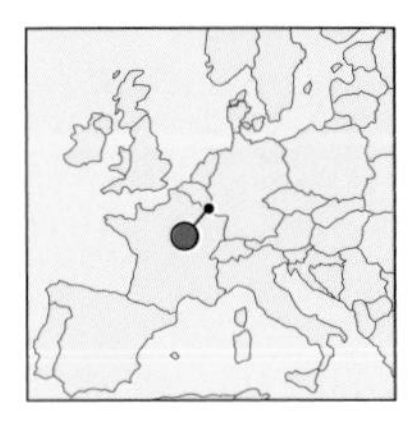

Luxembourg's flag was adopted in the 1840s. It uses colors taken from the coat of arms of the Grand Dukes. The flag is almost identical with that of the Netherlands. But the blue stripe is lighter and the flag itself is longer.

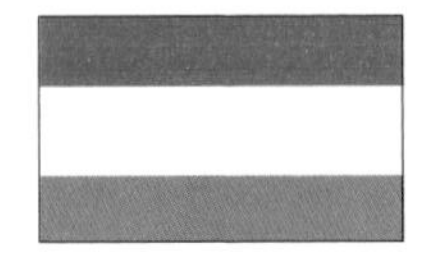

The Grand Duchy of Luxembourg is one of the smallest and oldest countries in Europe. The north is mountainous, belonging to an upland region which includes the Ardenne in Belgium and Luxembourg, and the Eifel highlands in Germany. This scenic region contains the country's highest point, at 1,854 ft [565 m] above sea level. The southern two-thirds of Luxembourg is a rolling plateau, called the Bon Pays. This region contains rich farmland, especially in the fertile Moselle and Sauer river valleys in the east.

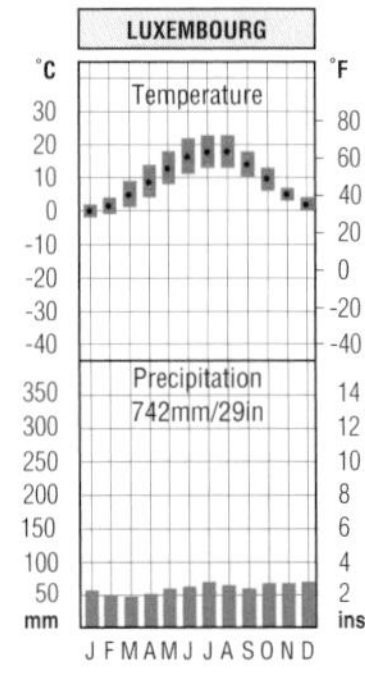

CLIMATE

Luxembourg has a temperate climate. The south of the country has warm summers and warm falls, when the grapes ripen in sheltered southeastern valleys. Winters can sometimes be severe, especially in the Ardenne region, where snow can cover the land for some weeks.

VEGETATION

Forests cover about a fifth of Luxembourg, mainly in the north, where deer and wild boar are found.

AREA 1,000 sq miles [2,590 sq km]
POPULATION 443,000
CAPITAL (POPULATION) Luxembourg (76,000)
GOVERNMENT Constitutional monarchy (Grand Duchy)
ETHNIC GROUPS Luxembourger 71%, Portuguese 10%, Italian 5%, French 3%, Belgian 3%, German 3%
LANGUAGES Luxembourgish (official), French, German
RELIGIONS Roman Catholic 95%, Protestant, Judaism, Islam
CURRENCY Euro = 100 cents

Farms cover about 25% of the land and pasture covers another 20%.

HISTORY AND POLITICS

Luxembourg became an independent state in AD 963 and a duchy (a territory ruled under a duke) in 1354. In the 1440s, Luxembourg came under the House of Burgundy and, in the early 16th century, under the rule of the Habsburgs. From 1684, it came successively under France (1684–97), Spain (1697–1714), and Austria until 1795, when it reverted to French rule.

In 1815, following the defeat of France, Luxembourg became a Grand Duchy under the Netherlands. This

was because the Grand Duke was also the king of the Netherlands. In 1890, when Wilhelmina became queen of the Netherlands, Luxembourg broke away and the Grand Duchy passed to Adolphus, Duke of Nassau-Weilburg.

Germany occupied Luxembourg in World War I (1914–18) and again in World War II (1939–45). In 1944–5, northern Luxembourg was the scene of the famous Battle of the Bulge.

In 1948, Luxembourg joined Belgium and the Netherlands in a union called Benelux. In the 1950s, it was one of the six founders of what is now the European Union. Since then, it has played an important part in European affairs. The capital Luxembourg contains the headquarters of several international agencies, including the European Coal and Steel Community and the European Court of Justice. The city is also a major financial center.

ECONOMY

Luxembourg has iron-ore reserves and is a major steel producer. It also has many high-technology industries, producing electronic goods and computers. Steel and other manufactures, including chemicals, rubber products, glass, and aluminum, dominate the country's exports. Other major activities include tourism and financial services.

Farmers raise cattle, sheep, pigs, and poultry. Major crops include oats, barley, fruits, including grapes for wine-making, potatoes, and wheat.

Luxembourg is an attractive country, with small towns and villages with beautiful old churches and romantic castles. The two stamps illustrated here, showing Bourscheid and Vianden castles, formed a set issued in 1982, entitled "Classified Monuments."

MACEDONIA (FYROM)

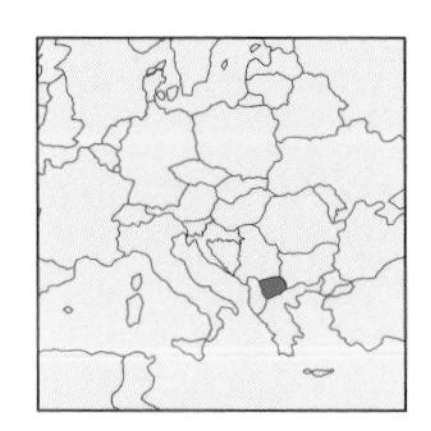

Macedonia's flag was introduced in August 1992. The emblem in the center of the flag was the device from the war-chest of Philip of Macedon. The Greeks claimed this symbol as their own. In 1995, Macedonia agreed to redesign their flag, as shown here.

The Republic of Macedonia is a country in southeastern Europe, which was once one of the six republics that made up the former Federal People's Republic of Yugoslavia. This landlocked country is largely mountainous or hilly. The highest point is Mount Korab, which reaches 9,071 ft [2,764 m] above sea level on the border with Albania. Most of the country is drained by the River Vardar and its many tributaries. In the southwest, Macedonia shares two large lakes – Ohrid and Prespa – with Albania and Greece.

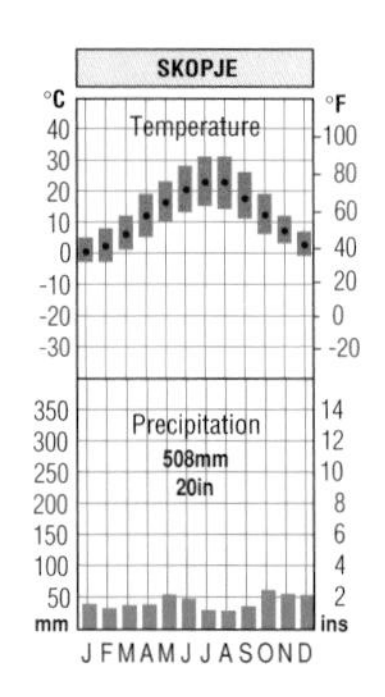

CLIMATE

Macedonia has hot summers, though highland areas are cooler. Winters are cold and snowfalls are often heavy. The climate is fairly continental in character. Rain occurs throughout the year. It is slightly heavier in early summer and the fall.

VEGETATION

Forests of beech, oak, and pine are common, especially in the mountains. Farmland, including pasture, covers about 30% of Macedonia.

AREA 9,927 sq miles [25,710 sq km]
POPULATION 2,046,000
CAPITAL (POPULATION) Skopje (541,000)
GOVERNMENT Multiparty republic
ETHNIC GROUPS Macedonian 67%, Albanian 23%, Turkish 4%, Romanian 2%, Serb 2%
LANGUAGES Macedonian and Albanian (both official), Turkish, Serbo-Croat
RELIGIONS Macedonian Orthodox, Islam
CURRENCY Dinar = 100 paras

HISTORY

Until the 20th century, Macedonia's history was closely tied to that of a larger area, also called Macedonia, which included parts of northern Greece and southwestern Bulgaria. The Macedonian Greek king, Alexander the Great, who built a world empire, came from this region.

Between the 14th and 19th centuries, Macedonia was ruled by the Turkish Ottoman empire. Between 1912 and 1913, the region was divided between Serbia, which controlled the north and center, Bulgaria which took a small area in the east, and Greece which gained the south. At the end of World War I (1914–18), Serbian Macedonia became part of

the Kingdom of the Serbs, Croats, and Slovenes, which was renamed Yugoslavia in 1929. After World War II (1939–45), a Communist government under President Josip Broz Tito took power in Yugoslavia.

POLITICS

Tito died in 1980 and, in the early 1990s, the country broke up into five separate republics. Macedonia declared its independence in September 1991. Greece objected to this territory using the name Macedonia, which it considered to be a Greek name. It also objected to a symbol on Macedonia's flag and a reference in the constitution to the desire to reunite the three parts of the old Macedonia.

Macedonia adopted a new clause in its constitution rejecting any Macedonian claims on Greek territory and, in 1993, the United Nations accepted the new republic as a member under the name of The Former Yugoslav Republic of Macedonia (FYROM).

By the end of 1993, all the countries of the European Union, except for Greece, were establishing diplomatic relations with the FYROM. In 1995, Macedonia agreed to redesign its flag. In 2001, fighting broke out in the north between Albanian-speaking Macedonians and government troops. To appease the rebels, the government agreed in 2002 to make Albanian an official language.

ECONOMY

According to the World Bank, Macedonia ranks as a "lower-middle-income" developing country. The leading sector of the economy is manufacturing, and manufactures dominate the country's exports. Macedonia mines coal, but has to import all its oil and natural gas. Chromium, copper, iron ore, lead, manganese, uranium, and zinc are also mined. Manufactures include cement, chemicals, iron and steel, textiles, and tobacco products. Agriculture employs less than 10% of the people. Major crops include cotton, fruits, maize, tobacco, and wheat. Livestock rearing and forestry are also important.

Eastern Orthodox churches contain many beautiful religious paintings called "icons." This Macedonian Christmas stamp uses an icon depicting the Nativity.

MADAGASCAR

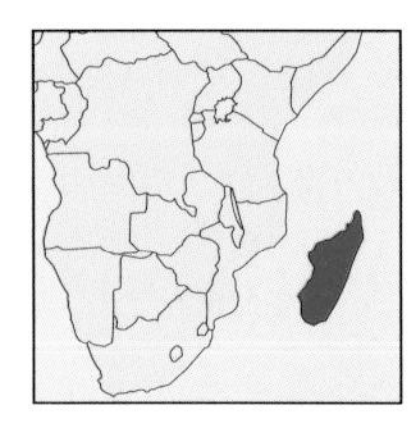

The colors on this flag are those used on historic flags in Southeast Asia. It was from this region that the ancestors of many Madagascans came around 2,000 years ago. This flag was adopted in 1958, when Madagascar became a self-governing republic under French rule.

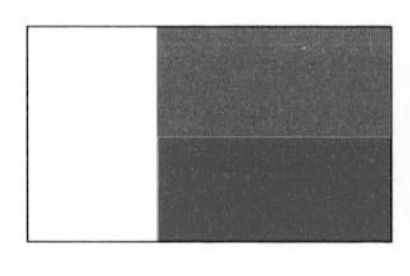

AREA 226,656 sq miles [587,040 sq km]

POPULATION 15,983,000

CAPITAL (POPULATION) Antananarivo (1,053,000)

GOVERNMENT Republic

ETHNIC GROUPS Merina 27%, Betsimisaraka 15%, Betsileo 11%, Tsimihety 7%, Sakalava 6%

LANGUAGES Malagasy and French (both official)

RELIGIONS Traditional beliefs 52%, Christianity 41%, Islam 7%

CURRENCY Malagasy franc = 100 centimes

The Democratic Republic of Madagascar, in southeastern Africa, is an island nation, which has a larger area than France. Behind the narrow coastal plains in the east lies a highland zone, mostly between 2,000 ft and 4,000 ft [610 m to 1,220 m] above sea level. Some volcanic peaks, such as Tsaratanana in the north, rise above this level. The highlands are Madagascar's most densely populated region. Western Madagascar is an area of broad plains bordering the Mozambique Channel.

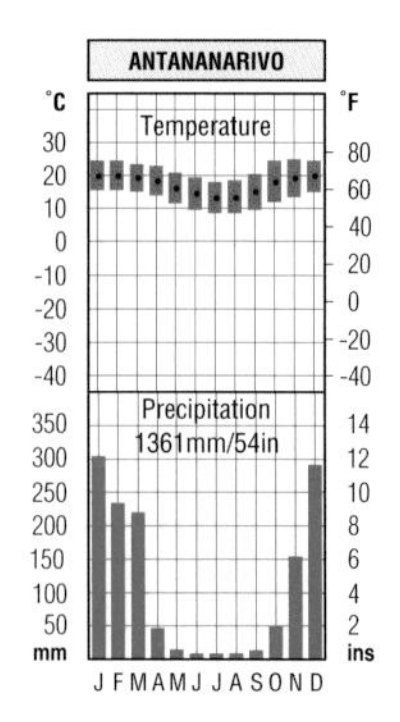

CLIMATE

Antananarivo lies in the tropics, but temperatures are moderated by the effects of altitude. The winters (from April to September) are dry, but heavy rain falls in the summer months. The eastern coastlands are warm and humid. The west is drier, and the south and southwest are hot and dry.

VEGETATION

Grassland and scrub grow in the south. Forest and savanna (grassland with patches of woodland) once covered much of the country, but large areas have been cleared for farming. This has seriously threatened Madagascar's wildlife.

HISTORY

People from Southeast Asia began to settle on Madagascar around 2,000 years ago. Other immigrants from Africa and Arabia settled on the coasts. The Malagasy language is of Southeast Asian origin, though it includes words from Arabic, Bantu languages, and European languages. The Asian immigrants introduced rice farming into Madagascar. Today, rice remains the main food crop.

The first Europeans to reach Madagascar were Portuguese. Later,

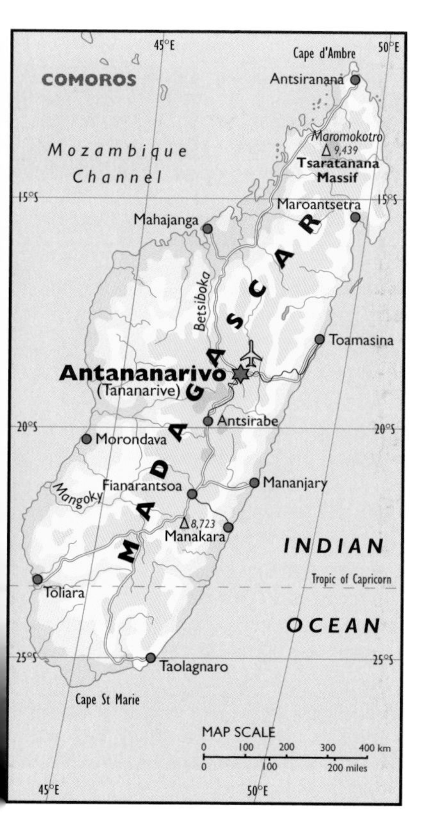

the island, which was ruled by powerful monarchs, became a haven for pirates. France made contacts with the island in the 1860s. Finally, French troops defeated the army of the King of the Merina in 1895. Madagascar was a French colony until 1960 when it achieved full independence as the Malagasy Republic.

POLITICS

In 1972, army officers took over the government. In 1975, under the leadership of Lt.-Commander Didier Ratsiraka, the country was renamed Madagascar. In 1977, elections were held, but Ratsiraka remained president of a one-party socialist state. In 2002, Ratsiraka refused to accept his defeat in a presidential election and the country nearly lurched into civil war. But after Ratsiraka fled the country, Marc Ravalomanana was installed as president.

ECONOMY

Madagascar is one of the world's poorest countries. The land has been badly eroded because of the cutting down of the forests and overgrazing of the grasslands. Farming, fishing, and forestry employ more than 80% of the people.

The country's food crops include bananas, cassava, rice, and sweet potatoes. Coffee is the leading export. Other exports include cloves, sisal, sugar, and vanilla. There are few manufacturing industries and mining is unimportant at present.

DID YOU KNOW

- that Madagascar is the world's fourth largest island after Greenland, New Guinea, and Borneo
- that Madagascar produces about two-thirds of the world's natural vanilla
- that the ancestors of many people in Madagascar came from Southeast Asia
- that Madagascar has many animals, including lemurs, which are not found in any other part of the world – more than 150,000 species of plants and animals are unique to the island

MALAWI

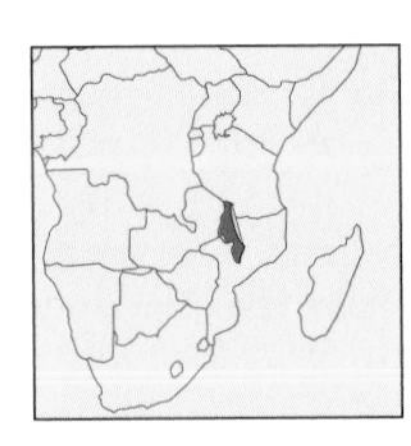

The colors in Malawi's flag come from the flag of the Malawi Congress Party, which was adopted in 1953. The symbol of the rising sun was added when Malawi became independent from Britain in 1964. It represents the beginning of a new era for Malawi and Africa.

The Republic of Malawi is a landlocked country in southern Africa. It includes part of Lake Malawi, which is drained in the south by the River Shire, a tributary of the River Zambezi. Lake Malawi, which is called Lake Nyasa in Tanzania and Mozambique, lies on the floor of the Great African Rift Valley. This huge gash in the Earth's crust extends from Mozambique to southwestern Asia. The land is mostly mountainous. The highest peak, Mulanje, reaches 9,846 ft [3,000 m] in the southeast.

AREA 45,745 sq miles [118,480 sq km]
POPULATION 10,548,000
CAPITAL (POPULATION) Lilongwe (395,000)
GOVERNMENT Multiparty republic
ETHNIC GROUPS Maravi (Chewa, Nyanja, Tonga, Tumbuka) 58%, Lomwe 18%, Yao 13%, Ngoni 7%
LANGUAGES Chichewa and English (both official)
RELIGIONS Protestant 55%, Roman Catholic 20%, Islam 20%
CURRENCY Kwacha = 100 tambala

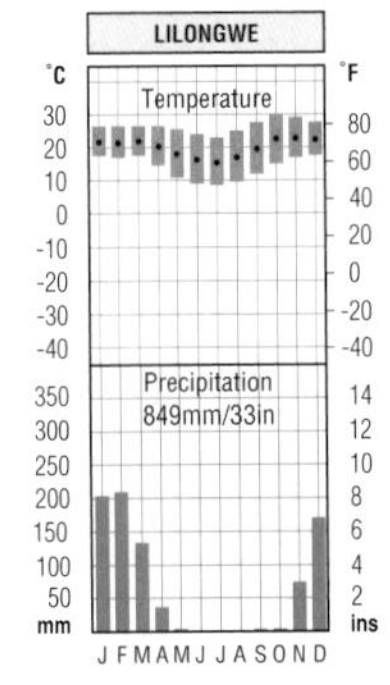

CLIMATE

While the low-lying areas of Malawi are hot and humid all year round, the uplands have a pleasant climate. Lilongwe, at about 3,609 ft [1,100 m] above sea level, has a warm and sunny climate. Frosts sometimes occur in July and August, in the middle of the long dry season.

VEGETATION

Grassland and savanna (tropical grassland with scattered trees) cover much of Malawi. Woodland flourishes in wet areas, such as river valleys. Commercial forests of softwoods have been planted in some areas.

HISTORY

In the early 19th century, the area now called Malawi was a center of the slave trade, which British missionaries helped to stamp out. In 1891, the area was made a British protectorate.

In 1953, Britain made the country, then called Nyasaland, part of a federation called the Federation of Rhodesia and Nyasaland. The federation included what are now Zambia and Zimbabwe. Most black Africans opposed this step. In Nyasaland, the opposition was led by Dr Hastings Kamuzu Banda. The federation was dissolved in 1963. In 1964, the coun-

try became independent as Malawi, with Banda as its prime minister. Banda became president when the country became a republic in 1966.

POLITICS

President Banda, who was made president for life in 1971, and his Malawi Congress Party, the sole political party, dominated national politics after independence. A multiparty system was restored in 1993, and, in elections in May 1994, Banda and his party were defeated. Bakili Muluzi was elected president. In 2002, Muluzi accepted parliament's decision that the number of presidential terms be limited to two.

ECONOMY

Malawi is one of the world's poorest countries. More than 80% of the people are farmers, but malnutrition is common when droughts ruin harvests. Food crops include cassava, maize, and rice. Tobacco, tea, sugar, cotton, and groundnuts are exported. Malawi lacks mineral resources and has few manufacturing industries.

This stamp, one of a set of four showing wild flowers, was issued to commemorate Christmas 1987. The name of the plant is Ochna macrocalyx.

MALAYSIA

This flag was adopted when the Federation of Malaysia was set up in 1963. The red and white bands date back to a revolt in the 13th century. The star and crescent are symbols of Islam. The blue represents Malaysia's role in the Commonwealth.

AREA 127,316 sq miles [329,750 sq km]
POPULATION 22,229,000
CAPITAL (POPULATION) Kuala Lumpur (1,145,000)
GOVERNMENT Federal constitutional monarchy
ETHNIC GROUPS Malay and other indigenous groups 58%, Chinese 27%, Indian 8%
LANGUAGES Malay (official), Chinese, English
RELIGIONS Islam 53%, Buddhism 17%, Chinese folk religionist 12%, Hinduism 7%, Christianity 6%
CURRENCY Ringgit (Malaysian dollar) = 100 cents

The Federation of Malaysia consists of two main parts. Peninsular Malaysia, which is joined to mainland Asia, contains about 80% of the population. The other main regions, Sabah and Sarawak, are in northern Borneo, an island which Malaysia shares with Indonesia.

Much of Malaysia is mountainous, with coastal lowlands bordering the rugged interior. The highest peak, Kinabalu, reaches 13,459 ft [4,101 m] in Sabah. The country's longest rivers are also in Sabah and Sarawak.

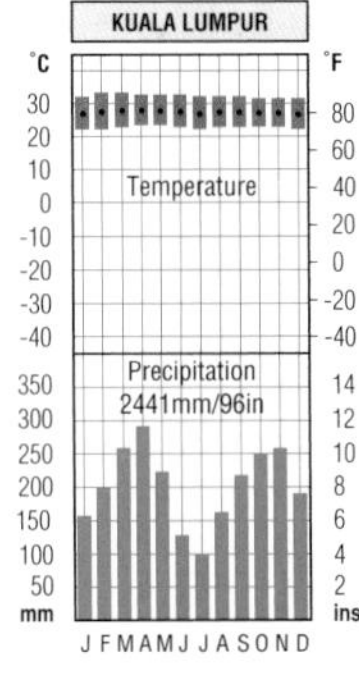

CLIMATE

Malaysia has a hot equatorial climate. The temperatures are high all through the year, though the mountains are much cooler than the lowland areas. The rainfall is heavy throughout the year. In the capital, Kuala Lumpur, rain falls, on average, on about 200 days every year.

VEGETATION

Dense rain forest, which is rich in plant species, covers about half of Malaysia. Farmland covers about 13% of the country, while scrub is the typical vegetation on formerly forested but unfarmed land.

HISTORY

In the early 16th century, Portuguese traders reached Melaka, a powerful Muslim kingdom on the southwest coast of Malaya (now Peninsular Malaysia). The Portuguese were succeeded by the Dutch and, later, the British.

In 1786, the British East India Company became established in Pinang and, in 1826, the British founded the Straits Settlement, which consisted of Pinang, Melaka, and Singapore. In 1867, the Straits Settlement became a British colony, while

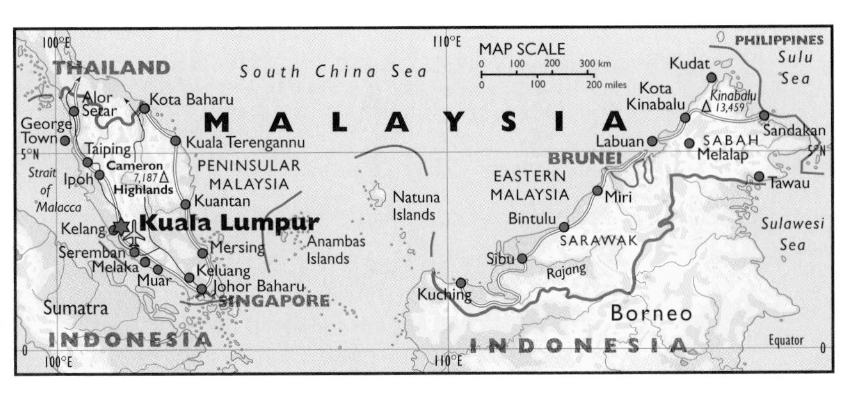

Sabah and Sarawak became a British protectorate in 1888.

Japan occupied Malaysia in World War II (1939–45) and, after the war, British troops fought a war against Communist guerrillas. In 1957, Malaya became independent. Malaysia was created in 1963, when Malaya, Singapore, Sabah, and Sarawak agreed to unite, but Singapore withdrew in 1965.

POLITICS

Ethnic tensions between the Chinese and the Muslim Malays have threatened Malaysia's stability. But no major clashes have occurred in recent years. Under the leadership of Dr Mahathir Mohamad, who became prime minister in 1981, Malaysia has achieved rapid economic progress, though it suffered a recession in the late 1990s. The country plays an important part in regional affairs, especially through its membership of ASEAN (Association of Southeast Asian Nations).

ECONOMY

The World Bank classifies Malaysia as an "upper-middle-income" developing country. Malaysia is a leading producer of palm oil, rubber, and tin. These products, together with oil and timber, underpin the economy. Agriculture remains important, and rice is the chief food crop. Other major crops include cocoa, coconuts, pepper, tobacco, pineapples, and tea.

Manufacturing is now important in the economy. Manufactures are diverse, including cars, chemicals, electronic goods, plastics, textiles, rubber, and food products.

Malaysia has a rich plant and animal life. Insects abound, including superbly colored butterflies, such as the Common Nawab shown on this stamp, which was one of a set of eight issued in 1970.

MALI

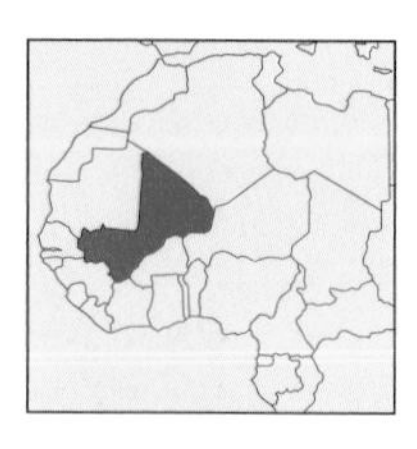

The colors on Mali's flag are those used on the flag of Ethiopia, Africa's oldest independent nation. They symbolize African unity. This flag was used by Mali's African Democratic Rally prior to the country becoming independent from France in 1960.

The Republic of Mali is a landlocked country in northern Africa. The land is generally flat, with the highest land in the Adrar des Iforhas on the border with Algeria. This highland region in the northern desert contains many wadis (dry valleys), where rivers once flowed. But today, the only permanent rivers and streams are in the south. The main rivers are the Sénégal, which flows westward to the Atlantic Ocean, and the Niger, which makes a large turn, called the Niger Bend, in south-central Mali.

AREA 478,837 sq miles [1,240,190 sq km]
POPULATION 11,009,000
CAPITAL (POPULATION) Bamako (810,000)
GOVERNMENT Multiparty republic
ETHNIC GROUPS Bambara 32%, Fulani (or Peul) 14%, Senufo 12%, Soninke 9%, Tuareg 7%, Songhai 7%, Malinke (Mandingo or Mandinke) 7%
LANGUAGES French (official), Voltaic languages
RELIGIONS Islam 90%, traditional beliefs 9%, Christianity 1%
CURRENCY CFA franc = 100 centimes

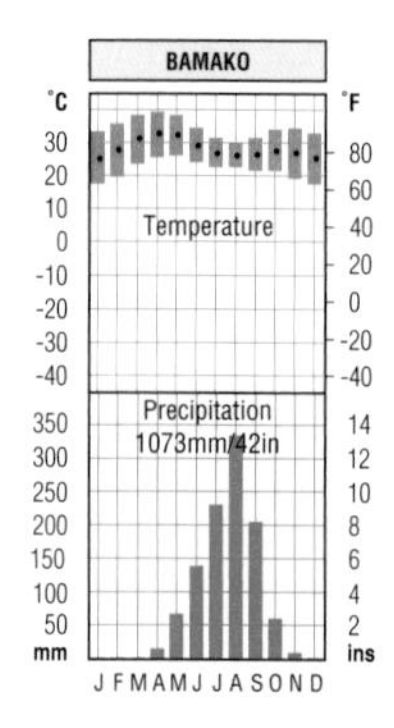

CLIMATE

Northern Mali is part of the Sahara, with a hot, practically rainless climate. But the south has enough rain for farming. Bamako, in the southwest of the country, has unpleasant weather when dry and dusty harmattan winds blow from the Sahara Desert. Bamako's rainy season is from May to October.

VEGETATION

The north is desert and plants are rare. The dry grasslands in central and southeastern Mali are part of a region called the Sahel. During long periods of drought, the northern Sahel dries up and becomes part of the Sahara. Southern Mali, which is covered by savanna (tropical grassland with scattered trees), is the most densely populated part of Mali.

HISTORY

From the 4th to the 16th centuries, Mali was part of three major black African cultures – ancient Ghana, Mali, and Songhai. Reports on these empires were made by Arab scholars who crossed the Sahara to visit them. One major center was Timbuktu (Tombouctou), in central Mali. In the 14th century, this town was a great center of learning in history, law, and the Muslim religion. It was also a trad-

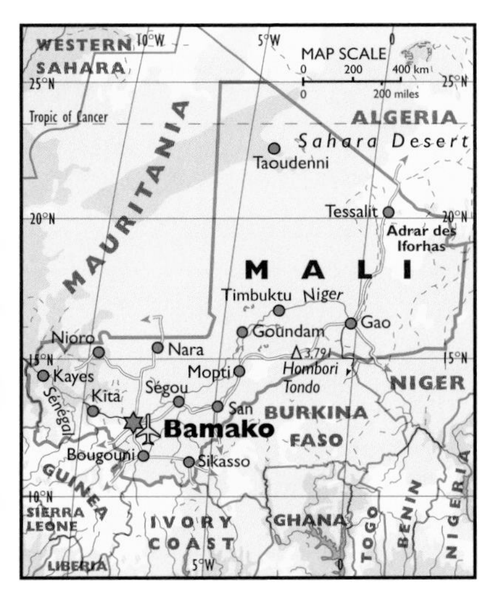

ing center and stopping point for Arabs and their camel caravans.

After initial resistance from the local people, France ruled the area, which they called French Sudan, between 1893 and 1960.

POLITICS

Mali's first president, Modibo Keita, introduced socialist policies. But he was overthrown in 1968 by an army group led by Moussa Traoré. Traoré's one-party rule was ended in 1991, when a military coup brought Lt.-Col. Amadou Touré to power. Multiparty democracy was restored in 1992. Elections were then held for the president and for the National Assembly. The new government agreed a pact providing for a special administration for the Tuaregs in the north.

ECONOMY

Mali is one of the world's poorest countries and 70% of the land is desert or semidesert. Only about 2% of the land is used for growing crops, while 25% is used for grazing animals.

However, agriculture, including nomadic livestock rearing, employs more than 80% of the people. Farming is hampered by water shortages, and the severe droughts in the 1970s and 1980s led to a great loss of animals and much human suffering.

The farmers in the south grow millet, rice, sorghum, and other food crops to feed their families. The chief cash crops are cotton, the main export, groundnuts, and sugarcane. Many of these crops are grown on land which is irrigated with river water. Fishing in the rivers is another important economic activity.

Mali has deposits of several minerals, but only salt and gold mining are of economic significance. Manufacturing is concerned mainly with processing farm products.

Stylized wood carvings are typical of black African art. This sculpture of two seated figures is in Mali's National Museum. The stamp was issued in 1979 to celebrate World Museums Day.

MALTA

Malta's flag was adopted when the country became independent from Britain in 1964. The colors are those of the Knights of Malta, who ruled the islands from 1530 to 1798. The George Cross, added in 1943, commemorates the heroism of the Maltese people in World War II.

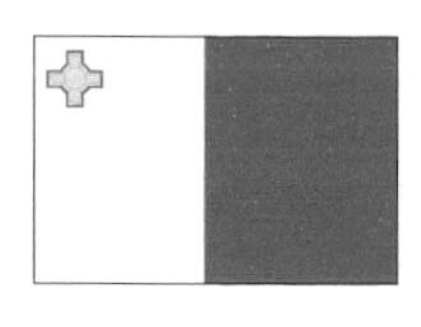

The Republic of Malta consists of two main islands, Malta and Gozo, a third, much smaller island called Comino lying between the two large islands, and two tiny islets. Malta island, which covers 95 sq miles [246 sq km], is largely made up of limestone. Because rainwater seeps down into underground caves through cracks in the limestone, the surface is mainly dry. Gozo island covers 26 sq miles [67 sq km]. Clay covers much of the land. The clay retains moisture and, as a result, Gozo's landscapes are much less arid.

AREA 122 sq miles [316 sq km]
POPULATION 395,000
CAPITAL (POPULATION) Valletta (102,000)
GOVERNMENT Multiparty republic
ETHNIC GROUPS Maltese 96%, British 2%
LANGUAGES Maltese and English (both official)
RELIGIONS Roman Catholic 99%
CURRENCY Maltese lira = 100 cents

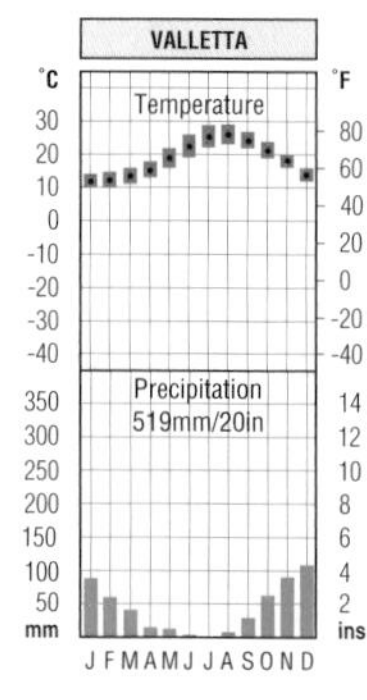

CLIMATE

Malta's climate is typical of Mediterranean lands. Summers are hot and dry, and the winters are mild and wet. A hot wind from North Africa, called the sirocco, may raise temperatures in spring. Northerly winds sometimes lower temperatures in the winter.

VEGETATION

With a low rainfall and many limestone landscapes, the vegetation is sparse on Malta island, though Gozo is greener. The country has no forest and farmland covers about 40% of the land.

HISTORY

From the Stone Age onward, Malta has been occupied by many people, and Stone Age homes and temples dating back about 4,000 years can still be seen. Phoenicians colonized Malta in about 850 BC. They were followed by the Carthaginians, Greeks, and Romans. In AD 395, Malta became part of the eastern Roman empire, but Arabs from North Africa invaded in 870. In 1091, the Norman king of Sicily, Roger I, restored Christian rule.

A succession of feudal lords ruled Malta until the early 16th century. In 1530 the Holy Roman Emperor gave Malta to the Hospitallers, who were also called the Knights of St John of Jerusalem, known for the part they played in the Crusades. The Knights held Malta against a Turkish Muslim siege in 1565. The French under Napoleon Bonaparte took Malta in 1798. With help from Britain, the French were driven out in 1800 and, in 1814, Malta became a British colony. Under Britain, Malta became a strategic base.

During World War I (1914–18) Malta was an important naval base. In World War II (1939–45), Italian and German aircraft bombed the islands. In recognition of the bravery of the Maltese, the British King George VI awarded the George Cross to Malta in 1942. In 1953, Malta became a base for NATO (North Atlantic Treaty Organization).

POLITICS

Malta became independent in 1964, and in 1974 it became a republic.

> **DID YOU KNOW**
> - that the Maltese language was not written until the 20th century
> - that Saint Paul was shipwrecked near Malta in AD 60; he spent three months there
> - that Malta has had strategic importance throughout history because of its position in the narrows between Sicily in Europe and Tunisia in Africa
> - that the Knights of Malta, also called the Knights of St John of Jerusalem, carried a special cross into battle during the Crusades; called the Maltese cross, it had eight points

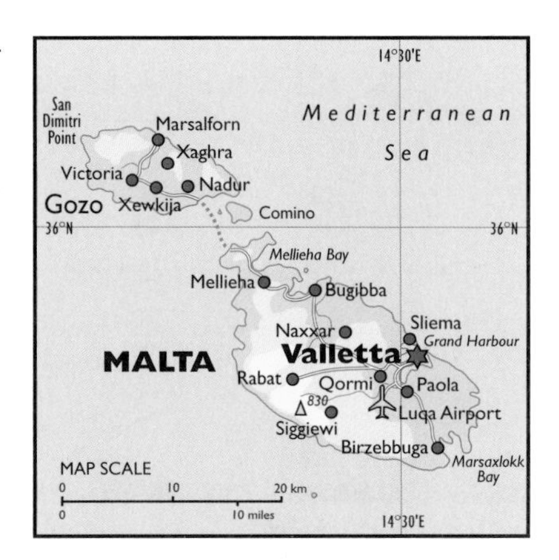

In 1979, Britain's military agreement with Malta expired, and Malta ceased to be a military base. In the 1980s, the people declared Malta to be a neutral country. Malta applied to join the European Union in the 1990s, but withdrew its bid between 1996 and 1998 when the Labor Party was in power. But, in 2002, it was invited to become a member in May 2004.

ECONOMY

The World Bank classifies Malta as an "upper-middle-income" developing country. It lacks natural resources, and most people work in the former naval dockyards, which are now used for commercial shipbuilding and repair, in manufacturing industries, and in the tourist industry.

Manufactures include chemicals, electronic equipment, processed food, and textiles. Farming is difficult, because of the rocky soils. The main crops are barley, fruits, potatoes and other vegetables, and wheat. The country also has a small fishing industry. But Malta produces only 20% of its food.

MAURITANIA

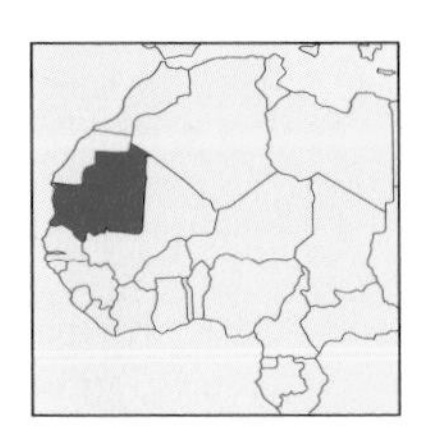

The Islamic Republic of Mauritania adopted its flag in 1959, the year before it became fully independent from France. It features a yellow star and crescent. These are traditional symbols of the national religion, Islam, as also is the color green.

The Islamic Republic of Mauritania in northwestern Africa is nearly twice the size of France. But France has more than 21 times as many people. Part of the world's largest desert, the Sahara, covers northern Mauritania and most Mauritanians live in the southwest. The land is mostly flat or gently rolling, with low rocky plateaux and areas of drifting sand dunes in the interior. The highest point, Kediet Ijill, reaches 3,003 ft [915 m] above sea level. The Kediet Ijill is rich in hematite (an iron ore).

AREA 397,953 sq miles [1,030,700 sq km]
POPULATION 2,747,000
CAPITAL (POPULATION) Nouakchott (735,000)
GOVERNMENT Multiparty Islamic republic
ETHNIC GROUPS Moor (Arab-Berber) 70%, Wolof 7%, Tukulor 5%, Soninke 3%, Fulani 1%
LANGUAGES Arabic and Wolof (both official), French
RELIGIONS Islam 99%
CURRENCY Ouguiya = 5 khoums

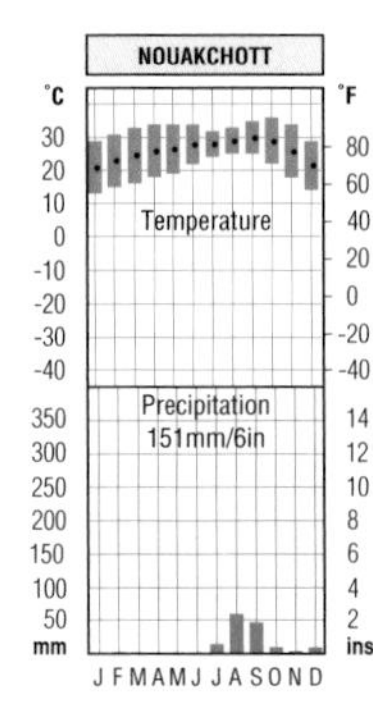

CLIMATE

The amount of rainfall and the length of the rainy season increase from north to south. Much of the land is desert, with dry northeast and easterly winds throughout the year. But southwesterly winds bring summer rain to the south.

VEGETATION

Savanna (tropical grassland with scattered trees) lies in the far south. Here, such trees as baobabs and palms dot the landscape. To the north, the savanna merges into the dry grasslands of the Sahel, where trees are rare. But most of Mauritania is desert and plants grow only around oases or in wadis (dry valleys), beneath which some water continues to flow.

HISTORY

From the 4th to the 16th centuries, parts of Mauritania belonged to two great African empires – ancient Ghana and Mali. Portuguese explorers arrived in the 1440s. But European contact did not begin in the area until the 17th century when trade in gum arabic, a substance obtained from an acacia tree, became important. Britain, France, and the Netherlands were all interested in this trade.

France set up a protectorate in

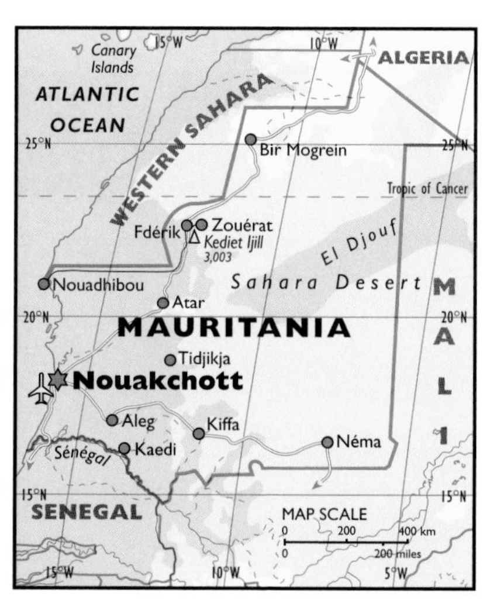

Mauritania in 1903 and the country became a territory of French West Africa and a French colony in 1920. French West Africa was a huge territory, which included present-day Benin, Burkina Faso, Guinea, Ivory Coast, Mali, Niger, and Senegal, as well as Mauritania. In 1958, Mauritania became a self-governing territory in the French Union and it became fully independent in 1960.

POLITICS

In 1961, Mauritania's political parties were merged into one by the president, Mokhtar Ould Daddah, who made the country a one-party state in 1965. In 1976, Spain withdrew from Spanish (now Western) Sahara, a territory bordering Mauritania to the north. Morocco occupied the northern two-thirds of this territory, while Mauritania took the rest. But Saharan guerrillas belonging to POLISARIO (the Popular Front for the Liberation of Saharan Territories) began an armed struggle for independence. In 1978, military leaders overthrew Ould Daddah and set up a military regime. In 1979, Mauritania withdrew from the southern part of Western Sahara, which was then occupied by Morocco.

In 1991, the country adopted a new constitution when the people voted to create a multiparty government. Elections were held in 1992. Maaouiya Ould Sidi Ahmed Taya, head of the military regime that governed between 1984 and 1992, was elected president. He was re-elected in 1997.

ECONOMY

The World Bank classifies Mauritania as a "low-income" developing country. Agriculture employs 38% of the people. Some are herders, who move around with herds of cattle and sheep, though droughts force many farmers to seek aid in cities. Farmers in the southeast grow beans, dates, millet, rice, and sorghum. The chief resource is iron ore, which is also the leading export. Other exports include livestock, processed fish, and gum arabic.

The savanna regions in the far south of Mauritania contain many wild animal species. This stamp, which is one of a set of 12 issued in 1963, shows a spotted hyena.

MEXICO

Mexico's flag dates from 1821. The stripes were inspired by the French tricolor. The emblem in the center contains an eagle, a snake, and a cactus. It is based on an ancient Aztec legend about the founding of their capital, Tenochtitlán (now Mexico City).

AREA 756,061 sq miles [1,958,200 sq km]
POPULATION 101,879,000
CAPITAL (POPULATION) Mexico City (15,643,000)
GOVERNMENT Federal republic
ETHNIC GROUPS Mestizo 60%, Amerindian 30%, White 9%
LANGUAGES Spanish (official)
RELIGIONS Roman Catholic 90%, Protestant 5%
CURRENCY New peso = 100 centavos

The United Mexican States, as Mexico is officially named, is the world's most populous Spanish-speaking country. Much of the land is mountainous. In the north, two ranges, the Sierra Madre Oriental and the Sierra Madre Occidental, enclose the plateau of Mexico, which contains most of Mexico's main cities. In the south, the plateau is dotted with volcanoes, one of which, Orizaba (or Citlaltépetl), is Mexico's highest point, reaching 18,707 ft [5,700 m] above sea level. Earthquakes are common in this area. Mexico contains two large peninsulas, Lower (or Baja) California in the northwest and the flat Yucatán peninsula in the southeast.

CLIMATE

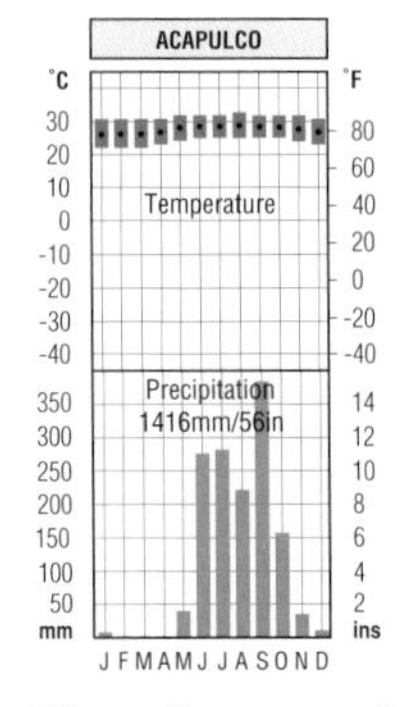

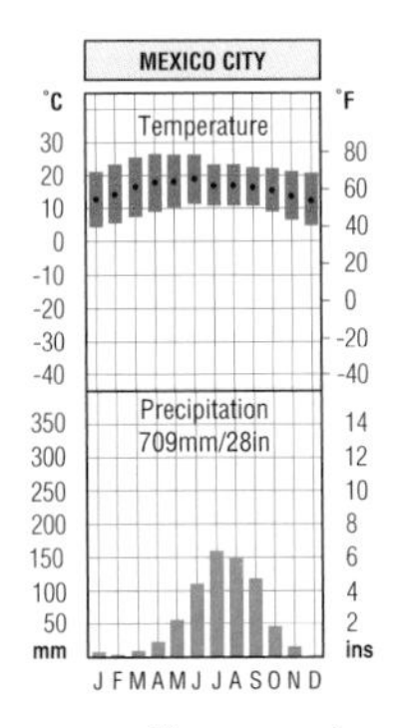

The climate varies according to the altitude. The resort of Acapulco on the southwest coast has a dry and sunny climate. Mexico City, at about 7,546 ft [2,300 m] above sea level, is much cooler. Most rain occurs between June and September. The rainfall decreases north of Mexico City and northern Mexico is mainly arid. In fact, about 70% of the country is desert or semidesert.

VEGETATION

The vegetation ranges from deserts in the north to rain forests in the south. Mexico also has areas of tropical grassland. The high mountains contain zones of vegetation which vary according to the altitude.

HISTORY AND POLITICS

From early times, Amerindians, including the Olmecs, Toltecs, Mayans, and Aztecs, built up major civiliz-

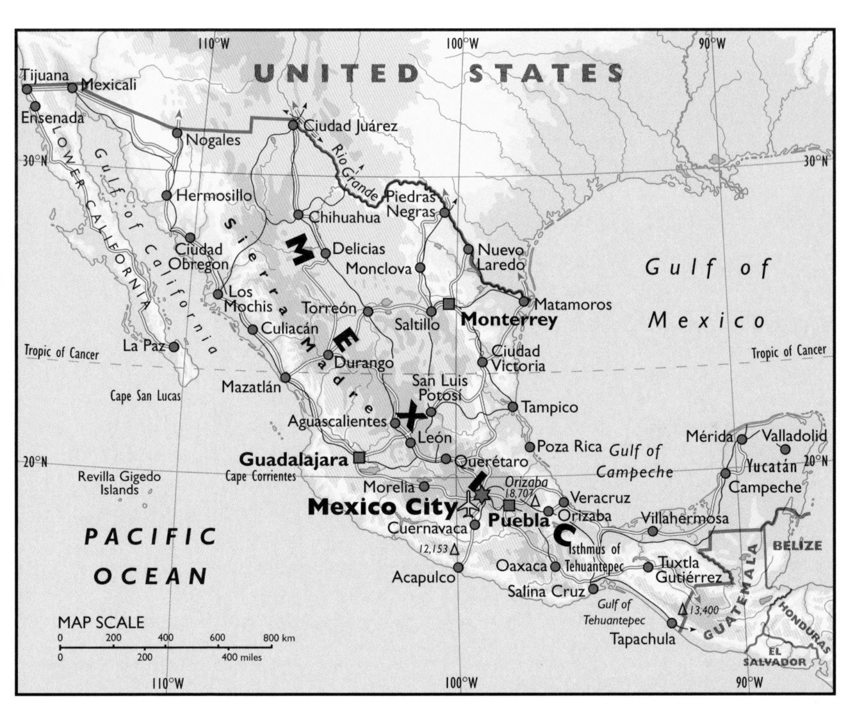

ations in Mexico. Spain conquered the Aztecs in 1519–21 and ruled until 1821. Mexico lost land to the United States in the mid-19th century, and revolutions created chaos from 1910 to 1921. In 1929, the Institutional Revolution Party (PRI) was formed. It ruled Mexico until opposition groups won parliamentary elections in 1997. The opposition leader, Vicente Fox, became Mexico's president in 2000.

Mexico's problems include rapid city growth, unemployment, demands by groups in the south for increased rights, and illegal emigration. Hope may lie in cooperation with Canada and the USA through NAFTA (North American Free Trade Association), which became operational in 1994.

ECONOMY

The World Bank classifies Mexico as an "upper-middle-income" developing country. Agriculture is important. Food crops include beans, maize, rice, and wheat, while cash crops include coffee, cotton, fruits, and vegetables. Beef cattle, dairy cattle, and other livestock are raised, and fishing is also important.

But oil and oil products are the chief exports, while manufacturing is the most valuable activity. Many factories near the northern border assemble goods, such as car parts and electrical products, for US companies. These factories are called *maquiladoras*. Other growing industries include forestry and tourism.

MOLDOVA

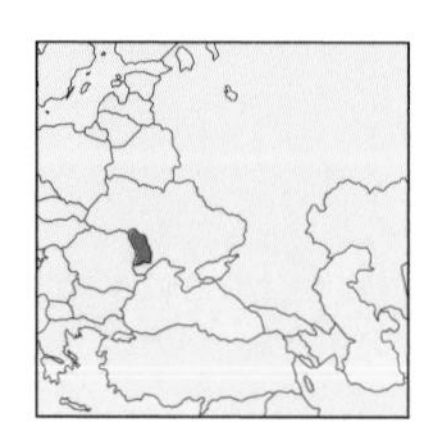

Moldova's flag, adopted in 1990, uses the colors and the eagle which appeared on the flag of Romania in pre-Communist times. The bull's head is a Moldovan emblem. The official description states that the tricolor represents "the past, present, and future" of Moldova.

The Republic of Moldova, or Moldavia as it is also called, is a small country sandwiched between Ukraine and Romania. It was formerly one of the 15 republics that made up the Soviet Union. Much of the land is hilly and the highest areas are near the center of the country, where the highest point, 1,408 ft [429 m] above sea level, is situated, northwest of the capital Chisinau. A large plain covers southern Moldova. The main river is the Dniester, which flows through the eastern part of the country.

AREA 13,010 sq miles [33,700 sq km]
POPULATION 4,432,000
CAPITAL (POPULATION) Chisinau (or Kishinev, 658,000)
GOVERNMENT Multiparty republic
ETHNIC GROUPS Moldovan/Romanian 65%, Ukrainian 14%, Russian 13%, Jewish 2%, Bulgarian 2%, Gagauz
LANGUAGES Moldovan/Romanian (official), Russian, Gagauz
RELIGIONS Eastern Orthodox, Judaism
CURRENCY Leu = 100 bani

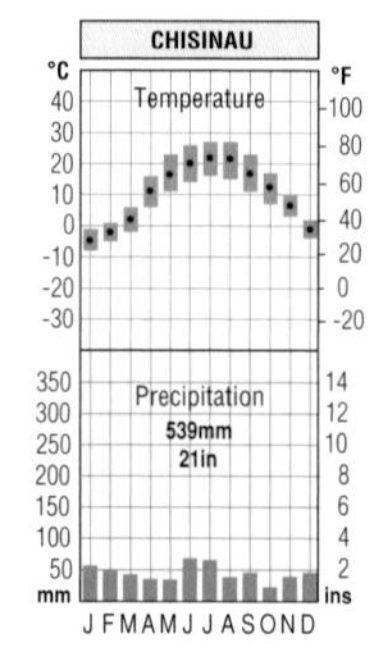

CLIMATE

Moldova has a moderately continental climate, with warm summers and fairly cold winters when temperatures dip below freezing point. Most of the rain comes in the warmer months, especially during heavy showers in the summer.

VEGETATION

Forests of hornbeam, oak, and other trees grow in northern and central Moldova. In the drier south, the land was originally grassy steppe, but most of the region is now used for farming. Rich pasture lies along the rivers.

HISTORY AND POLITICS

In the 14th century, the Moldavians formed a state called Moldavia. It included part of Romania and Bessarabia (now the modern country of Moldova). The Ottoman Turks took the area in the 16th century, but in 1812 Russia took over Bessarabia. In 1861, Moldavia and Walachia united to form Romania. Russia retook southern Bessarabia in 1878.

After World War I (1914–18), all of Bessarabia was returned to Romania, but the Soviet Union did not recognize this act. From 1944, the Moldovan Soviet Socialist Republic was part of the Soviet Union. In 1989,

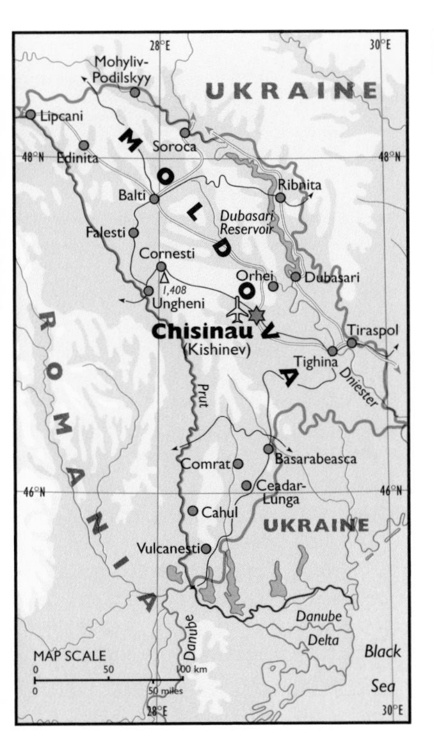

DID YOU KNOW

- that the country's name comes from the small Moldova River which is in Romania
- that Moldova is landlocked but has access to the Black Sea via a shoreline on the River Danube (less than 0.6 miles [1 km] long) and the River Dniester

the Moldovans asserted their independence by making Romanian the official language and, at the end of 1991, when the Soviet Union broke up, Moldova became an independent country.

In 1992, fighting occurred between Moldovans and Russians in Trans-Dniester, the land east of the River Dniester. Moldova held its first multiparty elections in 1994. In 2000, constitutional changes made Moldova a parliamentary republic instead of a semipresidential one.

ECONOMY

Moldova has a developing, "lower-middle-income" economy. Agriculture is important and major products include fruits, grapes for wine-making, maize, sugar beet, sunflower seeds, tobacco, vegetables, and wheat. Farmers also raise livestock, including dairy cattle and pigs.

Moldova has few natural resources and imports materials and fuels for its industries. Manufactures include processed food, farm machinery, refrigerators, television sets, and washing machines. Exports include food, wine, tobacco, and textiles.

Moldova issued a set of stamps to commemorate the 25th anniversary of the first Moon landing by Apollo 11 *in 1969. One of the stamps shows the Saturn V rocket taking off from Earth, and another depicts the lunar module which landed astronauts on the Moon.*

MONGOLIA

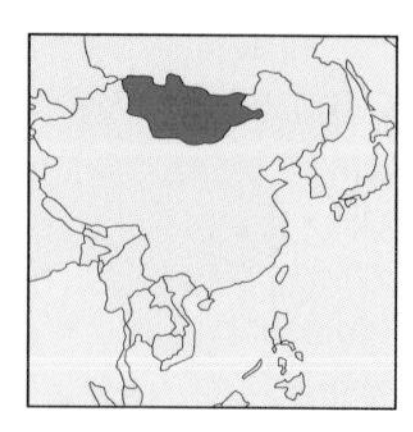

Mongolia's flag contains blue, the national color, together with red for Communism. The traditional Mongolian golden *soyonbo* symbol represents freedom. Within this, the flame is seen as a promise of prosperity and progress.

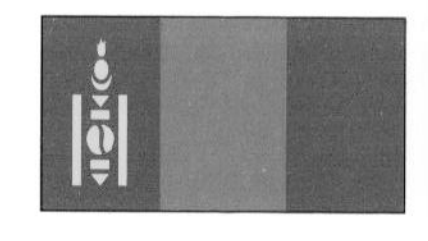

The State of Mongolia, which is sandwiched between China and Russia, is the world's largest landlocked country. High plateaux cover most of Mongolia. The highest plateaux are in the west, between the Altai Mountains (or Aerhta Shan) and the Hangayn Mountains (or Hangayn Nuruu).

The Altai Mountains contain the country's highest peaks, which reach 14,316 ft [4,362 m] above sea level. The land descends toward the east and south, where part of the huge Gobi Desert is situated.

AREA 604,826 sq miles [1,566,500 sq km]

POPULATION 2,655,000

CAPITAL (POPULATION) Ulan Bator (673,000)

GOVERNMENT Multiparty republic

ETHNIC GROUPS Khalkha Mongol 85%, Kazakh 6%

LANGUAGES Khalkha Mongolian (official), Turkic, Russian

RELIGIONS Tibetan Buddhist Lamaism 96%, Islam, Shamanism, Christianity

CURRENCY Tugrik = 100 möngös

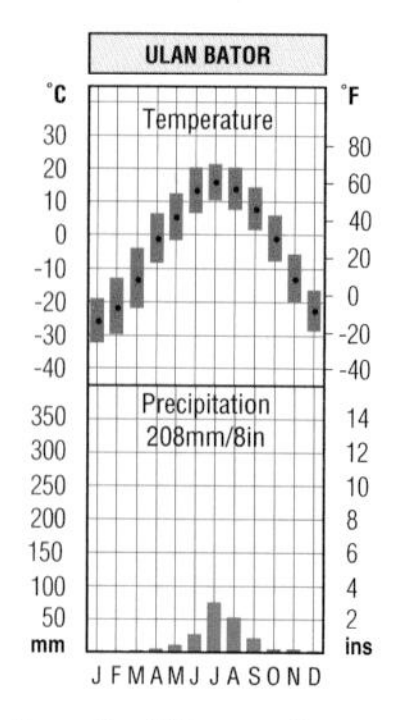

CLIMATE

Ulan Bator lies on the northern edge of a desert plateau in the heart of Asia. It has bitterly cold winters. In the summer months, the temperatures are moderated by the height of the land above sea level. The rain, which occurs mainly in summer, decreases to the south.

VEGETATION

Forests, found mainly in the northern mountains, contain such trees as birch, cedar, larch, pine, and spruce. Mongolia also has large areas of steppe grassland, but plants become increasingly sparse to the south.

HISTORY

In the 13th century, Genghis Khan united the Mongolian peoples and built up a great empire. Under his grandson, Kublai Khan, the Mongol empire extended from Korea and China to eastern Europe and Mesopotamia (now Iraq).

The Mongol empire broke up in the late 14th century. In the early 17th century, Inner Mongolia came under Chinese control, and by the late 17th century Outer Mongolia had become a Chinese province. In 1911, the Mongolians drove the Chinese out of Outer Mongolia and made the area a Buddhist kingdom. But in 1924, under Russian influence, the Communist Mongolian People's Republic was set up and, in the 1930s, the government worked to suppress religious worship. From the 1950s, Mongolia supported the Soviet Union in its disputes with China.

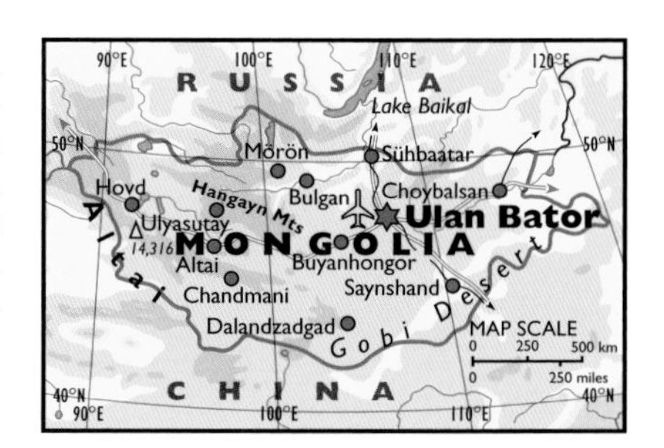

POLITICS

In the late 1980s, Mongolia's leaders were influenced by reforms occurring in the Soviet Union. In 1990, the people demonstrated for more freedom. Free elections in June 1990 resulted in victory for the Mongolian People's Revolutionary Party (MPRP), composed of Communists. Although the MPRP started to move away from Communist policies, it was defeated in 1996. However, the MPRP won a landslide victory in 2000.

ECONOMY

The World Bank classifies Mongolia as a "lower-middle-income" developing country. Most people were once nomads, who moved around with their herds of sheep, cattle, goats, and horses. Under Communist rule, most people were moved into permanent homes on government-owned farms. But livestock and animal products remain leading exports.

The Communists also developed industry, especially the mining of coal, copper, gold, molybdenum, tin, and tungsten, and manufacturing, especially processing food and animal products. Minerals and fuels now account for around half of Mongolia's exports.

DID YOU KNOW

- that Mongolia is one of the highest countries in the world; practically all of the country is over 3,280 ft [1,000 m] high
- that the traditional home of the nomadic people is the *gher* or *yurt*, a tent-like structure made of wood and felt
- that Ulan Bator is probably the coldest capital in the world. The average temperature is below freezing for six months of the year

Camels are used for transport and as beasts of burden in Mongolia. A set of seven stamps issued in 1978 celebrated the country's Bactrian camels. The stamp shows a camel race.

MOROCCO

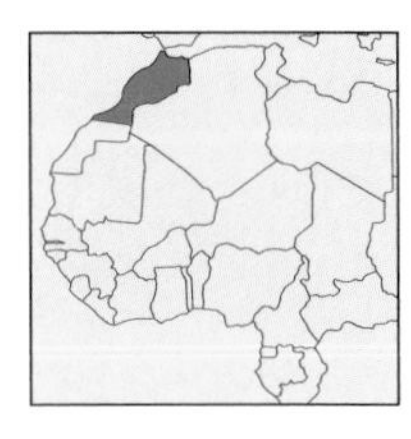

Morocco has flown a red flag since the 16th century. The green pentagram (five-pointed star), called the Seal of Solomon, was added in 1915. This design was retained when Morocco gained its independence from French and Spanish rule in 1956.

The Kingdom of Morocco lies in northwestern Africa. Its name comes from the Arabic *Maghreb-el-Aksa*, meaning "the farthest west." The country has varied scenery. The High (*Haut*) Atlas contains the highest peak, Djebel Toubkal, at 13,669 ft [4,165 m]. Other ranges include the Anti Atlas in the south, the Middle (*Moyen*) Atlas, and the Rif Atlas (or Er Rif) in the far north.

Between the Atlas Mountains and the narrow western coastal plain lies a broad plateau. To the east, the land slopes down to the Sahara Desert.

AREA 172,413 sq miles [446,550 sq km]
POPULATION 30,645,000
CAPITAL (POPULATION) Rabat (1,220,000)
GOVERNMENT Constitutional monarchy
ETHNIC GROUPS Arab 70%, Berber
LANGUAGES Arabic (official), Berber, French
RELIGIONS Islam 99%
CURRENCY Moroccan dirham = 100 centimes

CLIMATE

The Atlantic coast of Morocco is cooled by the Canaries Current. Inland, summers are hot and dry. The winters are mild. In winter, between October and April, south-westerly winds from the Atlantic Ocean bring moderate rainfall, and snow often falls on the High Atlas Mountains.

VEGETATION

The Sahara Desert is barren, but cedars, firs, and junipers swathe the mountain slopes. Cork oaks are common, with wild olives on the high plateaux. Scrub and grasses grow in dry areas.

HISTORY AND POLITICS

The original people of Morocco were the Berbers. But in the 680s, Arab invaders introduced Islam and the Arabic language. By the early 20th century, France and Spain controlled Morocco, but Morocco became an independent kingdom in 1956.

Although Morocco is a constitutional monarchy, King Hassan II ruled the country in an authoritarian manner from his accession to the throne in 1961 to his death in 1999. His son and successor Mohamed VI faced several problems, including the future of Western Sahara which Hassan II had claimed for Morocco.

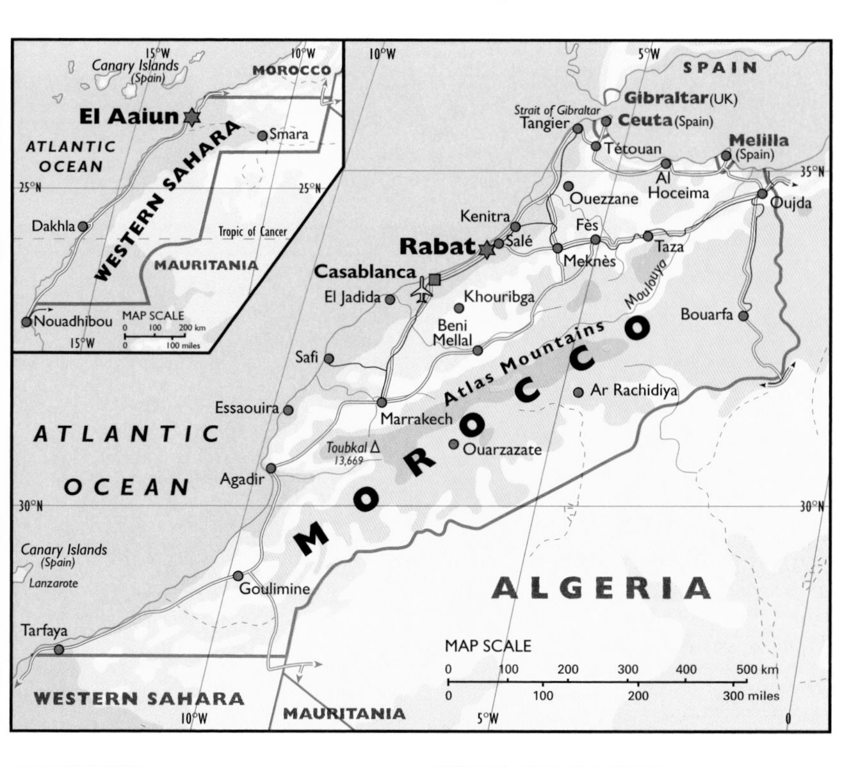

ECONOMY

The United Nations classifies the Kingdom of Morocco as a "lower-middle-income" developing country. Morocco's main resource is phosphate rock, which is used to make fertilizers. Morocco is the world's third largest producer. One of the reasons why Morocco wants to keep Western Sahara is that it, too, has large phosphate reserves. Farming employs 46% of the people. The main crops include barley, beans, citrus fruits, maize, olives, sugar beet, and wheat. Fishing is also important. Processed phosphates are exported, but most of Morocco's manufactures are for home consumption.

WESTERN SAHARA

Western Sahara, a former Spanish possession, covers 102,700 sq miles [266,000 sq km], with a population of 251,000. In 1976, Spain withdrew. Morocco took the northern two-thirds of Western Sahara and Mauritania the rest. Local people in POLISARIO (Popular Front for the Liberation of Saharan Territories) began a guerrilla war. Mauritania withdrew, and Morocco occupied the entire territory. But the war continued.

Throughout the 1990s, the United Nations failed to resolve the issue. In 2002, King Mohamed VI rejected UN plans for a referendum to decide the future of the territory.

MOZAMBIQUE

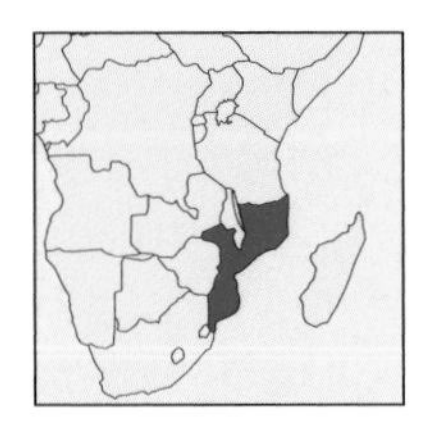

Mozambique's flag was adopted when the country became independent from Portugal in 1975. The green stripe represents fertile land, the black stands for Africa, and the yellow for mineral wealth. The badge on the red triangle contains a rifle, a hoe, a cogwheel, and a book.

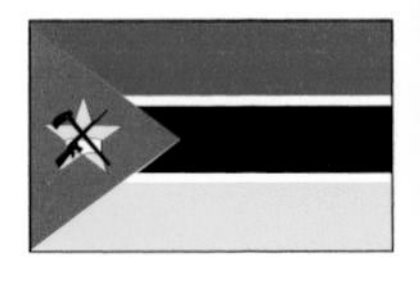

The Republic of Mozambique borders the Indian Ocean in southeastern Africa. The coastal plains are narrow in the north but broaden in the south. The coastal plains make up nearly half of the country. Inland lie plateaux and hills, which make up another two-fifths of Mozambique, with highlands along the borders with Zimbabwe, Zambia, Malawi, and Tanzania. The coast is fringed by sand dunes and swamps, with coral reefs offshore. The only natural harbor is at the capital, Maputo, in the far south.

AREA 309,494 sq miles [801,590 sq km]
POPULATION 19,371,000
CAPITAL (POPULATION) Maputo (2,000,000)
GOVERNMENT Multiparty republic
ETHNIC GROUPS Indigenous tribal groups (Shangaan, Chokwe, Manyika, Sena, Makua, others) 99%
LANGUAGES Portuguese (official), many others
RELIGIONS Traditional beliefs 48%, Roman Catholic 31%, Islam 20%
CURRENCY Metical = 100 centavos

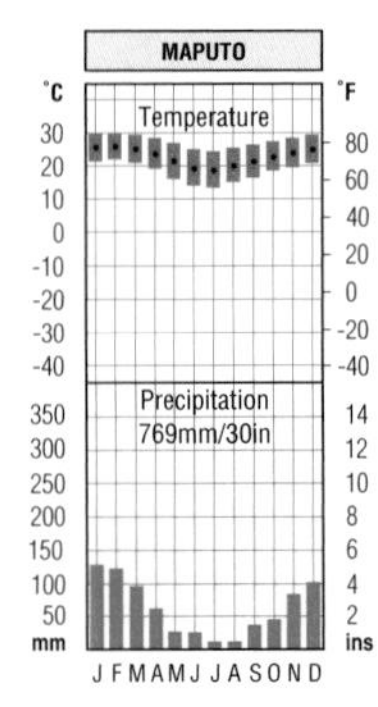

CLIMATE

Mozambique has a tropical climate, though Maputo lies just outside the tropics. The warm southward-flowing Mozambique Current gives Maputo hot and humid summers, though the winters are mild and fairly dry. Inland, the temperatures and the humidity are a little lower.

VEGETATION

Palm trees are common along the coast. Inland are patches of rain forest, containing such trees as ebony and ironwood. But savanna (tropical grassland with scattered trees) is the most widespread type of vegetation.

HISTORY AND POLITICS

Arab traders began to operate in the area in the 9th century AD, and Portuguese explorers arrived in 1497. The Portuguese set up trading stations in the early 16th century and the area became a source of slaves.

In 1885, when the European powers divided Africa, Mozambique was recognized as a Portuguese colony. But black African opposition to European rule gradually increased. In 1961, the Front for the Liberation of Mozambique (FRELIMO) was founded to oppose Portuguese rule. In 1964, FRELIMO launched a guerrilla

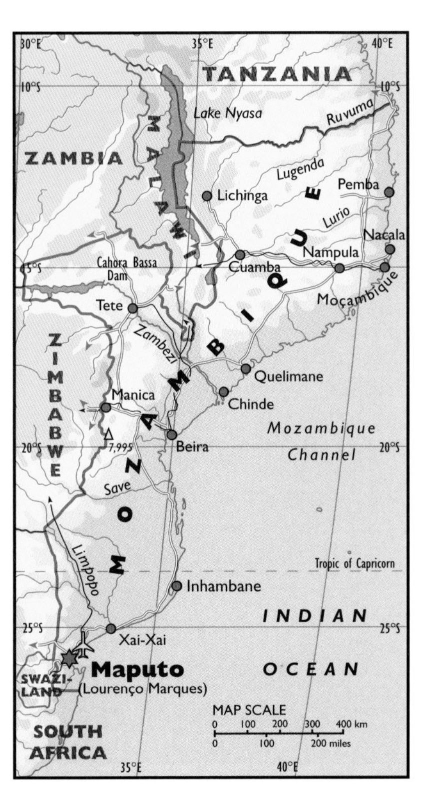

war, which continued for ten years. Mozambique became independent in 1975, when FRELIMO, which followed Marxist-Leninist policies, took over the government.

After independence, Mozambique became a one-party state. Its government aided African nationalists in Rhodesia (now Zimbabwe) and South Africa. But the white governments of these countries helped an opposition group, the Mozambique National Resistance Movement (RENAMO), to lead an armed struggle against Mozambique's government. The civil war, combined with severe droughts, caused much human suffering in the 1980s. In 1989, FRELIMO declared that it had dropped its Communist policies and ended one-party rule. The war officially ended in 1992 and multiparty elections in 1994 heralded more stable conditions. In 1995 Mozambique became the 53rd member of the Commonwealth, joining its English-speaking allies in southern Africa.

ECONOMY

In the early 1990s, the UN rated Mozambique as one of the world's poorest countries. Civil war, droughts, and floods had caused an economic collapse. However, the second half of the 1990s saw a surge in economic growth, although about 80% of the people remain poor. Agriculture is the main activity and crops include cassava, cotton, cashew nuts, fruits, maize, rice, sugarcane, and tea. Manufacturing is small scale, though electricity generated at the Dahora Bassa Dam on the River Zambezi is exported to South Africa.

Mozambique's colonial past is recalled by this stamp issued in 1969 to celebrate the 400th anniversary of Portugal's national poet, Luis de Camoes, author of a great epic poem, "The Lusiads." The portrait on the stamp was painted in 1581.

NAMIBIA

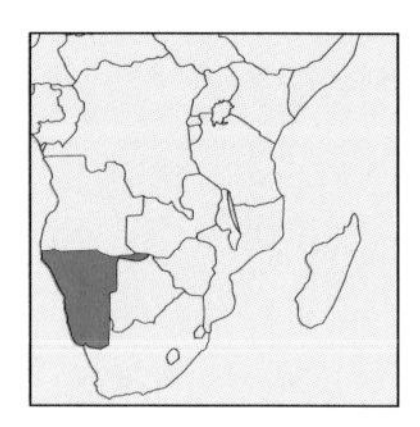

Namibia adopted this flag in 1990 when it gained its independence from South Africa. The red diagonal stripe and white borders are symbols of Namibia's human resources. The green and blue triangles, and the gold sun represent the country's resources.

The Republic of Namibia was formerly ruled by South Africa, who called it South West Africa. The country became independent in 1990. The coastal region contains the arid Namib Desert, which is virtually uninhabited. Inland is a central plateau, mostly between 2,950 ft and 6,560 ft [900 m to 2,000 m] above sea level, which is bordered by a rugged spine of mountains stretching from north to south.

Eastern Namibia contains part of the Kalahari, a semidesert area which extends into Botswana. The Orange River forms Namibia's southern border, while the Cunene and Okavango rivers form parts of the northern borders.

AREA 318,434 sq miles [825,414 sq km], including Walvis Bay, a former South African territory

POPULATION 1,798,000

CAPITAL (POPULATION) Windhoek (126,000)

GOVERNMENT Multiparty republic

ETHNIC GROUPS Ovambo 50%, Kavango 9%, Herero 7%, Damara 7%, White 6%, Nama 5%

LANGUAGES English (official), Ovambo, Afrikaans, German

RELIGIONS Christianity 90% (Lutheran 51%)

CURRENCY Namibian dollar = 100 cents

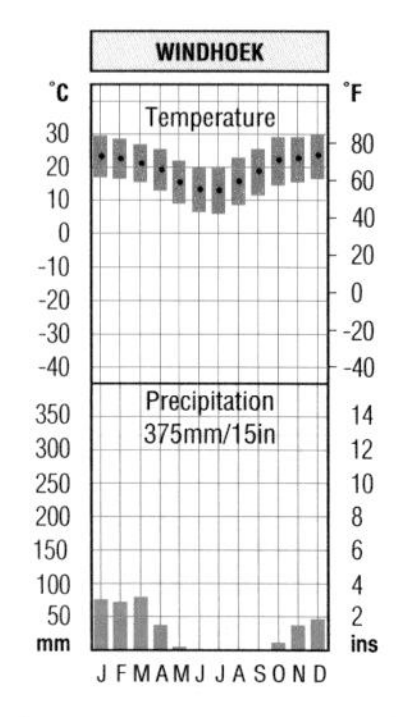

CLIMATE

Namibia is a warm and mostly arid country. Lying at 5,500 ft [1,700 m] above sea level, Windhoek has an average annual rainfall of about 15 inches [370 mm], often occurring during thunderstorms in the hot summer months. Most other parts of the country are even drier. The north is the rainiest part of Namibia, with 20 inches [500 mm] of rain per year.

VEGETATION

The Namib Desert is one of the bleakest in the world, with many high sand dunes. One strange plant, the *Welwitschia*, lives up to 1,000 years or more. It has two large leaves, growing from a short, woody stem. Windblown sand often shreds the leaves into long threads.

Grassland and shrub cover much of the interior. The wetter north has large numbers of wild animals. Etosha National Park in the north is a magnificent wildlife reserve.

HISTORY AND POLITICS

The earliest people in Namibia were the San (also called Bushmen) and the

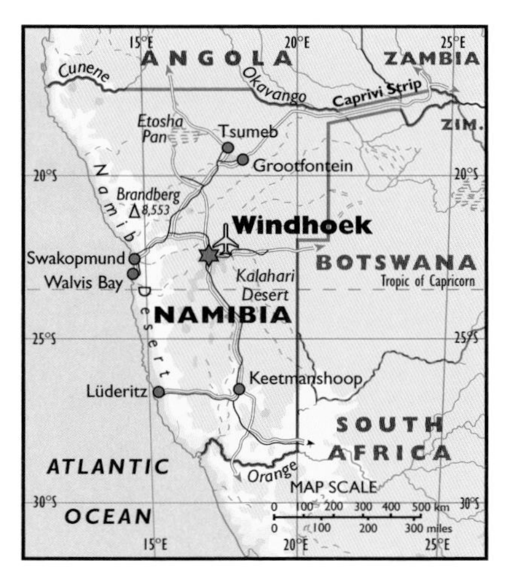

Damara (Hottentots). Later arrivals were people who spoke Bantu languages. These include the Ovambo, Kavango, and Herero. Germany took over the country in 1884. But in 1915, during World War I, South Africa occupied the country which was then called South West Africa.

After World War II (1939–45), many people challenged the right of South Africa to govern the country. A civil war began in the 1960s between African guerrillas, opposed to South Africa's racial policies, and South African troops. In 1971, the International Court of Justice declared that South African rule over the country was illegal.

A ceasefire in Namibia's long-running civil war was agreed in 1989 and the country became independent in 1990. A small area on Namibia's coast, called Walvis Bay (Walvisbaai), remained part of South Africa. But South Africa handed this area over to Namibia in 1994. After winning independence, the government pursued a successful policy of "national reconciliation." In 1999, a rebellion was put down by Lozi people demanding independence for the Caprivi Strip.

ECONOMY

Namibia is rich in mineral reserves, including diamonds, uranium, zinc, and copper. Minerals make up 90% of the exports. But farming employs about two out of every five Namibians. The main activities are cattle and sheep farming, because most of the country is too dry for crop farming. But droughts in the last 20 years have greatly reduced the number of farm animals. The chief food products are maize, millet, and vegetables.

Fishing in the Atlantic Ocean is also important, though overfishing has reduced the yields of Namibia's fishing fleet. The country has few manufacturing industries.

This stamp was issued in 1986, when Namibia was still ruled by South Africa under the name South West Africa (SWA). The stamp shows petrified sand dunes (sand dunes which have been turned into hard rock) in the Namib Desert between Windhoek and Walvis Bay.

NETHERLANDS

The flag of the Netherlands, one of Europe's oldest, dates from 1630, during the long struggle for independence from Spain which began in 1568. The tricolor became a symbol of liberty which inspired many other revolutionary flags around the world.

The Kingdom of the Netherlands is one of the "Low Countries." The others are Belgium and Luxembourg. The Netherlands lies at the western end of the North European Plain, which extends to the Ural Mountains in Russia. Except for the far southeastern corner, the Netherlands is flat and about 40% lies below sea level at high tide. To prevent flooding, the Dutch have built dykes (sea walls) to hold back the waves. Large areas which were once under the sea, but which have been reclaimed, are called *polders*.

AREA 16,033 sq miles [41,526 sq km]
POPULATION 15,981,000
CAPITAL (POPULATION) Amsterdam (1,115,000); The Hague (seat of government, 700,000)
GOVERNMENT Constitutional monarchy
ETHNIC GROUPS Dutch 95%, Indonesian, Turkish, Moroccan, Surinamese
LANGUAGES Dutch (official), Frisian
RELIGIONS Roman Catholic 34%, Protestant 21%, Islam 4%
CURRENCY Euro = 100 cents

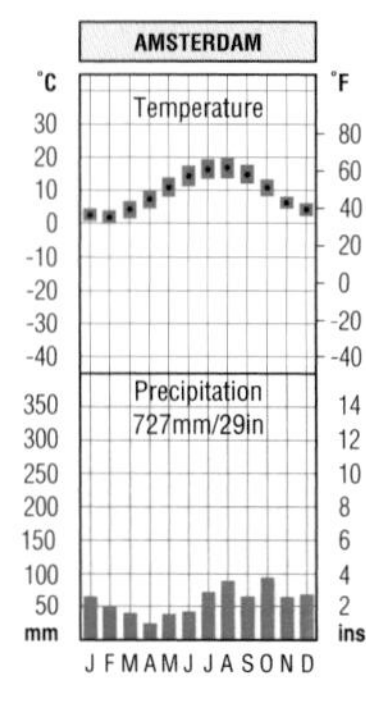

CLIMATE

Because of its position on the North Sea, the Netherlands has a temperate climate. The winters are mild, with rain coming from the Atlantic depressions which pass over the country. In the winter months, the southwesterly winds warm the country. In summer, they have a cooling effect.

VEGETATION

The Netherlands is one of the most crowded countries in the world, and the plants and animals have been greatly modified by human activity. About two-thirds of the land is farmed or used for grazing, though some interesting salt and water plants grow on the coast, and woodland covers around 9% of the land.

HISTORY AND POLITICS

Before the 16th century, the area that is now the Netherlands was under a succession of foreign rulers, including the Romans, the Germanic Franks, the French, and the Spanish. The Dutch declared their independence from Spain in 1581 and their status was finally recognized by Spain in 1648. In the 17th century, the Dutch built up a great overseas empire, especially in Southeast Asia. But in the early 18th century, the Dutch lost control of the seas to England.

France controlled the Netherlands from 1795 to 1813. In 1815, the Netherlands, then containing Belgium and Luxembourg, became an independent kingdom. Belgium broke away in 1830 and Luxembourg followed in 1890.

The Netherlands was neutral in World War I (1914–18), but was occupied by Germany in World War II (1939–45). After the war, the Netherlands Indies became independent as Indonesia. The Netherlands became active in West European affairs. With Belgium and Luxembourg, it formed a customs union called Benelux in 1948. In 1949, it joined NATO (the North Atlantic Treaty Organization), and the European Coal and Steel Community (ECSC) in 1953. In 1957, it became a founder member of the European Economic Community (now the European Union), and its economy prospered. In 2002, a right-wing, anti-immigration party came second in parliamentary elections, but its support collapsed in elections in 2003.

ECONOMY

The Netherlands is a highly industrialized country, and industry and commerce are the most valuable activities. Its resources include natural gas, some oil, salt, and china clay. But the Netherlands imports many of the materials needed by its industries and it is, therefore, a major trading country. Industrial products are wide-ranging, including aircraft, chemicals, electronic equipment, machinery, textiles, and vehicles. Agriculture employs only 3% of the people, but scientific methods are used and yields are high. Dairy farming is the leading farming activity. Major products include barley, flowers and bulbs, potatoes, sugar beet, and wheat.

DID YOU KNOW

- that the Netherlands is often called Holland, though this name refers correctly only to one part of the country
- that, according to an old Dutch saying, "God created the world, but the Dutch created Holland"
- that the Dutch parliament meets in The Hague and not in the capital, Amsterdam
- that the Netherlands includes two areas in the West Indies: Aruba and Netherlands Antilles

NEW ZEALAND

New Zealand's flag was designed in 1869 and adopted as the national flag in 1907 when New Zealand became an independent dominion. The flag includes the British Blue Ensign and four of the five stars in the Southern Cross constellation.

New Zealand lies about 994 miles [1,600 km] southeast of Australia. It consists of two main islands and several other small ones. Much of North Island is volcanic. Active volcanoes include Ngauruhoe and Ruapehu. Hot springs and geysers are common, and steam from the ground is used to produce electricity. The north also contains fertile peninsulas and basins.

South Island contains the Southern Alps, where Aoraki Mount Cook, New Zealand's highest peak at 12,313 ft [3,753 m], is situated. South Island also has some large, fertile plains.

AREA 103,737 sq miles [268,680 sq km]

POPULATION 3,864,000

CAPITAL (POPULATION) Wellington (329,000)

GOVERNMENT Constitutional monarchy

ETHNIC GROUPS New Zealand European 74%, New Zealand Maori 10%, Polynesian 4%

LANGUAGES English and Maori (both official)

RELIGIONS Anglican 24%, Presbyterian 18%, Roman Catholic 15%, Methodist 5%, Baptist 2%

CURRENCY New Zealand dollar = 100 cents

CLIMATE

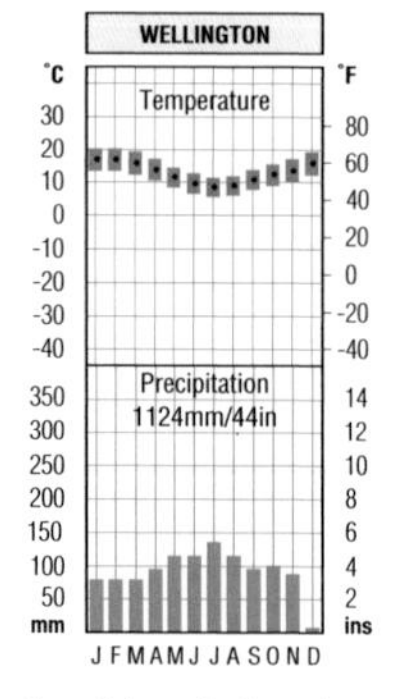

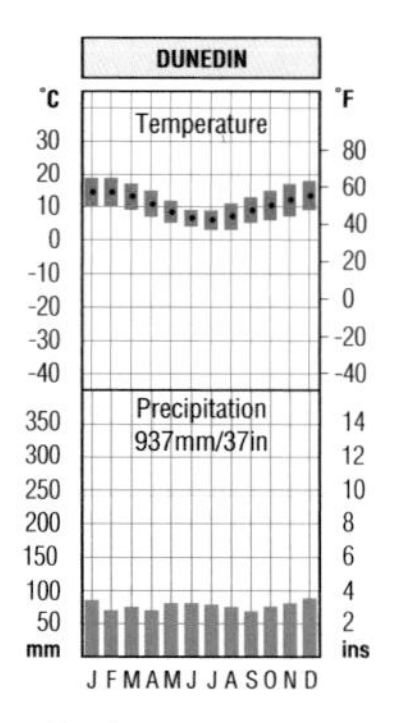

Auckland in the north has a warm, humid climate throughout the year. Wellington has cooler summers, while in Dunedin, in the southeast, temperatures sometimes dip below freezing in winter. The rainfall is heaviest on the western highlands.

VEGETATION

Much of the original vegetation has been destroyed and only small areas of the original pine-like kauri forests survive. But beech forests grow on the cooler highlands, and large plantations of the foreign radiata pine and other species are grown by the timber industry.

HISTORY

Evidence suggests that early Maori settlers arrived in New Zealand more than 1,000 years ago. The Dutch

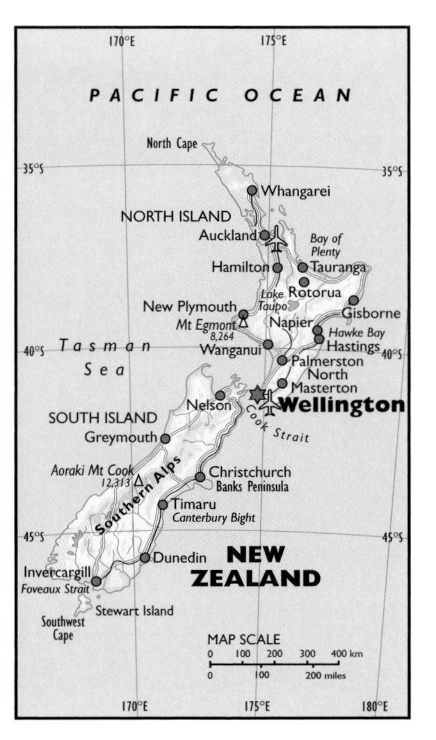

navigator Abel Tasman reached New Zealand in 1642, but his discovery was not followed up. In 1769, the British Captain James Cook rediscovered the islands. In the early 19th century, British settlers arrived in New Zealand. In 1840, under the Treaty of Waitangi, Britain took possession of the country. Clashes occurred with the Maoris in the 1860s, but from the 1870s the Maoris were gradually integrated into society.

POLITICS

In 1907, New Zealand became a self-governing dominion in the British Commonwealth. The country's economy developed quickly and the people became increasingly prosperous. However, after Britain joined the European Economic Community in 1973, New Zealand's exports to Britain shrank and the country had to reassess its economic and defense strategies, and seek new markets. The world economic recession also led the government to cut back on its spending on welfare services in the 1990s. Maori rights and the preservation of Maori culture are other major political issues.

ECONOMY

New Zealand's economy has traditionally depended on agriculture, but manufacturing now employs twice as many people as agriculture. Meat and dairy products are the most valuable items produced on farms. The country has more than 45 million sheep, 4.3 million dairy cattle, and 4.6 million beef cattle. Crops include barley, fruits, potatoes, and wheat. Processed foods are the chief manufactures, but there is a wide range of other industries. Tourism is also important.

New Zealand has much spectacular scenery, including hundreds of waterfalls, one of which, Sutherland Falls, is the world's fifth highest. This stamp, showing Bridal Veil Falls, is one of a set of four illustrating waterfalls issued in 1976.

NICARAGUA

Nicaragua's flag was adopted in 1908. It was the flag of the Central American Federation (1823–39), which included Costa Rica, El Salvador, Guatemala, Honduras, and Nicaragua. It resembles the flag of El Salvador except for the shading of the blue and the central motif.

The Republic of Nicaragua is the second largest country in Central America. In the east is a broad plain bordering the Caribbean Sea. The plain is drained by rivers that flow from the Central Highlands, which rise to 6,443 ft [nearly 2,000 m] in the Cordillera Isabella. The western Pacific region is in an unstable part of the Earth's crust. It contains about 40 volcanoes, many of which are active. Earthquakes are common in this mostly fertile area. The Pacific region contains two large lakes, Managua and Nicaragua.

AREA 50,193 sq miles [130,000 sq km]

POPULATION 4,918,000

CAPITAL (POPULATION) Managua (864,000)

GOVERNMENT Multiparty republic

ETHNIC GROUPS Mestizo 69%, White 17%, Black 9%, Amerindian 5%

LANGUAGES Spanish (official), Misumalpan

RELIGIONS Roman Catholic 85%

CURRENCY Córdoba oro (gold córdoba) = 100 centavos

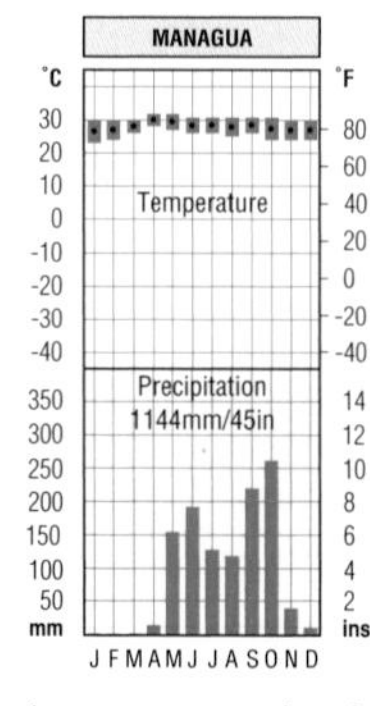

CLIMATE

Nicaragua has a tropical climate. Managua is hot throughout the year and there is a marked rainy season from May to October. The Central Highlands are cooler and wetter. But the wettest part of the country is the Caribbean region, which has about 165 inches [4,200 mm] of rain a year.

VEGETATION

Rain forests cover large areas in the east. The main trees include cedars, mahogany, and walnut. Savanna (tropical grassland with scattered trees) is common in the west.

HISTORY AND POLITICS

Christopher Columbus reached Nicaragua in 1502 and claimed the land for Spain. Spain ruled Nicaragua until 1821. In the mid-19th century, Nicaragua was ravaged by civil war. By the early 20th century, the United States had considerable influence in the country, and in 1912 US forces entered the country in order to protect US interests. From 1927 to 1933, rebels under General Augusto César Sandino tried to drive the US forces out of the country. In 1933, the US marines set up a Nicaraguan army, called the National Guard, to help in defeating the rebels. Its leader, Anastasio Somoza Garcia, had Sandino murdered in 1934. From 1937, Somoza ruled Nicaragua as a dictator.

In the mid-1970s, many people began to protest against Somoza's rule. Many joined a guerrilla force, called the Sandinista National Liberation Front, named after General Sandino. The rebels defeated the Somoza regime in 1979. In the 1980s, the US-supported forces, called the "Contras," launched a campaign against the Sandinista government. The US government opposed the Sandinista regime, under Daniel José Ortega Saavedra, claiming that it was a Communist dictatorship.

A coalition, the National Opposition Union, defeated the Sandinistas in elections in 1990. In 1996 and again in 2001, the Sandinista leader, Daniel José Ortega Saavedra, was defeated in presidential elections.

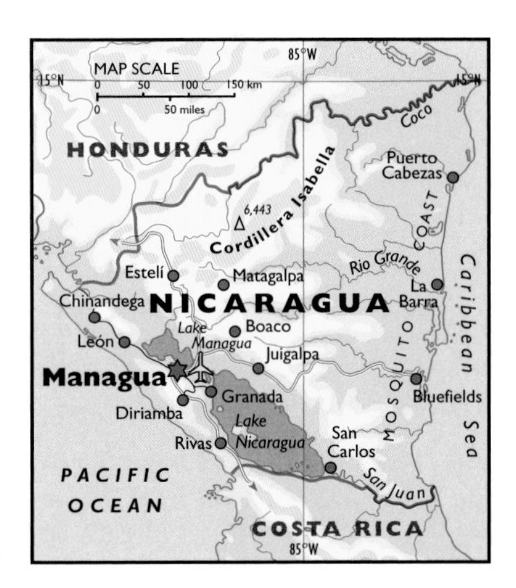

ECONOMY

In the early 1990s, Nicaragua faced many problems in rebuilding its shattered economy. Agriculture is the main activity, employing nearly half of the people. Coffee, cotton, sugar, and bananas are grown for export, while rice is the main food crop. The country has some copper, gold, and silver, but mining is underdeveloped. Most manufacturing industries are based in and around Managua.

DID YOU KNOW

- that Dutch, English, and French pirates once used hideouts on Nicaragua's Caribbean coast
- that earthquakes are common in Nicaragua; Managua was severely damaged in 1931 and 1972
- that Nicaragua was named after an Amerindian chief, Nicarao
- that in October 1998, Hurricane Mitch caused massive devastation, with floods and mudslides causing an estimated 1,800 deaths, with many other people unaccounted for

Nicaragua's rich plant life was celebrated by a set of nine stamps issued in 1974, showing wild flowers, such as the malva on this stamp, and cacti.

NIGER

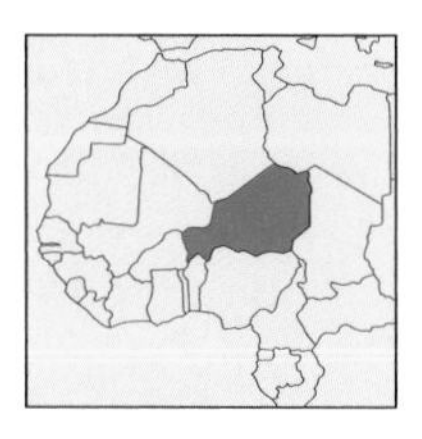

This flag was adopted shortly before Niger became independent from France in 1960. The orange stripe represents the Sahara in the north, and the green the grasslands in the south. Between them, the white stripe represents the River Niger, with a circle for the sun.

The Republic of Niger is a landlocked nation in north-central Africa. The northern plateaux lie in the Sahara Desert. Central Niger contains the rugged, partly volcanic Aïr Mountains, which reach a height of 6,636 ft [2,022 m] above sea level.

The south consists of broad plains. The Lake Chad basin lies in southeastern Niger on the borders with Chad and Nigeria. The only permanent rivers are the Niger and its tributaries in the southwest. The narrow Niger Valley is the country's most fertile and densely populated region.

AREA 489,189 sq miles [1,267,000 sq km]
POPULATION 10,355,000
CAPITAL (POPULATION) Niamey (398,000)
GOVERNMENT Multiparty republic
ETHNIC GROUPS Hausa 56%, Djerma 22%, Tuareg 8%, Fula 8%
LANGUAGES French (official), Hausa, Djerma
RELIGIONS Islam 98%
CURRENCY CFA franc = 100 centimes

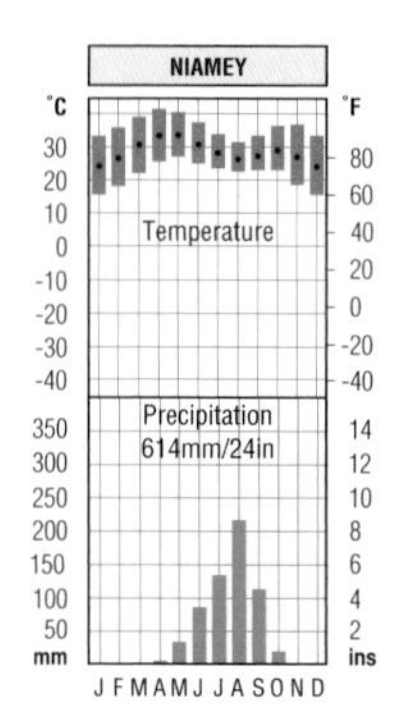

CLIMATE

Niamey has a tropical climate, with a rainy season from June to September. The hottest months are March to May, when the hot, dry harmattan wind blows from the Sahara. The rainfall decreases in both quantity and reliability from south to north. Northern Niger is practically rainless.

VEGETATION

The far south of the country consists of savanna (tropical grassland with scattered trees). Savanna animals, such as buffaloes, elephants, giraffes, and lions, can be seen in the "W" national park, which Niger shares with Benin and Burkina Faso.

But most of southern Niger lies in the Sahel, a region of dry grassland where long droughts sometimes turn the land into desert. The Aïr Mountains have enough rain to support grasses and scrub, but the northern deserts are generally barren. Windblown sand dunes cover large areas.

HISTORY

Nomadic Tuaregs settled in the Aïr Mountains about 1,000 years ago, and by the 15th century they had built up an empire based on Agadez. At around the same time, the Zerma-Songhai people founded the Songhai empire along the River Niger. The

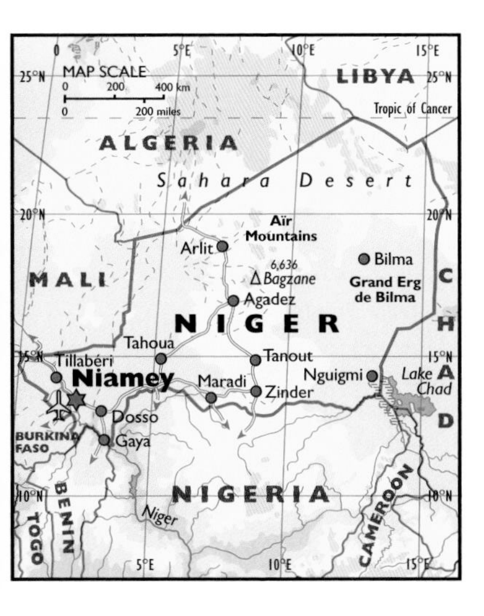

Songhai conquered the Tuaregs, but in the late 16th century Songhai was defeated by a Moroccan army.

Later on, the Hausa and then the Fulani set up kingdoms in the area. France became involved in Niger in the late 19th century. From 1900 to 1958, France ruled Niger as part of French West Africa. In 1958, the country became an autonomous republic, and in 1960 it became fully independent.

POLITICS

Since independence, Niger has been badly hit by droughts, causing great suffering. Food shortages and the destruction of the traditional nomadic lifestyle of some of Niger's people have caused political instability. In 1974, a group of army officers, led by Lt.-Col. Seyni Kountché, overthrew the country's first president, Hamani Diori, and seized control of the government, suspending the constitution. Civilian rule was restored in 1989. A new constitution was adopted in 1992, and elections held in 1993. The military took power in 1996 and the coup leader was elected president. But another coup in 1999 led to the restoration of democracy and the election of President Tandja Mamadou.

ECONOMY

Niger's chief resource is uranium and it is the fourth largest producer in the world. Some tin and tungsten are also mined, though other mineral resources are largely untouched.

Despite its resources, Niger is one of the world's poorest countries. Farming employs 76% of the population, though only 3% of the land can be used for crops and 7% for grazing. Food crops include beans, cassava, millet, rice, and sorghum. Groundnuts and cotton are major cash crops.

African countries are becoming increasingly aware that wildlife is a major resource and that tourism can bring in much needed foreign currency. Niger issued this stamp in 1959 as part of a "Protection of Wild Animals" set.

NIGERIA

Nigeria's flag was adopted in 1960 when Nigeria became independent from Britain. It was selected after a competition to find a suitable design. The green represents Nigeria's forests. The white in the center stands for peace.

The Federal Republic of Nigeria is the most populous nation in Africa. The country's main rivers are the Niger and Benue, which meet in central Nigeria. North of the two river valleys are high plains and plateaux. The Lake Chad basin is in the northeast, with the Sokoto plains in the northwest. Southern Nigeria contains hilly uplands and broad coastal plains, including the swampy Niger delta. Highlands form the border with Cameroon.

CLIMATE

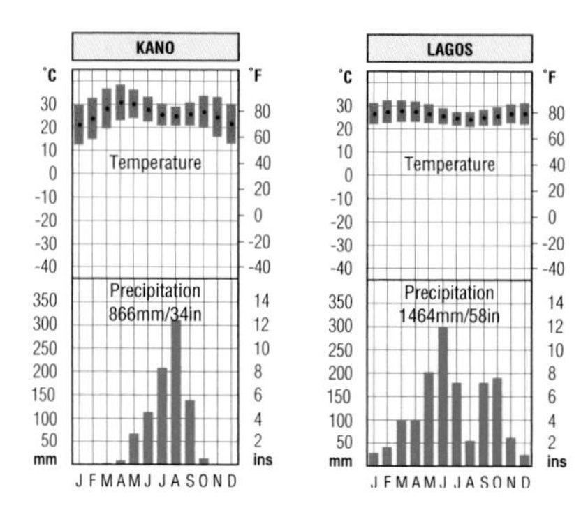

Lagos has a tropical climate, with high temperatures and rain throughout the year. The north of the country is drier and often hotter than the south, though the highlands are cooler. Kano, in north-central Nigeria, has a marked dry season from October to April.

AREA 356,668 sq miles [923,770 sq km]

POPULATION 126,636,000

CAPITAL (POPULATION) Abuja (339,000)

GOVERNMENT Federal multiparty republic

ETHNIC GROUPS Hausa and Fulani 29%, Yoruba 21%, Ibo (or Igbo) 18%, Ijaw 10%, Kanuri 4%

LANGUAGES English (official), Hausa, Yoruba, Ibo

RELIGIONS Islam 50%, Christianity 40%, traditional beliefs 10%

CURRENCY Naira = 100 kobo

VEGETATION

Behind the mangrove swamps along the coast are rain forests, though large areas have been cleared by farmers. The north contains large areas of savanna (tropical grassland with scattered trees) with forests along the rivers. Open grassland and semi-desert occur in drier areas.

HISTORY AND POLITICS

Among the civilizations that grew up in Nigeria were the Nok (500 BC to AD 200), Ife, which developed around 1,000 years ago, and Benin, which flourished between the 15th and 17th centuries. These cultures are famous for their beautiful sculptures.

Nigeria was a center of the slave

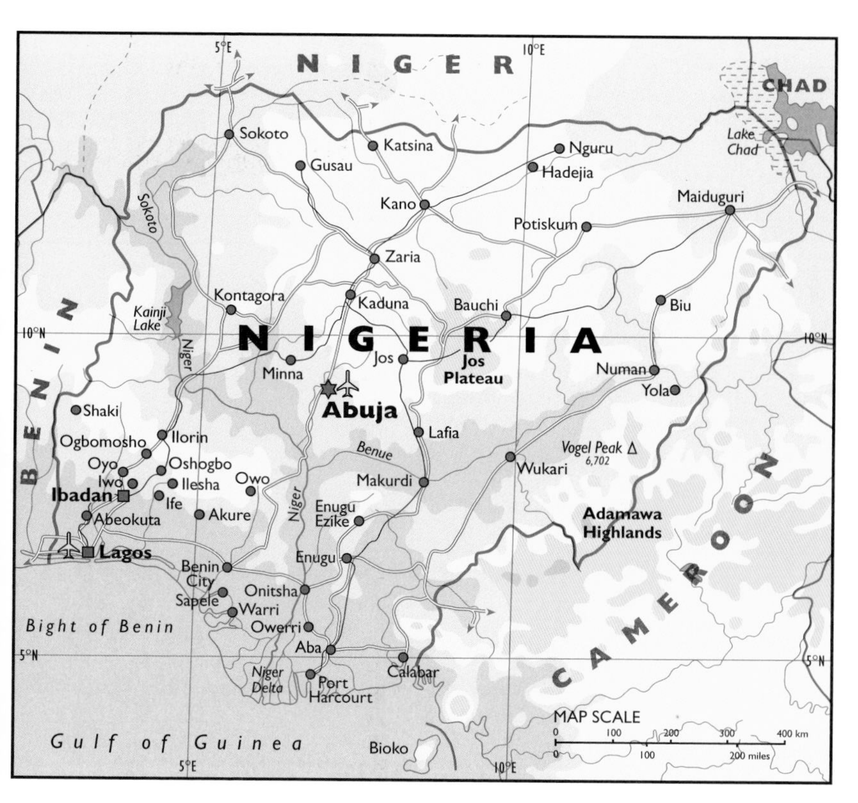

trade from the 16th century until the 19th century. In the second half of the 19th century, Britain gradually extended its influence over Nigeria. By 1914, it ruled the entire country.

In 1960, Nigeria became independent, and in 1963 it became a federal republic. The division of the country into states reflected the fact that Nigeria contains more than 250 ethnic and language groups, as well as religious ones. Civil war occurred between 1967 and 1970, when people of the southeast fought unsuccessfully to make their area a separate country, called Biafra.

In recent times, military regimes have ruled Nigeria for much of the time. But civilian government was restored in 1999 when Olesegun Obasanjo became president. However, in the early 2000s problems arose when ethnic and religious clashes occurred.

ECONOMY

Nigeria is a developing country, whose greatest resource is oil, which accounts for most of its exports. But agriculture employs 43% of the people. Nigeria is a major producer of cocoa, palm oil and kernels, groundnuts, and rubber. Manufacturing is becoming important in the economy.

NORWAY

This flag became the national flag of Norway in 1898, although merchant ships had used it since 1821. The design is based on the Dannebrog, the flag of Denmark, the country which ruled Norway from the 14th century until the early 19th century.

The Kingdom of Norway forms the western part of the mountainous Scandinavian peninsula. Norway has one of the world's most indented coastlines. The deep inlets along the coast were worn out by glaciers during the Ice Age. Ice has also sculpted the inland mountain scenery.

Norway's main lowlands are around Oslo and around Trondheim. About 150,000 islands lie off the coast. Two distant island groups, Svalbard and the Jan Mayen islands, are Norwegian possessions.

AREA 125,050 sq miles [323,900 sq km]
POPULATION 4,503,000
CAPITAL (POPULATION) Oslo (502,000)
GOVERNMENT Constitutional monarchy
ETHNIC GROUPS Norwegian 97%
LANGUAGES Norwegian (official), Lappish, Finnish
RELIGIONS Evangelical Lutheran 88%
CURRENCY Krone = 100 ore

CLIMATE

The warm waters off the coast of Norway give the country a mild climate. Oslo has mild winters and cool summers. Nearly all Norway's seaports are always ice-free. Inland, winters are colder, but summers are warmer. Snow covers the land for at least three months every year.

VEGETATION

Large areas of the rugged mountains are bare rock. The country also has some icefields. Forests and woodland cover about 27% of the country. The main trees include birch, pine, and spruce.

HISTORY AND POLITICS

Norway has a long tradition of seafaring. Vikings from Norway raided parts of Europe for about 300 years after AD 870. Between 1380 and 1814, Norway was united with Denmark, but it was under Danish rule.

Under a treaty in 1814, Denmark handed Norway over to Sweden, but it kept Norway's colonies – Greenland, Iceland, and the Faroe Islands. Norway briefly became independent, but Swedish forces defeated the Norwegians and Norway had to accept Sweden's king as its ruler.

The union between Norway and Sweden ended in 1903. During World War II (1939–45), Germany occupied

This stamp, which was issued in 1975, commemorates a World Scout Jamboree held in Lillehammer, north of Oslo. The Guide emblem is on the left and the Scout emblem is on the right.

Norway. Norway's economy developed quickly after the war and the country now enjoys one of the world's highest standards of living. In 1960, Norway, together with six other countries, formed the European Free Trade Association (EFTA). In November 1994, the Norwegians voted against joining the European Union.

ECONOMY

Norway's chief resources and exports are oil and natural gas, which come from wells under the North Sea. Farmland covers only 3% of the land.

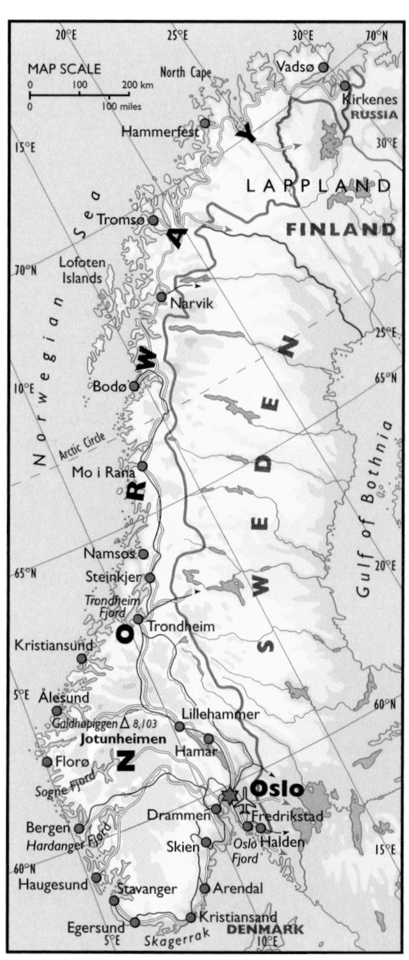

Dairy farming and meat production are the chief activities. But Norway has to import food. Using cheap hydroelectric power, Norway has set up many industries. Manufactures include petroleum products, chemicals, aluminum, wood pulp and paper, machinery, clothing, and furniture.

DID YOU KNOW

- that Norway has the world's fifth largest merchant fleet
- that the sun does not set at Tromsø in northern Norway for three months every year
- that skiing is Norway's national sport
- that Norway produces more hydroelectricity per person than any other country in the world

OMAN

The former Sultanate of Muscat and Oman had a plain red flag, the traditional color of the local people. A new design, including white and green panels, was adopted in 1970 when the country was renamed Oman. Further slight alterations were made in 1995. The state arms include a sword and a dagger.

The Sultanate of Oman occupies the southeastern corner of the Arabian peninsula. It also includes the tip of the Musandam peninsula, overlooking the strategic Strait of Hormuz. This area is separated from the rest of Oman by the United Arab Emirates. A narrow, fertile plain, the Al Batinah, borders the Gulf of Oman in the north. The plain is backed by the Al Hajar Mountains. In the south, most of the land is barren and rocky. It includes part of the Rub' al Khali, a bleak desert also known as the "Empty Quarter."

AREA 82,031 sq miles [212,460 sq km]

POPULATION 2,622,000

CAPITAL (POPULATION) Muscat (350,000)

GOVERNMENT Monarchy with a consultative council

ETHNIC GROUPS Omani Arab 74%, Pakistani 21%

LANGUAGES Arabic (official), Baluchi, English

RELIGIONS Islam 86%, Hinduism

CURRENCY Omani rial = 100 baizas

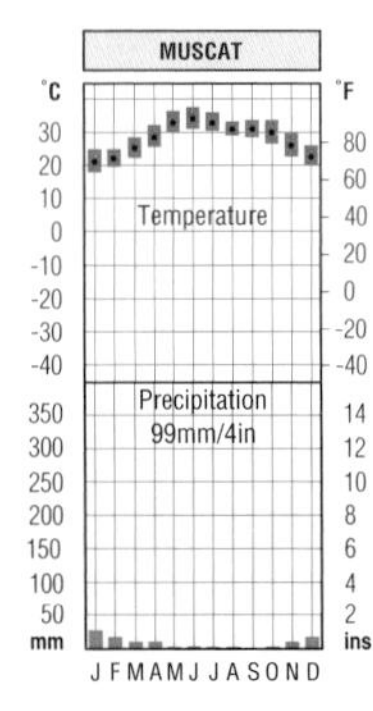

CLIMATE

Oman has a hot tropical climate. In Muscat, temperatures may rise as high as 117°F [47°C] in the summer. Parts of the northern mountains have an average yearly rainfall of 16 inches [400 mm], but most of the country has less than 5 inches [125 mm].

VEGETATION

Date palms grow on the northern coastal plain and also around oases in the interior. Grassy pasture occurs on the Al Hajar Mountains and also along the southern coasts. But 95% of the land is desert, with huge, lifeless sand dunes in the largely unexplored Rub' al Khali.

HISTORY

In ancient times, Oman became an important trading area on the main route between The Gulf and the Indian Ocean. Islam was introduced in the 7th century AD and the Arab-Muslim culture remains the unifying force in Oman.

Portuguese sailors captured several seaports in Oman, including Muscat, in 1507, but Arabs expelled the Portuguese in 1650. Oman set up trading posts in East Africa, including Zanzibar in 1698, and they held this territory until the 1860s.

In the 1740s, the Al Bu Said family

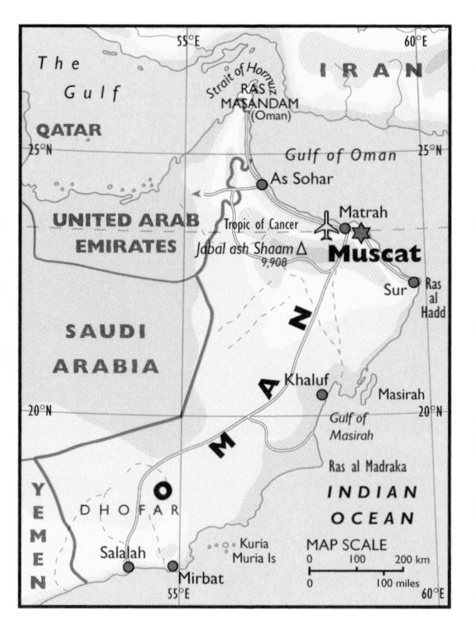

took power in Oman, and this family has ruled Oman ever since. In 1798, Oman signed the first of several treaties of friendship with Britain.

In the late 19th and early 20th centuries, war broke out between the Sultan's forces and people from the interior who wanted a religious leader to take over the government. Conflict broke out again in the 1950s and again, in the south, between 1964 and 1975, but the rebels were put down with the aid of British troops.

POLITICS

Oil was discovered in 1964 and the country became fully independent in 1971. Since 1970, when Sultan Said bin Taimur was deposed by his son, Qaboos bin Said, the government has sought to modernize the country. Qaboos is an absolute ruler, but Oman has a consultative parliament. In the 2000 elections, two women were among the successful candidates. Ties with Britain are close and, from 1980, Oman has been an ally of the USA, allowing it to use air and naval bases during emergencies. Oman works with other countries in eastern Arabia through the Gulf Cooperation Council.

ECONOMY

The World Bank classifies Oman as an "upper-middle-income" developing country. Its economy is based on oil production and oil accounts for more than 90% of Oman's export revenues. But agriculture still provides a living for half of the people. Major crops include alfalfa, bananas, coconuts, dates, limes, tobacco, vegetables, and wheat. Some farmers raise camels and cattle, and fishing, especially for sardines, is also important. But Oman has to import food to feed its people. Oil refining and the processing of copper are among Oman's few manufacturing industries.

Since 1970, Oman's armed forces have played an important part in putting down revolts and maintaining stability. This stamp, which is one of two issued in 1982 to commemorate "Armed Forces Day," shows a mounted army band. In 1980, Oman negotiated a defense agreement with the United States.

PACIFIC OCEAN

AMERICAN SAMOA

Officially an "unincorporated territory" of the United States. **AREA** 77 sq miles [200 sq km]; **POPULATION** 67,000; **CAPITAL** Pago Pago.

FIJI

The Republic of Fiji contains two large islands, Vanua Levu and Viti Levu, and over 800 small ones. They were British between 1874 and 1970. Conflict between native Fijians and descendants of Indian workers has marred progress since independence. **AREA** 7,054 sq miles [18,270 sq km]; **POPULATION** 844,000; **CAPITAL** Suva.

FRENCH POLYNESIA

A French overseas territory consisting of five scattered island groups. Tahiti is the main island. **AREA** 1,520 sq miles [3,941 sq km]; **POPULATION** 254,000; **CAPITAL** Papeete (population 26,000).

GUAM

An "unincorporated territory" of the US. **AREA** 212 sq miles [549 sq km]; **POPULATION** 158,000; **CAPITAL** Agana (population 4,000).

KIRIBATI

The Republic of Kiribati contains the former Gilbert Islands, Banaba Island, eight of the Phoenix Islands, and 11 of the Line Islands. The Gilbert and Ellice Islands became British in 1892. The Ellice Islands broke away in 1975 and Kiribati became fully independent in 1979. Copra is exported. **AREA** 281 sq miles [728 sq km]; **POPULATION** 94,000; **CAPITAL** Tarawa (population 20,000).

MARSHALL ISLANDS

A former US territory, the Republic of the Marshall Islands became fully independent in 1991. Agriculture and tourism are important. **AREA** 70 sq miles [181 sq km]; **POPULATION** 71,000; **CAPITAL** Dalap-Uliga-Darrit, on Majuro island.

MICRONESIA, FED. STATES OF

A former US territory, the Federated States of Micronesia became fully independent in 1991. Copra and phosphates are exported. **AREA** 272 sq miles [705 sq km]; **POPULATION** 135,000; **CAPITAL** Palikir.

NAURU

The Republic of Nauru, which achieved full independence in 1968, became prosperous by mining its phosphate rock, which is likely to be exhausted by the early 21st century. **AREA** 8 sq miles [21 sq km]; **POPULATION** 12,000; **CAPITAL** Yaren.

NEW CALEDONIA

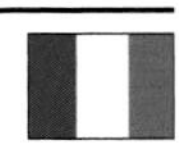

New Caledonia, a French overseas territory since 1958, has rich deposits of nickel. **AREA** 7,174 sq miles [18,580 sq km]; **POPULATION** 205,000; **CAPITAL** Nouméa (population 98,000).

NORTHERN MARIANAS

The Northern Mariana Islands are a Commonwealth in union with the United States, consisting of all 17 Mariana Islands except the most southerly one, Guam. **AREA** 184 sq miles [477 sq km]; **POPULATION** 75,000; **CAPITAL** Saipan (population 39,000).

PALAU

The Republic of Palau (or Belau) became fully independent in October 1994. Its main products are cassava, coconuts, copra, and fish. **AREA** 177 sq miles [458 sq km]; **POPULATION** 19,000; **CAPITAL** Koror.

PITCAIRN

A British dependency, Pitcairn is famous as the place where mutineers from HMS *Bounty* settled in 1790. **AREA** 19 sq miles [48 sq km]; **POPULATION** 47; **CAPITAL** Adamstown.

SAMOA

Samoa (formerly Western Samoa) was administered by New Zealand from 1920 until becoming an independent monarchy in 1962. The islands are volcanic. Samoa exports coconut oil and cocoa. **AREA** 1,097 sq miles [2,840 sq km]; **POPULATION** 179,000; **CAPITAL** Apia (population 32,000).

SOLOMON ISLANDS

A British protectorate from the 1890s, the islands became independent in 1978. Farming, forestry, and fishing are important. **AREA** 10,954 sq miles [28,370 sq km]; **POPULATION** 480,000; **CAPITAL** Honiara.

TONGA

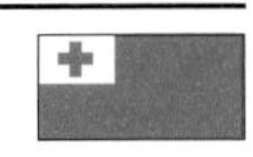

Tonga became an independent monarchy in 1970. Products include coconut oil, copra, fruits, and fish. **AREA** 290 sq miles [750 sq km]; **POPULATION** 104,000; **CAPITAL** Nuku'alofa (population 29,000).

TUVALU

Tuvalu, formerly the Ellice Islands, became an independent monarchy in 1978. Copra is the only major export. **AREA** 9 sq miles [24 sq km]; **POPULATION** 11,000; **CAPITAL** Fongafale.

VANUATU

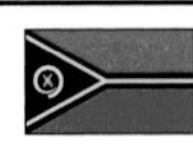

The Republic of Vanuatu, formerly the New Hebrides, became independent in 1980. It produces copra and cocoa. **AREA** 4,707 sq miles [12,190 sq km]; **POPULATION** 193,000; **CAPITAL** Port-Vila (population 20,000).

WALLIS AND FUTUNA ISLANDS

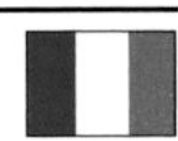

A French overseas territory. **AREA** 77 sq miles [200 sq km]; **POPULATION** 15,000; **CAPITAL** Mata-Utu.

PACIFIC OCEAN

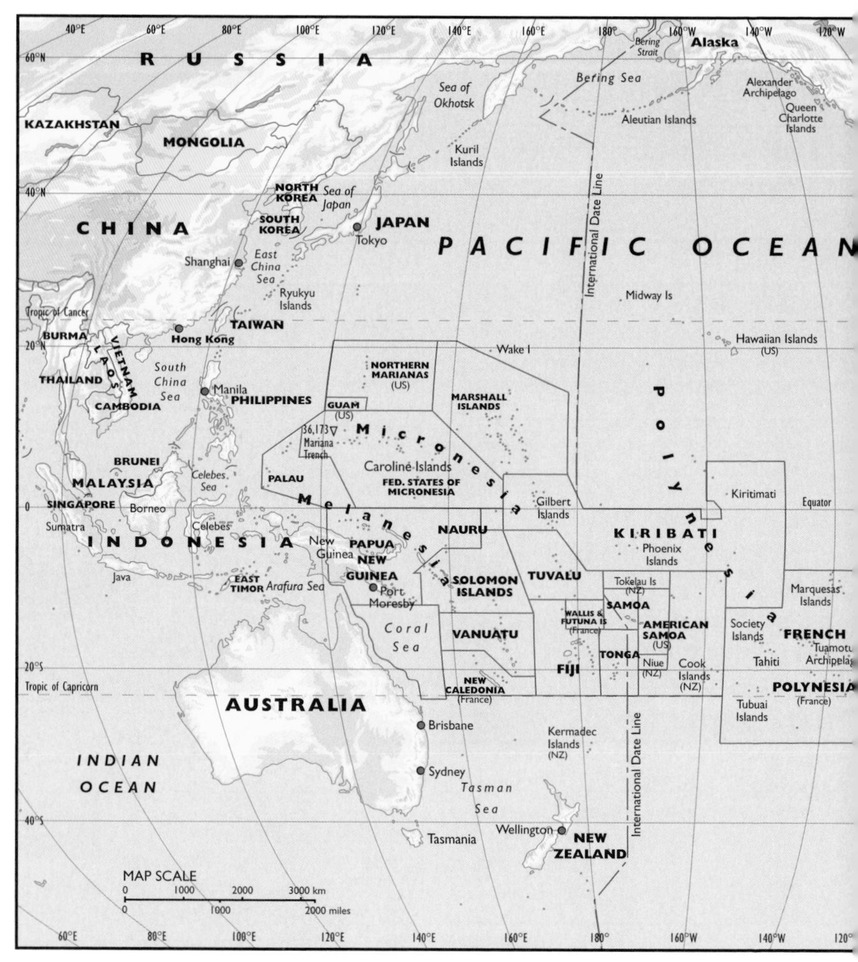

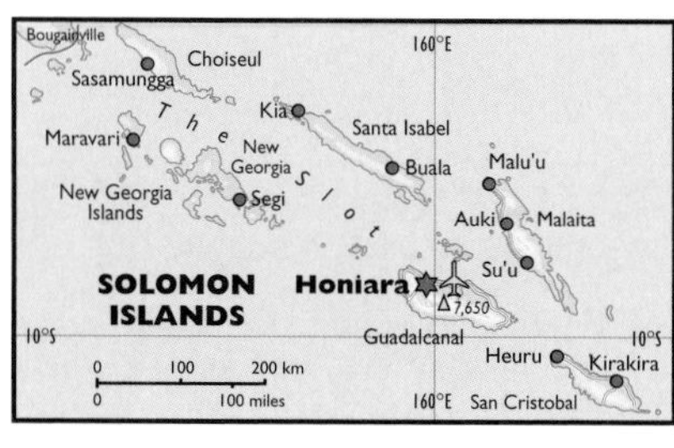

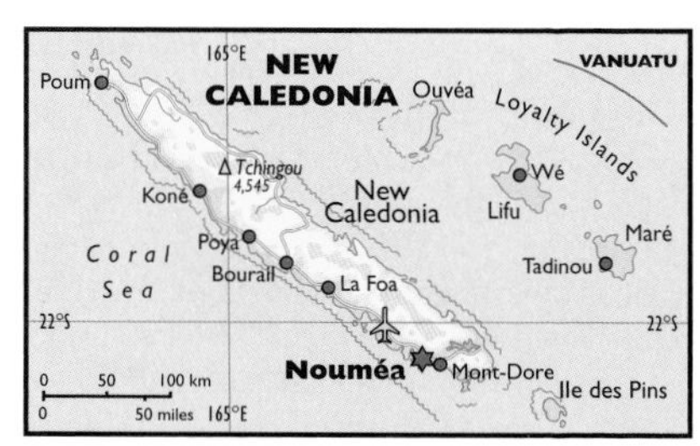

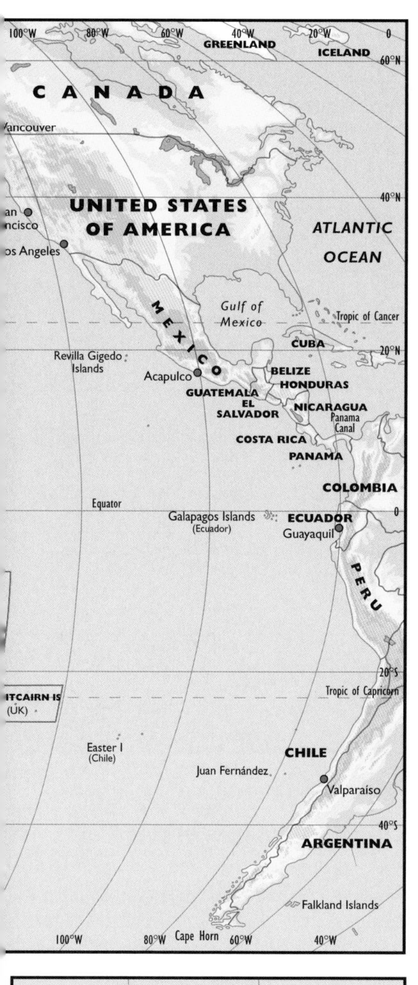
CANADA
Vancouver
UNITED STATES OF AMERICA
Los Angeles
ATLANTIC OCEAN
GREENLAND
ICELAND
MEXICO
Gulf of Mexico
Tropic of Cancer
CUBA
Revilla Gigedo Islands
Acapulco
BELIZE
HONDURAS
GUATEMALA
EL SALVADOR
NICARAGUA
Panama Canal
COSTA RICA
PANAMA
COLOMBIA
Equator
Galapagos Islands (Ecuador)
ECUADOR
Guayaquil
PERU
PITCAIRN IS (UK)
Tropic of Capricorn
Easter I (Chile)
CHILE
Juan Fernández
Valparaíso
ARGENTINA
Falkland Islands
Cape Horn

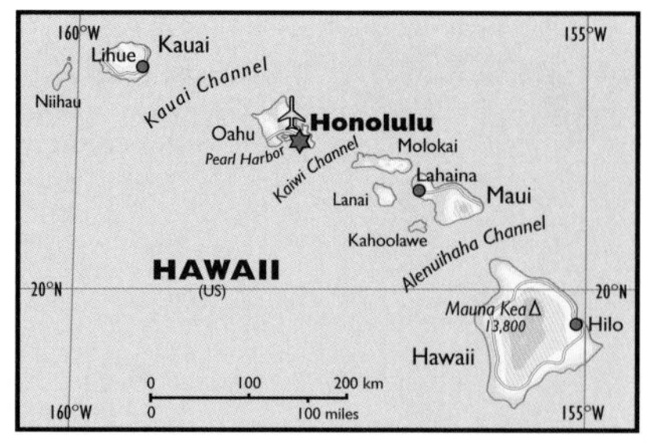
Kauai
Lihue
Niihau
Kauai Channel
Oahu
Honolulu
Pearl Harbor
Kaiwi Channel
Molokai
Lahaina
Lanai
Maui
Kahoolawe
Alenuihaha Channel
HAWAII (US)
Mauna Kea 13,800
Hilo
Hawaii

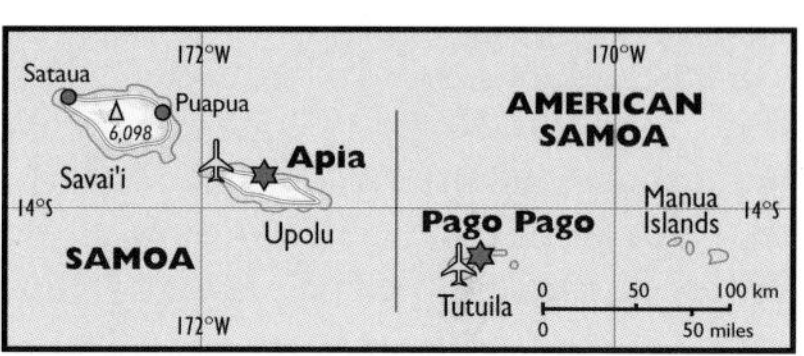
Sataua
Puapua
6,098
Savai'i
Apia
Upolu
SAMOA
AMERICAN SAMOA
Pago Pago
Tutuila
Manua Islands

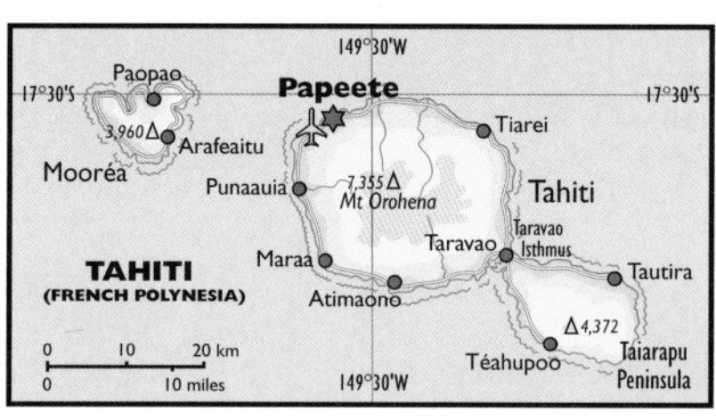
Paopao
3,960
Arafeaitu
Mooréa
Papeete
Tiarei
Punaauia
7,355 Mt Orohena
Tahiti
Taravao
Taravao Isthmus
Tautira
Maraa
Atimaono
TAHITI (FRENCH POLYNESIA)
4,372
Téahupoo
Taiarapu Peninsula

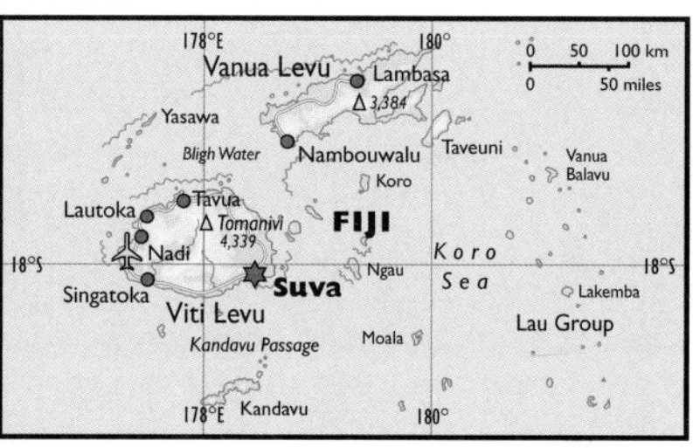
Vanua Levu
Lambasa
3,384
Yasawa
Bligh Water
Nambouwalu
Taveuni
Vanua Balavu
Koro
Lautoka
Tavua
Tomanivi 4,339
Nadi
FIJI
Koro Sea
Ngau
Singatoka
Suva
Viti Levu
Kandavu Passage
Moala
Lakemba
Lau Group
Kandavu

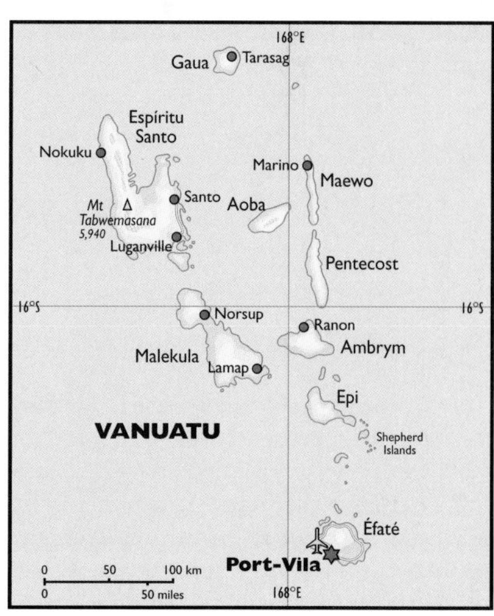
Gaua
Tarasag
Espíritu Santo
Nokuku
Marino
Maewo
Santo
Aoba
Mt Tabwemasana 5,940
Luganville
Pentecost
Norsup
Ranon
Ambrym
Malekula
Lamap
Epi
VANUATU
Shepherd Islands
Éfaté
Port-Vila

PAKISTAN

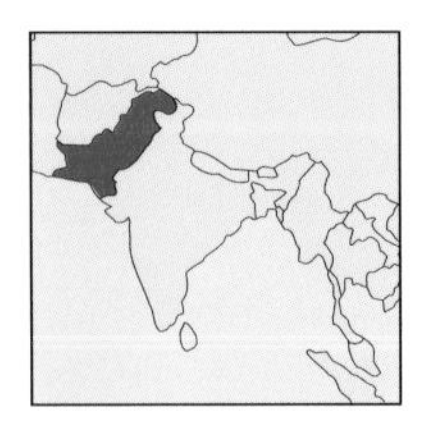

Pakistan's flag was adopted in 1947, when the country became independent from Britain. The color green, the crescent Moon, and the five-pointed star are all traditional symbols of Islam. The white stripe represents the other religions in Pakistan.

The Islamic Republic of Pakistan contains high mountains, fertile plains and rocky deserts. The Karakoram range, which contains K2, the world's second highest peak, lies in the northern part of Jammu and Kashmir, which is occupied by Pakistan but claimed by India. Other mountains rise in the west.

Plains, drained by the River Indus and its tributaries, occupy much of eastern Pakistan. The Thar Desert is in the southeast. The dry Baluchistan plateau is in the southwest.

AREA 307,374 sq miles [796,100 sq km]
POPULATION 144,617,000
CAPITAL (POPULATION) Islamabad (525,000)
GOVERNMENT Military regime
ETHNIC GROUPS Punjabi 60%, Sindhi 12%, Pushtun 13%, Baluch, Muhajir
LANGUAGES Urdu (official), many others
RELIGIONS Islam 97%, Christianity, Hinduism
CURRENCY Pakistan rupee = 100 paisa

CLIMATE

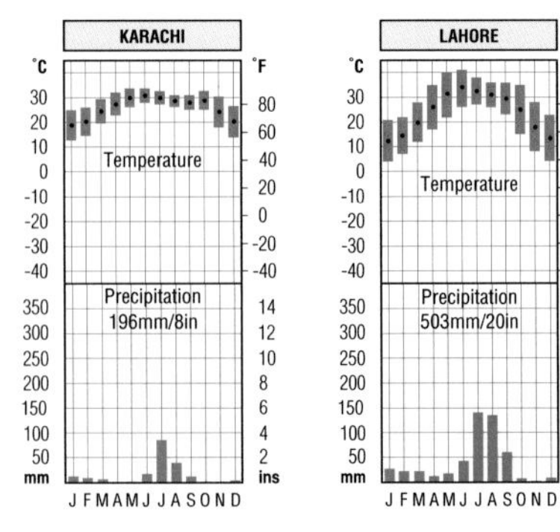

The mountains have cold, snowy winters. But most of Pakistan has hot summers and cool winters. The rainfall is sparse throughout much of the country. Most of it comes between the months of July and September, when the southwest monsoon winds blow.

VEGETATION

Forests grow on mountain slopes, but most of Pakistan is covered by dry grassland and low bushes, with only occasional trees.

HISTORY AND POLITICS

Pakistan was the site of the Indus Valley civilization, which developed about 4,500 years ago. But Pakistan's modern history dates from 1947, when British India was divided into India and Pakistan. Muslim Pakistan was divided into two parts: East and West Pakistan, but East Pakistan broke away in 1971 to become Bangladesh. In 1948–9, 1965, and 1971, Pakistan and India clashed over the disputed territory of Kashmir.

Pakistan has been subject to several periods of military rule. In 1999, after

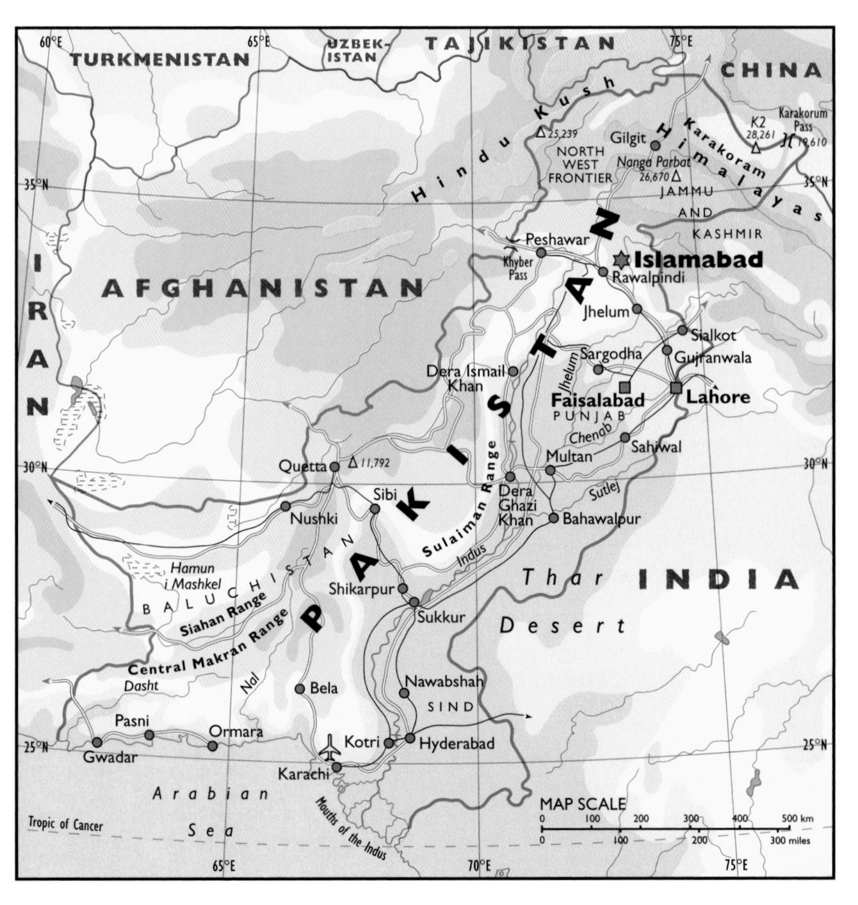

a military coup, General Pervez Musharraf became Pakistan's leader. Musharraf supported the West in its assault on the Taliban government of Afghanistan in 2001. Elections were held in 2002, but Musharraf remained president and army chief, with the power to dismiss parliament.

ECONOMY

According to the World Bank, Pakistan is a "low-income" developing country. The economy is based on farming or rearing goats and sheep. Agriculture employs about 40% of the people. Major crops include cotton, fruits, rice, sugarcane, and, most important of all, wheat. Much of the farmland is irrigated.

Manufactures include clothing, processed food, and textiles. Carpets, embroidered goods, pottery, and wood products are made by small-scale craft industries.

PANAMA

This flag dates from 1903, when Panama, a former province of Colombia, became independent. The blue stands for the Conservative Party, the red for the Liberal Party, and the white for the hope of peace. The red star represents law and order, and the blue star "public honesty."

The Republic of Panama links Central and South America. It is an example of an isthmus – a narrow strip of land joining two large land areas. The narrowest part of Panama is less than 37 miles [60 km] wide. The Panama Canal, 50.7 miles [81.6 km] long, which cuts across the isthmus, has made Panama a major transport center. Most Panamanians live within 12 miles [20 km] of this major waterway. Most of the land between the Pacific and Caribbean coastal plains is mountainous, rising to 11,405 ft [3,475 m] at the volcano Barú.

AREA 29,761 sq miles [77,080 sq km]
POPULATION 2,846,000
CAPITAL (POPULATION) Panama City (452,000)
GOVERNMENT Multiparty republic
ETHNIC GROUPS Mestizo 70%, Black and Mulatto 14%, White 10%, Amerindian 6%
LANGUAGES Spanish (official)
RELIGIONS Roman Catholic 84%, Protestant 5%
CURRENCY Balboa = 100 centésimos; US dollar = 100 cents

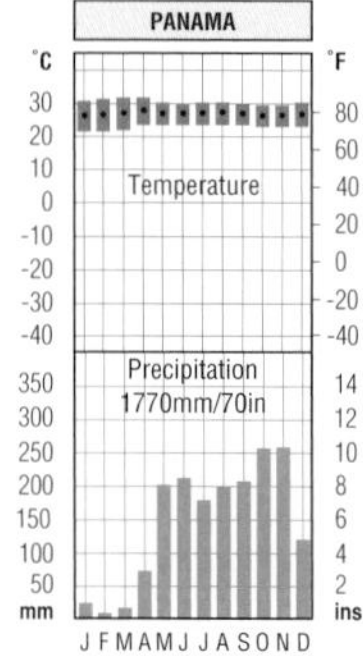

CLIMATE

Panama has a tropical climate. Temperatures are high, though the mountains are much cooler than the coastal plains. The main rainy season is between May and December. The Caribbean side of the country has about twice as much rain as the Pacific side.

VEGETATION

Tropical forests cover about half of Panama, with mangrove swamps along the coast. Subtropical woodland grows on the mountains, while savanna (tropical grassland with scattered trees) occurs along the Pacific coast.

HISTORY AND POLITICS

Christopher Columbus landed in Panama in 1502 and Spain soon took control of the area. In 1821, Panama became independent from Spain and it became a province of Colombia.

In 1903, Colombia refused a request by the United States to build a canal. Panama then revolted against Colombia, and became independent. The United States then began to build the canal, which was opened in 1914. The United States administered the Panama Canal Zone, a strip of land along the canal. But many Panamanians resented US influence and, in

1979, the Canal Zone was returned to Panama. Control of the canal itself was handed over by the USA to Panama on December 31, 1999.

POLITICS

Panama's government has changed many times since independence, and there have been periods of military dictatorships. In 1983, General Manuel Antonio Noriega became Panama's leader. In 1988, two US grand juries in Florida indicted Noriega on charges of drug trafficking. In 1989, Noriega was apparently defeated in a presidential election, but the election was declared invalid. After the killing of a US marine, US troops entered Panama and arrested Noriega, who was later convicted of drug offences in 1992. Revenues from the Panama Canal rose in the early 21st century, but the economy slowed and public discontent grew.

ECONOMY

The World Bank classifies Panama as a "lower-middle-income" developing country. The Panama Canal is an important source of revenue and it generates many jobs in commerce, trade, manufacturing, and transport.

Away from the Canal, the main activity is agriculture, which employs 16% of the people. Rice is the main food crop, while bananas are grown for export. Other exports include shrimps, sugar, clothing, and coffee.

DID YOU KNOW

- that the opening of the Panama Canal shortened the sea journey from New York City to San Francisco by about 7,830 miles [12,600 km]
- that ships from many countries sail under Panama's flag; this is because of the low taxes and wages in Panama
- that an average of 33 ships pass through the Panama Canal every day; it takes eight to nine hours to get through
- that the first European to cross Panama and see the Pacific Ocean was a Spaniard, Vasco Núñez de Balboa, in 1513

The forests of Panama are rich in wildlife, including many birds with magnificently coloured plumage. A set of six stamps entitled "Wild Birds," that was issued in 1967, included this stamp, which shows a red-necked aracari, a kind of toucan.

PAPUA NEW GUINEA

Papua New Guinea's flag was first adopted in 1971, four years before the country became independent from Australia. It includes a local bird of paradise, the *kumul*, in flight, together with the stars of the Southern Cross. The colors are those often used by local artists.

Papua New Guinea is an independent country in the Pacific Ocean, north of Australia. It is part of a Pacific island region called Melanesia. Papua New Guinea includes the eastern part of New Guinea, the Bismarck Archipelago, the northern Solomon Islands, the D'Entrecasteaux Islands, and the Louisiade Archipelago. The land is largely mountainous. The highest peak, Mount Wilhelm, reaches 14,795 ft [4,508 m] in eastern New Guinea. Eastern New Guinea also has some large coastal lowlands.

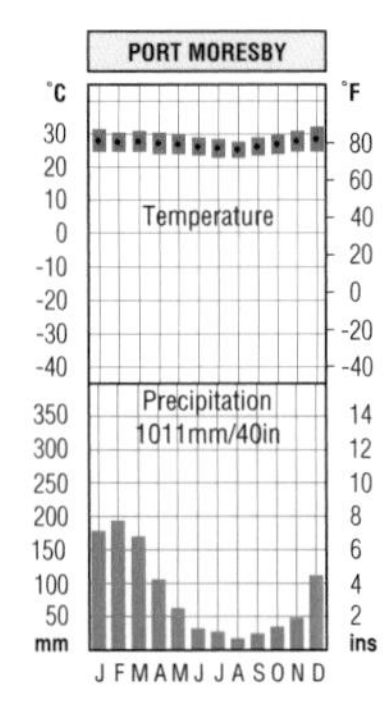

CLIMATE

Papua New Guinea has a tropical climate, with high temperatures throughout the year. Most of the rain occurs during the monsoon season (from December to April), when the northwesterly winds blow. Winds blow from the southeast during the dry season.

VEGETATION

Forests cover more than 70% of the land and the dominant kind of vegetation is lowland rain forest. Mangrove swamps border the coast, with lowland grasslands inland, while "cloud" forest and tussock grasses are found on the higher peaks.

AREA 178,703 sq miles [462,840 sq km]

POPULATION 5,049,000

CAPITAL (POPULATION) Port Moresby (174,000)

GOVERNMENT Constitutional monarchy

ETHNIC GROUPS Papuan, Melanesian

LANGUAGES English (official), about 800 others

RELIGIONS Traditional beliefs 34%, Roman Catholic 22%, Lutheran 16%

CURRENCY Kina = 100 toea

HISTORY

Little is known about the early history of the Melanesian peoples of Papua New Guinea. The first European contact was made by the Portuguese in 1526 and the region was later explored by Spanish, Dutch, French, and British navigators. But no settlements were established until the last quarter of the 19th century.

The Dutch took western New Guinea (now part of Indonesia) in 1828, but it was not until 1884 that Germany took northeastern New Guinea and Britain the southeast. In 1906, Britain handed the southeast over to Australia. It then became

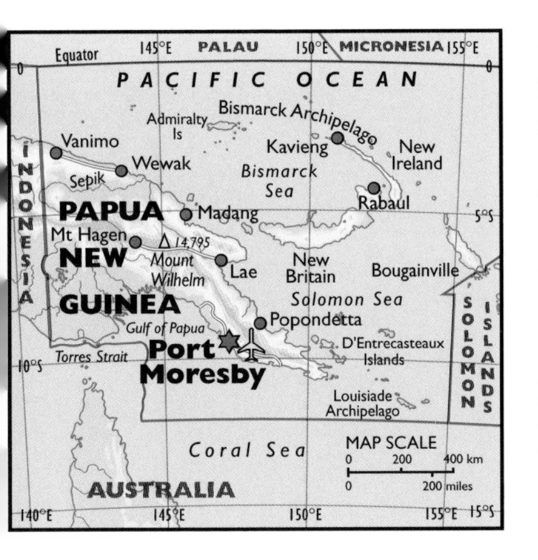

known as the Territory of Papua. When World War I broke out in 1914, Australia took German New Guinea and, in 1921, the League of Nations gave Australia a mandate to rule the area, which was named the Territory of New Guinea.

Japan invaded New Guinea in 1942, but the Allies reconquered the area in 1944. In 1949, Papua and New Guinea were combined into the Territory of Papua and New Guinea. Papua New Guinea became fully independent in 1975.

POLITICS

Since independence, the government of Papua New Guinea has worked to develop its mineral reserves. One of the most valuable mines was on Bougainville, in the northern Solomon Islands. But the people of Bougainville demanded a larger share in the profits of the mine. Conflict broke out, the mine was closed and the Bougainville Revolutionary Army proclaimed the island independent. But their attempted secession was not recognized internationally and the Papua New Guinea government expressed its intention to keep the island. An agreement ending the conflict was signed in 1998, and the island was granted local autonomy in 2000.

ECONOMY

The World Bank classifies Papua New Guinea as a "lower-middle-income" developing country. Agriculture employs three out of every four people, many of whom produce little more than they need to feed their families. But minerals, notably copper and gold, are the most valuable exports. Other exports include coffee, timber, palm oil, cocoa, and lobsters. Manufacturing industries process farm, fish, and forest products.

Papua New Guinea contains 33 of the 42 known species of the bird of paradise. Their feathers were used by local people to decorate costumes worn in ceremonies. The stamps show two species: Astrapia mayeri, *the ribbon-tailed bird of paradise, and* Astrapia stephaniae, *the Princess Stephanie bird of paradise.*

PARAGUAY

The front (obverse) side of Paraguay's tricolor flag, which evolved in the early 19th century, contains the state emblem, which displays the May Star, commemorating liberation from Spain in 1811. The reverse side shows the treasury seal – a lion and staff.

The Republic of Paraguay is a landlocked country in South America. Rivers form most of its borders. They include the River Paraná in the south and east, the Pilcomayo (Brazo Sur) in the southwest, and the Paraguay in the northeast. West of the River Paraguay is a region known as the Chaco, which extends into Bolivia and Argentina. The Chaco is mostly flat, but the land rises to the northwest. East of the Paraguay is a region of plains, hills, and, in the east, the Paraná plateau region.

AREA 157,046 sq miles [406,750 sq km]

POPULATION 5,734,000

CAPITAL (POPULATION) Asunción (945,000)

GOVERNMENT Multiparty republic

ETHNIC GROUPS Mestizo 90%, Amerindian 3%

LANGUAGES Spanish and Guaraní (both official)

RELIGIONS Roman Catholic 96%, Protestant 2%

CURRENCY Guaraní = 100 céntimos

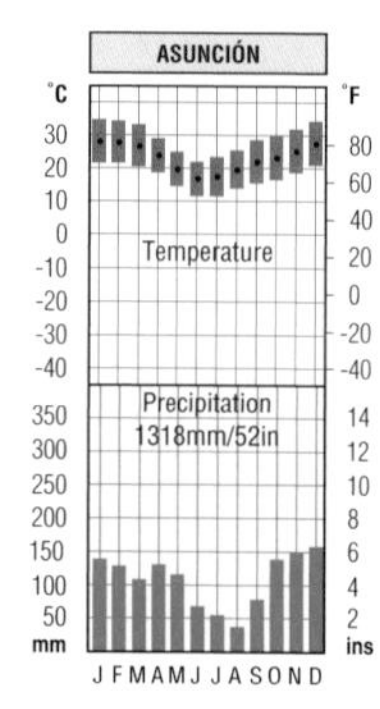

CLIMATE

Northern Paraguay lies in the tropics, while the south is subtropical. Most of the country has a warm, humid climate. The Chaco is the driest and hottest part of the country. The rainfall increases to the Paraná plateau in the southeast.

VEGETATION

The Chaco is a region of coarse grasses, shrubs, scrub forests, and, in places, some patches of hardwood forests. Trees include quebrachos, which are a source of tannin, used to process leather. Marshes occur near some rivers. Eastern Paraguay has grassy plains, with forests on the hills.

HISTORY

The earliest known inhabitants of Paraguay were the Guaraní, an Amerindian people. Spanish and Portuguese explorers reached the area in the early 16th century. In 1537, a Spanish expedition built a fort at Asunción, which later became the capital of Spain's colonies in southeastern South America. From the late 16th century, Jesuit missionaries worked to protect the Guaraní and convert them to Christianity. But many colonists opposed them. They wanted to use the people as laborers

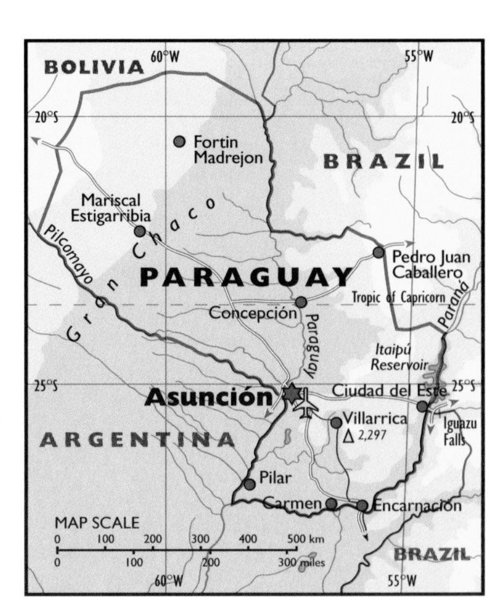

on their farms. Finally, in 1767, the Spanish king expelled the Jesuits from Paraguay. In 1776, Paraguay became part of a large colony called the Viceroyalty of La Plata, with Buenos Aires as the capital. Paraguayans opposed this move and the country declared its independence in 1811.

For many years, Paraguay was torn by internal strife and conflict with its neighbors. A war against Brazil, Argentina, and Uruguay (1865–70) led to the deaths of more than half of Paraguay's population, and a great loss of territory. Some territory was regained in the Chaco War against Bolivia between 1929 and 1935, but a civil war in 1947 was another setback for the country.

POLITICS

General Alfredo Stroessner took power in 1954 and ruled as a dictator. His government imprisoned many opponents. Stroessner was overthrown in 1989. Free multiparty elections held in 1993 resulted in the restoration of a civilian president, but democracy in Paraguay remains fragile because of rivalries between politicians and military leaders.

DID YOU KNOW

- that the world's most massive dam by volume is the Itaipú Dam on the River Paraná on the border between Paraguay and Brazil
- that Asunción was named because work on the original fort on the site began on the Feast of the Assumption (*Asunción* in Spanish), August 15, 1537
- that Paraguay tea, also called *maté* or *yerba maté*, is made from the leaves and roots of a holly tree
- that German-speaking people of the Mennonite faith have set up several farming communities in Paraguay

ECONOMY

The World Bank classifies Paraguay as a "lower-middle-income" developing country. Agriculture and forestry are the leading activities, employing 32% of the population. The country has large cattle ranches, while many crops are grown in the fertile soils of eastern Paraguay. Major exports include cotton, soya beans, timber, vegetable oils, coffee, tannin, and meat products.

The country has abundant hydroelectricity and it exports power to Argentina and Brazil. Its factories produce such things as cement, processed food, leather goods, and textiles. The country has no major mineral or fossil fuel resources.

PERU

Peru's flag was adopted in 1825. The colors are said to have been inspired by a flock of red and white flamingos which the Argentine patriot General José de San Martín saw flying over his marching army when he arrived in 1820 to liberate Peru from Spain.

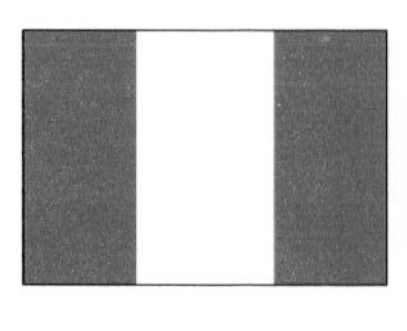

The Republic of Peru lies in the tropics in western South America. A narrow coastal plain borders the Pacific Ocean in the west. Inland are ranges of the Andes Mountains, which rise to 22,213 ft [6,768 m] at Mount Huascarán, an extinct volcano. The Andes also contain active volcanoes, windswept plateaux, broad valleys, and, in the far south, part of Lake Titicaca, the world's highest navigable lake. To the east the Andes descend to a hilly region and a huge plain. Eastern Peru is part of the Amazon basin.

AREA 496,223 sq miles [1,285,220 sq km]
POPULATION 27,484,000
CAPITAL (POPULATION) Lima (Lima-Callao, 6,601,000)
GOVERNMENT Transitional republic
ETHNIC GROUPS Amerindian 45%, Mestizo 37%, White 15%, Black, Japanese, Chinese
LANGUAGES Spanish and Quechua (both official), Aymara
RELIGIONS Roman Catholic 90%
CURRENCY New sol = 100 centavos

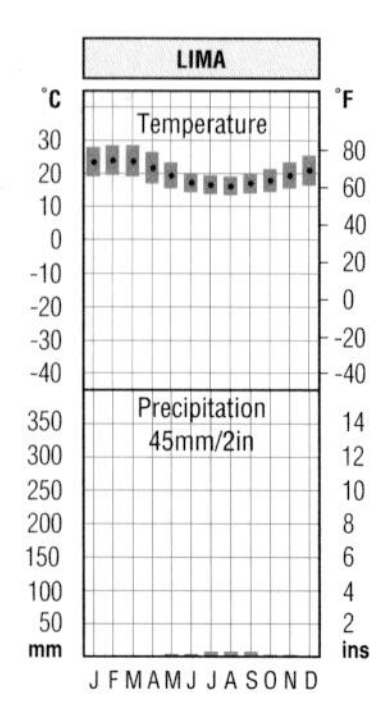

CLIMATE

Lima, on the coastal plain, has an arid climate. The coastal region is chilled by the cold, offshore Humboldt Current. The rainfall increases inland and many mountains in the high Andes are snow-capped. Eastern Peru has a hot, rainy climate.

VEGETATION

The coastal plain is a desert, though trees and farmland are found along rivers that flow from the mountains. The vegetation in the Andes varies with the altitude, with large areas of mountain grassland on the higher slopes. Eastern Peru is a region of *selva* – rain forest with many valuable trees, such as rosewood and rubber.

HISTORY

Amerindian people probably reached the area about 12,000 years ago. Several civilizations developed in the Andes region. By about AD 1200, the Inca were established in southern Peru. In 1500, their empire extended from Ecuador to Chile. The Spanish adventurer Francisco Pizarro visited Peru in the 1520s. Hearing of Inca riches, he returned in 1532. By 1533, he had conquered most of Peru.

In 1820, the Argentinian José de San Martín led an army into Peru

and declared the country to be independent. But Spain still held large areas. In 1823, the Venezuelan Simón Bolívar led another army into Peru and, in 1824, one of his generals defeated the Spaniards at Ayacucho. The Spaniards surrendered in 1826. Peru suffered much instability throughout the 19th century.

POLITICS

Instability continued in the 20th century. In 1980, when civilian rule was restored, a left-wing group called the *Sendero Luminoso*, or the "Shining Path," began guerrilla warfare against the government. In 1990, Alberto Fujimori, son of Japanese immigrants, became president. In 1992, he suspended the constitution and dismissed the legislature. The guerrilla leader, Abimael Guzmán, was arrested in 1992, but instability continued. A new constitution introduced in 1993 gave more power to the president. A border dispute with Ecuador dating back to 1942 was resolved in 1998.

ECONOMY

The World Bank classifies Peru as a "lower-middle-income" developing country. Agriculture employs 33% of the people. Major food crops include beans, maize, potatoes, and rice. Sugarcane, coffee, and cotton are exported. Some farmers grow coca, which is used to make cocaine. However, the most valuable export is copper.

Peru also produces iron ore, lead, oil, silver, and zinc. Its manufactures include chemicals, clothing and textiles, furniture, paper products, processed food, and steel.

Peru is one of the world's top ten fishing nations. This stamp, issued in 1952, was one of a set depicting Peruvian animals, industries, and monuments. It shows a tuna fishing boat, together with several local fish species.

PHILIPPINES

This flag was adopted in 1946, when the country won its independence from the United States. The eight rays of the large sun represent the eight provinces which led the revolt against Spanish rule in 1898. The three smaller stars stand for the three main island groups.

The Republic of the Philippines is an island country in south-eastern Asia. It includes about 7,100 islands, of which 2,770 are named and about 1,000 are inhabited. Luzon and Mindanao, the two largest islands, make up more than two-thirds of the country. The land is mainly mountainous and lacks large lowlands. The country lies in an unstable region and earthquakes are common. The islands also have several active volcanoes, one of which is the highest peak, Mount Apo, at 9,695 ft [2,954 m] above sea level.

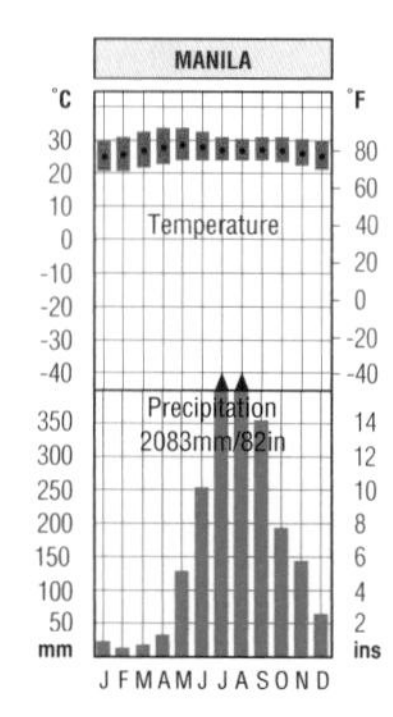

CLIMATE

The country has a tropical climate, with high temperatures all through the year. The dry season runs from December to April. The rest of the year is wet. The high rainfall is associated with typhoons which periodically strike the east coast.

VEGETATION

Forests cover large areas, especially in the mountains, with mangrove swamps along many coasts. Coarse grasses grow in areas where the forest has been cleared. The islands are rich in flowering plants and ferns.

AREA 115,300 sq miles [300,000 sq km]

POPULATION 82,842,000

CAPITAL (POPULATION) Manila (8,594,000)

GOVERNMENT Multiparty republic

ETHNIC GROUPS Christian Malay 92%, Muslim Malay 4%, Chinese, other

LANGUAGES Filipino (based on Tagalog) and English (both official), Spanish, many others

RELIGIONS Roman Catholic 83%, Protestant 9%, Islam 4%

CURRENCY Philippine peso = 100 centavos

HISTORY

Around 5,000 years ago, Malay people, the ancestors of modern Filipinos, arrived in the Philippines. They displaced earlier peoples, including the ancestors of the Negritos who survive in some remote areas. Islam was introduced in the late 14th century. But the arrival of Spaniards in the 16th century prevented the spread of Islam throughout the region.

The first European to reach the Philippines was the Portuguese navigator Ferdinand Magellan in 1521.

Spanish explorers claimed the region in 1565 when they established a settlement on Cebu. The Spaniards ruled the country until 1898, when the United States took over at the end of the Spanish-American War. Japan invaded the Philippines in 1941, but US forces returned in 1944. The country became fully independent as the Republic of the Philippines in 1946.

POLITICS

Since independence, the country's problems have included armed uprisings by left-wing guerrillas demanding land reform, and Muslim separatist groups, crime, corruption, and unemployment. The dominant figure in recent times was Ferdinand Marcos, who ruled in a dictatorial manner from 1965 to 1986. His successors restored democracy, but they were faced by conflict with the Muslim guerrillas in the south. Despite help from the United States, this conflict continued into the 21st century.

ECONOMY

The Philippines is a developing country with a lower-middle-income economy. Agriculture employs 36% of the people. The main foods are rice and maize, while such crops as bananas, cocoa, coconuts, coffee, sugarcane, and tobacco are grown commercially. Farm animals include water buffaloes, cattle, and pigs. Many farm products are exported.

Forestry and fishing are also important, but manufacturing now plays an increasing role in the economy. Food processing is the main industry. Chemicals, electronic goods, textiles, and timber products are other important manufactures. The government is working to diversify the economy.

DID YOU KNOW

- that the Philippines was named after Philip II of Spain (1527–98)
- that the national sport in the Philippines is basketball
- that the Philippines has more than 20 active volcanoes
- that the Philippines is the only large country in Asia where Christianity is the religion of the majority of the people

POLAND

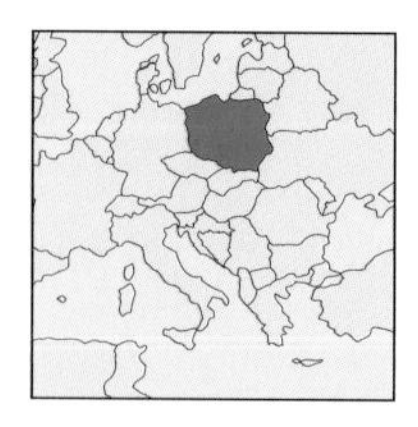

Poland's flag was adopted when the country became a republic in 1919. Its colors were taken from the 13th-century coat of arms of a white eagle on a red field. This coat of arms still appears on Poland's merchant flag.

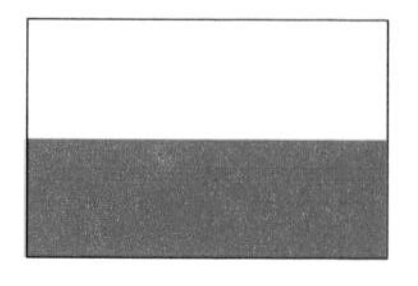

The Republic of Poland faces the Baltic Sea in north-central Europe. Behind the lagoon-fringed coast is a broad plain. Much of the soil is infertile, being made up of stony moraine (rock deposited by ice sheets during the Ice Age). The plains of central Poland are more fertile.

The land rises to a plateau region in the southeast of the country. The Sudeten Highlands straddle the border with the Czech Republic. Part of the Carpathian range lies on the southeastern border with the Slovak Republic.

AREA 120,726 sq miles [312,680 sq km]
POPULATION 38,643,000
CAPITAL (POPULATION) Warsaw (or Warszawa, 1,626,000)
GOVERNMENT Multiparty republic
ETHNIC GROUPS Polish 98%, Ukrainian 1%, German 1%
LANGUAGES Polish (official)
RELIGIONS Roman Catholic 94%, Orthodox 2%
CURRENCY Zloty = 100 groszy

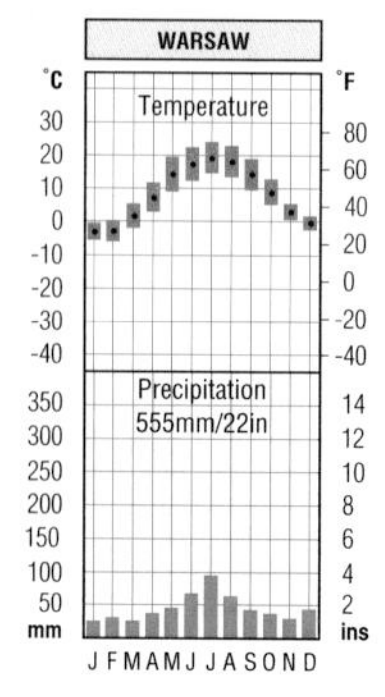

CLIMATE

Poland's climate is influenced by its position in Europe. Warm, moist air masses come from the west, while cold air masses come from the north and east. Summers are warm, but winters are cold and snowy. The climate of the coast is much milder than the southern highlands.

VEGETATION

Forests cover about 28% of Poland. Mixed forests of beech, fir, and oak grow on the lowlands, with spruce on the higher slopes. But about 60% of the land is farmed or used as pasture.

HISTORY AND POLITICS

Poland's boundaries have changed several times in the last 200 years. It disappeared from the map in the late 18th century, when a Polish state called the Grand Duchy of Warsaw was set up. But in 1815, the country was partitioned, between Austria, Prussia, and Russia. Poland became independent in 1918, but in 1939 it was divided between Germany and the Soviet Union. The country again became independent in 1945, when it lost land to Russia but gained some from Germany.

Communists took over the government of Poland in 1948. But opposition to Communism mounted and eventually became focused through

an organization called Solidarity.

Solidarity was led by a trade unionist, Lech Walesa. A coalition government was formed between Solidarity and the Communists in 1989. In 1990, the Communist party was dissolved and Walesa became president. But Walesa faced many problems in turning Poland toward a market economy. Throughout the 1990s, Poland pursued westward-looking policies and, in 1999, it achieved its aim of joining NATO. In 2002, the European Union offered membership to Poland in May 2004.

ECONOMY

Poland has large reserves of coal and deposits of various minerals which are used in its factories. Manufactures include chemicals, processed food, machinery, ships, steel, and textiles. Major crops include barley, potatoes, rye, sugar beet, and wheat. Machinery, metals, chemicals, and fuels are the country's leading exports.

PORTUGAL

Portugal's colors, which were adopted in 1910 when the country became a republic, represent the soldiers who died in the war (red), and hope (green). The armillary sphere – an early navigational instrument – reflects Portugal's leading role in world exploration.

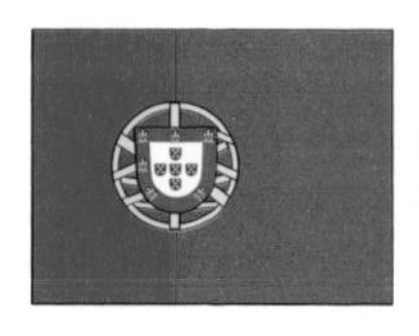

The Republic of Portugal shares the Iberian peninsula with Spain. It is the most westerly of Europe's mainland countries. The land rises from the coastal plains on the Atlantic Ocean to the western edge of the huge plateau, or Meseta, which occupies most of the Iberian peninsula. In central Portugal, the Serra da Estrela contains Portugal's highest point, at 6,534 ft [1,991 m]. Portugal also contains two autonomous regions, the Azores and Madeira island groups (*see Atlantic Ocean, pages 20 and 23*).

AREA 35,670 sq miles [92,390 sq km]
POPULATION 9,444,000
CAPITAL (POPULATION) Lisbon (or Lisboa, 2,561,000)
GOVERNMENT Multiparty republic
ETHNIC GROUPS Portuguese 99%, Cape Verdean, Brazilian, Spanish, British
LANGUAGES Portuguese (official)
RELIGIONS Roman Catholic 95%, other Christians 2%
CURRENCY Euro = 100 cents

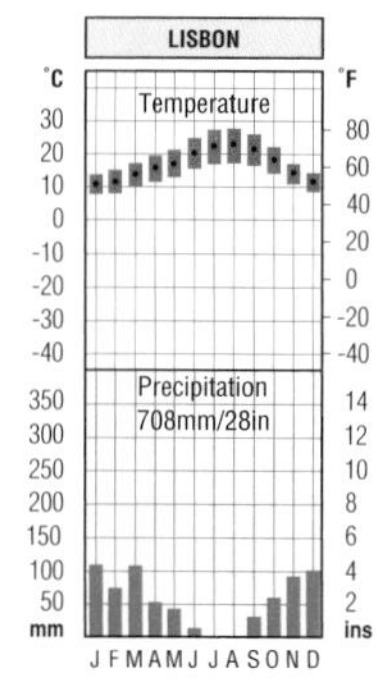

CLIMATE

The climate of the western part of the Iberian peninsula is moderated by winds blowing from the Atlantic Ocean. As a result, summers are cooler and the winters milder than in other Mediterranean lands. But most rain occurs in the winter.

VEGETATION

Forests cover about two-fifths of Portugal. Pine and oak are common trees. Portugal is the world's leading producer of cork, which is made from the bark of one species, the cork oak. Olive trees are also common, with almond, carob, and fig trees in the far south.

HISTORY AND POLITICS

Portugal became a separate country, independent of Spain, in 1143. In the 15th century, Portugal led the "Age of European Exploration." This led to the growth of a large Portuguese empire, with colonies in Africa, Asia, and, most valuable of all, Brazil in South America. Portuguese power started to decline in the 16th century and, between 1580 and 1640, the country was ruled by Spain. In 1822, Portugal lost Brazil.

In 1910, Portugal became a republic, but political instability marred its progress. Army officers seized power in 1926, and in 1928 they chose

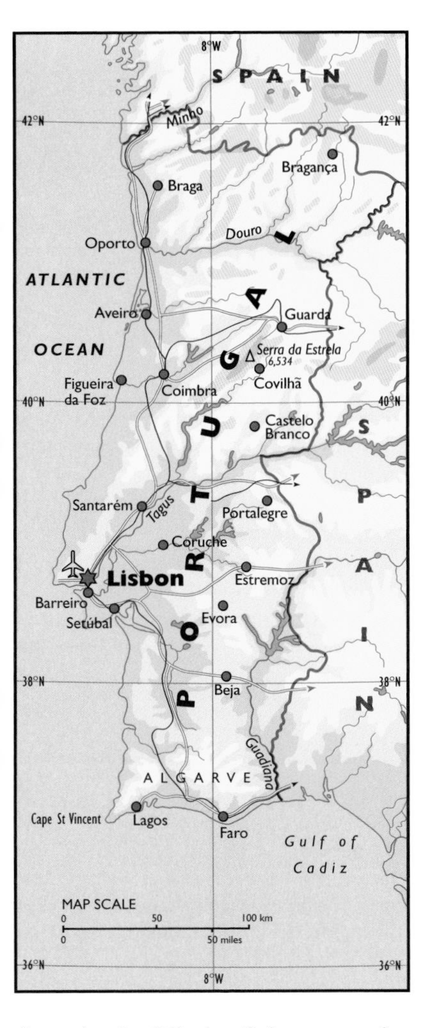

Antonio de Oliveira Salazar as minister of finance. Salazar became prime minister in 1932 and, from 1933, he ruled as a dictator.

Salazar ruled until 1968, but his successor, Marcello Caetano, was overthrown in 1974 by a group of army officers. The new government made most of Portugal's remaining colonies independent. Free elections were held in 1978. Portugal joined the European Union in 1986 and, despite its relatively small economy, by EU standards, it adopted the euro, the single currency of the EU, on January 1, 1999.

ECONOMY

Agriculture and fishing were the mainstays of the economy until the mid-20th century. But manufacturing is now the most valuable sector. Textiles, processed food, paper products, and machinery are important manufactures. Major crops include grapes for wine-making, olives, potatoes, rice, maize, and wheat. Cattle and other livestock are raised, and fishing catches include cod, sardines, and tuna.

Portugal's plants and animals are a fascinating mixture of Atlantic, Mediterranean, and African species. This stamp shows a lizard and the "little wild carnation," which belongs to a group of flowering plants called Dianthus. *It was issued in 1976 as one of a set of four to commemorate Portucale '77, a stamp exhibition held in Oporto.*

ROMANIA

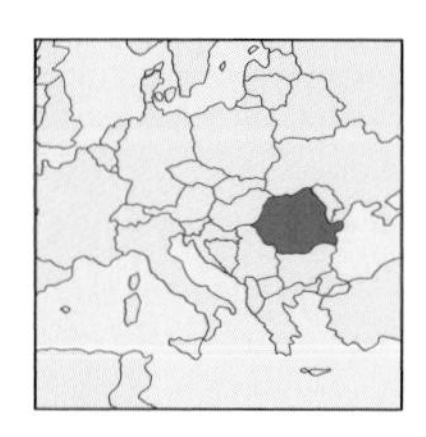

Romania's flag, adopted in 1948, uses colors from the arms of the provinces, which united in 1861 to form Romania. A central coat of arms, added in 1965, was deleted in 1990 after the fall of the Communist regime under the dictator Nicolae Ceaucescu.

Romania, which is sometimes spelled Rumania, is a country on the Black Sea in eastern Europe. Eastern and southern Romania form part of the Danube river basin. The delta region, near the mouths of the Danube, where the river flows into the Black Sea, is one of Europe's finest wetlands. The southern part of the coast contains several resorts. The heart of the country is called Transylvania. It is ringed in the east, south, and west by scenic mountains which are part of the Carpathian mountain system.

AREA 91,699 sq miles [237,500 sq km]

POPULATION 22,364,000

CAPITAL (POPULATION) Bucharest (or Bucuresti, 2,028,000)

GOVERNMENT Multiparty republic

ETHNIC GROUPS Romanian 89%, Hungarian 7%, Roma 2%

LANGUAGES Romanian (official), Hungarian, German

RELIGIONS Romanian Orthodox 70%, Protestant 6%, Roman Catholic 3%

CURRENCY Romanian leu = 100 bani

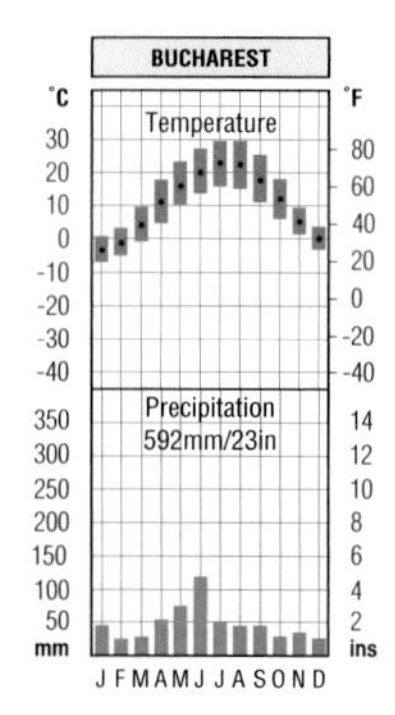

CLIMATE

Romania has hot summers and cold winters. The rainfall is heaviest in spring and early summer, when thundery showers are common. But with more than 2,000 hours of sunshine every year, Romania is one of the sunniest places in Europe.

VEGETATION

Farms and pasture make up nearly two-thirds of Romania and forests 28%. Oaks are the main trees at lower levels, with beeches and conifers on the higher slopes. Treeless mountain pastures occur at the highest levels.

HISTORY

Around 2,300 years ago, Romania was called Dacia, after the local Dacian people. But after the Romans conquered the area in AD 106, the Dacians embraced Roman culture and the Latin language so completely that the region became known as Romania. After the fall of the Roman empire, Romania was invaded many times. The first step toward the creation of the modern state occurred in the 14th century when two principalities (states ruled by princes) were formed: Walachia (or Valachi) in the south and Moldavia in the east. But they were conquered by the Ottoman Turks around 1500.

From the late 18th century, the

Turkish empire began to break up. The modern history of Romania began in 1861 when Walachia and Moldavia united. After World War I (1914–18), Romania, which had fought on the side of the victorious Allies, obtained large areas, including Transylvania, where most people were Romanians. This almost doubled the country's size and population. In 1939, Romania lost territory to Bulgaria, Hungary, and the Soviet Union. Romania fought alongside Germany in World War II, but Soviet troops occupied the country in 1944. In 1945, Hungary returned northern Transylvania to Romania, but Bulgaria and the Soviet Union kept former Romanian territory.

In 1947, Romania officially became a Communist country. Communist rule continued until 1989, when the country's Communist leader, Nicolae Ceaucescu, and his wife were executed.

POLITICS

In 1990, Romania held its first free elections since the end of World War II. The National Salvation Front, led by Ion Iliescu and containing many former Communist leaders, won a large majority. A new constitution, approved in 1991, made the country a democratic republic. Elections held under this constitution in 1992 again resulted in victory for Ion Iliescu, whose party was renamed the Party of Social Democracy (PDSR) in 1993.

The center-right Democratic Convention defeated the PDSR in 1996. Ion Iliescu regained the presidency in 2000, while the PDSR was the largest party in parliament.

DID YOU KNOW

- that Romania's Transylvanian Alps are the legendary home of Count Dracula
- that Romania means "land of the Romans"; the Romanian language has developed from Latin
- that the River Danube delta has been designated a wetland of international importance; it contains about 300 bird species
- that Bucharest (Bucuresti) is said to have been founded in the 15th century by a shepherd named Bucur

ECONOMY

According to the World Bank, Romania is a "lower-middle-income" economy. Under Communist rule, industry, including mining and manufacturing, became more important than agriculture. Oil and natural gas are the main mineral resources, while manufactures include cement, iron and steel, processed food, petroleum products, and wood products. Major crops include fruits, maize, potatoes, sugar beet, and wheat. Sheep and other livestock are reared.

RUSSIA

In August 1991, Russia's traditional flag, which had first been used in 1699, was restored as Russia's national flag. It uses colors from the flag of the Netherlands. This flag was suppressed when Russia was part of the Soviet Union.

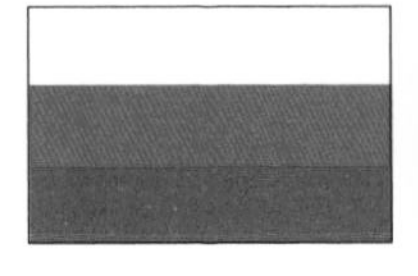

Russia, or the Russian Federation as it is also known, is the world's largest country. About 25% of the country lies west of the low Ural Mountains in European Russia. This area contains about 80% of Russia's population and is a vast plain, but the land rises in the far south to the Caucasus Mountains where Russia's highest peak, Elbrus, at 18,481 ft [5,633 m], is found. Asian Russia, which is also called Siberia, is a land of plains and plateaux, with mountains in the east and south. The Kamchatka peninsula in the far east has many active volcanoes.

Russia has some of the world's longest rivers, including the Amur, Lena, Ob, Volga, and Yenisei (or Yenisey). It also includes part of the world's largest lake, the Caspian Sea, and Lake Baikal (or Baykal), the world's deepest lake.

AREA 6,592,800 sq miles [17,075,000 sq km]

POPULATION 145,470,000

CAPITAL (POPULATION) Moscow (or Moskva, 8,405,000)

GOVERNMENT Federal multiparty republic

ETHNIC GROUPS Russian 82%, Tatar 4%, Ukrainian 3%, Chuvash 1%, more than 100 other nationalities

LANGUAGES Russian (official), many others

RELIGIONS Mainly Russian Orthodox, with Roman Catholic and Protestant minorities, Islam, Judaism

CURRENCY Russian rouble = 100 kopeks

MOSCOW
Temperature
Precipitation 624mm/25in

CLIMATE

Moscow has a continental climate with cold and snowy winters and warm summers. Krasnoyarsk in south-central Siberia has a harsher, drier climate, but it is not as severe as parts of northern Siberia. Vladivostok in the far southeast has bitterly cold winters.

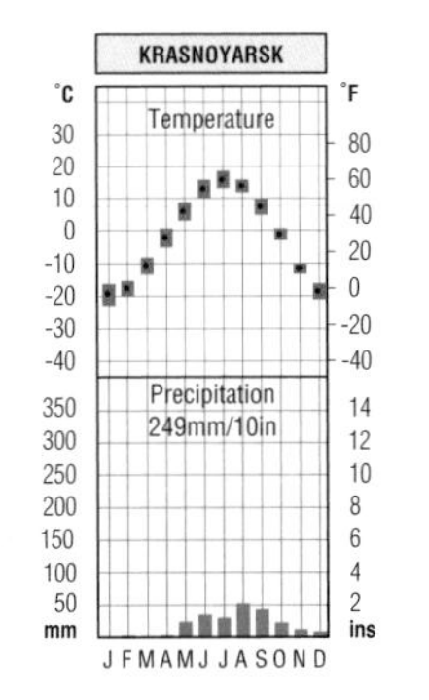

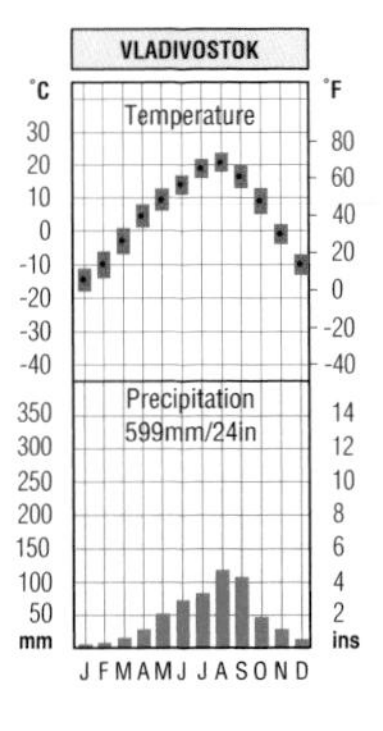

VEGETATION

The far north of the country is a bleak, treeless region, the tundra, where mosses, lichens, and stunted shrubs grow during the short summer. To the south is the taiga, with such coniferous trees as fir, pine, and spruce. In the west and east are mixed forests of conifers and such trees as oak and beech. South-central Russia contains large areas of former steppe (dry grassland), but most of it is now farmed.

This stamp was issued in 1985 by the USSR (or CCCP in the Russian alphabet), which consisted of Russia and 14 other republics. The stamp, one of a set of five entitled "Protected Animals," shows a Satunin's jerboa.

HISTORY

In the 9th century AD, a state called Kievan Rus was formed by a group of people called the East Slavs. Kiev, now capital of Ukraine, became a major trading center, but, in 1237, Mongol armies conquered Russia and destroyed Kiev. Russia was part of the Mongol empire until the late 15th century. Under Mongol rule, Moscow became the leading Russian city.

In the 16th century, Moscow's grand prince was retitled "tsar." The first tsar, Ivan the Terrible, expanded Russian territory. In 1613, after a period of civil war, Michael Romanov became tsar, founding a dynasty which ruled until 1917. In the early 18th century, Tsar Peter the Great began to westernize Russia and, by 1812, when Napoleon failed to conquer the country, Russia was a major European power. But during the 19th century, many Russians demanded reforms and discontent was widespread.

In World War I (1914–18), the Russian people suffered great hardships and, in 1917, Tsar Nicholas II was forced to abdicate. In November 1917, the Bolsheviks seized power under Vladimir Lenin. In 1922, the Bolsheviks set up a new nation, the Union of Soviet Socialist Republics (also called the Soviet Union).

From 1924, Joseph Stalin introduced a socialist economic program, suppressing all opposition. In 1939, the Soviet Union and Germany signed a non-aggression pact, but Germany invaded the Soviet Union in 1941. Soviet forces pushed the Germans back, occupying eastern Europe. They reached Berlin in May 1945. From the late 1940s, tension between the Soviet Union and its allies and Western nations developed into a "Cold War."

POLITICS

The Soviet Union collapsed in 1991 because its economic policies had failed. During the 1990s, Boris Yeltsin worked to develop democratic systems and reform the economy. In ill-health, Yeltsin retired in 1999 and was succeeded by Vladimir Putin, who won a resounding victory in

RUSSIA

2000. Russia continued to combat Muslim separatist guerrillas in the Chechen Republic in an attempt to maintain national unity. After the attacks on the USA on September 11, 2001, Putin and US President Bush found common cause in the campaign against international terrorism.

ECONOMY

Russia's economy was thrown into disarray after the collapse of the Soviet Union, and in the early 1990s the World Bank described Russia as a "lower-middle-income" economy. Industry is the most valuable activity. But under Communist rule, manufac-

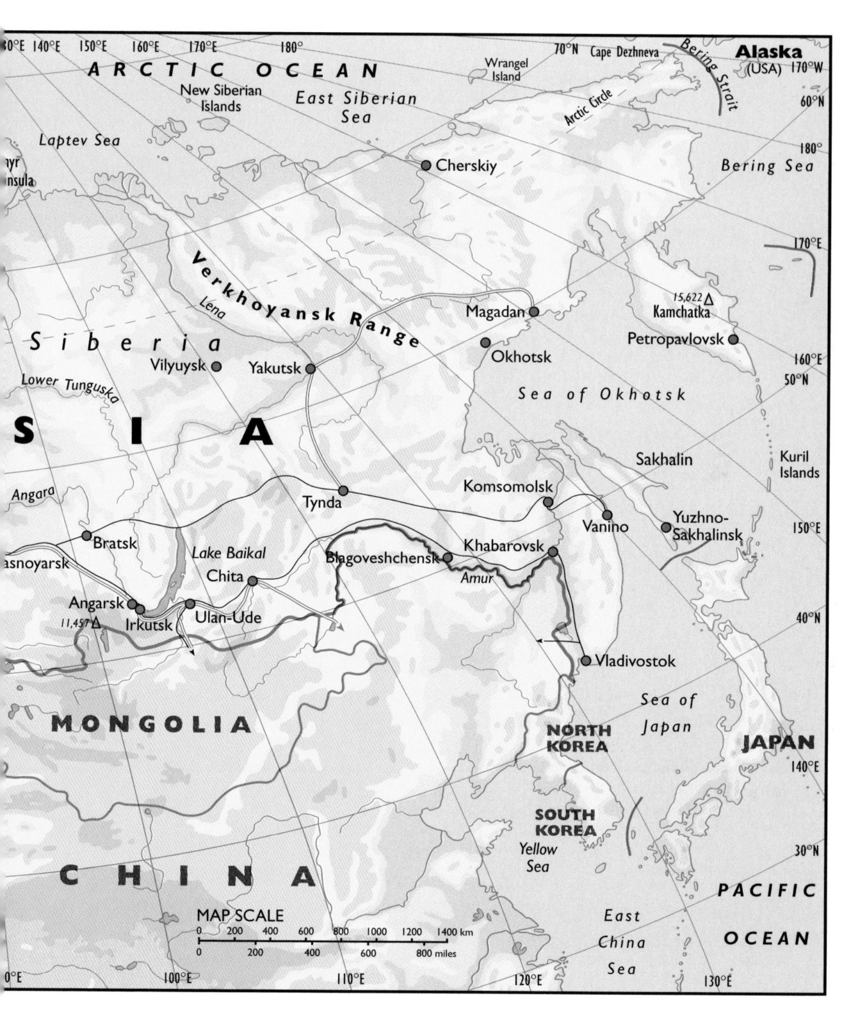

turing was less efficient than in the West and the emphasis was on heavy industry. Today light industries, many producing consumer goods, are also becoming important. Russia is rich in resources, including oil and natural gas, coal, timber, metal ores, and hydroelectric power.

Most farmland is still government-owned or run as collectives. Russia is a major producer of farm products, though it imports grains. Major crops include barley, flax, fruits, oats, rye, potatoes, sugar beet, sunflower seeds, vegetables, and wheat. Livestock farming is also important.

RWANDA

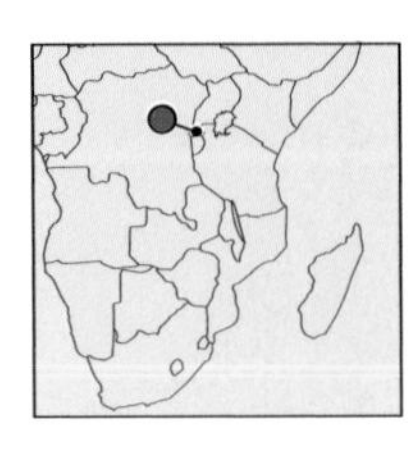

Rwanda's flag was adopted in 2002. The color blue is used to symbolize peace and tranquility. Yellow represents wealth, while green denotes prosperity, work, and productivity. The 24-ray golden sun symbolizes new hope.

The Republic of Rwanda is a landlocked African country, bordered by Uganda, Tanzania, Burundi, and Congo (Dem. Rep.). Lake Kivu and the River Ruzizi form the boundary in the west. These two features lie in the Great African Rift Valley. Overlooking the valley in western Rwanda are high, volcanic mountains reaching 14,792 ft [4,507 m] above sea level in the northwest. Eastern Rwanda consists of a series of plateaux. These plateaux slope to the east where there are several small lakes.

AREA 10,170 sq miles [26,340 sq km]
POPULATION 7,313,000
CAPITAL (POPULATION) Kigali (235,000)
GOVERNMENT Republic
ETHNIC GROUPS Hutu 84%, Tutsi 15%, Twa 1%
LANGUAGES French, English and Kinyarwanda (all official)
RELIGIONS Roman Catholic 57%, Protestant 26%, Adventist 11%, Islam 4%
CURRENCY Rwanda franc = 100 centimes

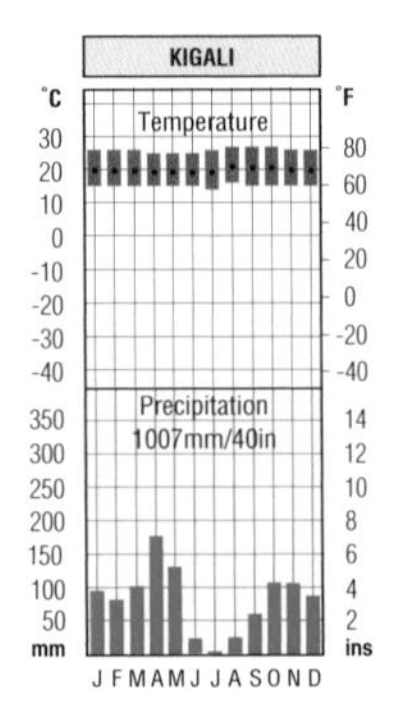

CLIMATE

Kigali stands on the central plateau of Rwanda. Here, temperatures are moderated by the altitude. The rainfall is abundant, but much heavier rain falls on the western mountains. The floor of the Great Rift Valley is warmer and drier than the rest of the country.

VEGETATION

Rain forests occur in the west. These forests contain many plant and animal species, including mountain gorillas. Forests once also covered eastern Rwanda, but they have been largely cut down to create farmland. The cleared land has been badly damaged by the heavy rain which washes away the soil.

HISTORY AND POLITICS

The Twa, a pygmy people, were the first known people to live in Rwanda. About 1,000 years ago, a farming people, the Hutu, settled in the area, gradually displacing the Twa.

From the 15th century, a cattle-owning people from the north, the Tutsi, began to dominate the Hutu, who had to serve the Tutsi overlords.

Germany conquered the area, called Ruanda-Urundi, in the 1890s. But Belgium occupied the region during World War I (1914–18) and ruled it until 1961, when the people of Ruanda voted for their country to become a republic, called Rwanda. This decision followed a rebellion by the majority Hutu people against the Tutsi monarchy. About 150,000 deaths resulted from this conflict. Many Tutsis fled to Uganda, where they formed a rebel army. Urundi became independent as a monarchy, though it became a republic in 1966.

Relations between the Hutus and Tutsis continued to cause friction. In 1994, civil war broke out when the presidents of Rwanda and Burundi were killed in a plane crash. In 1996, the conflict spread into Zaïre (now Congo [Dem. Rep.]). In the early 21st century, prosecutions began in both Belgium and Tanzania of Rwandans accused of genocide.

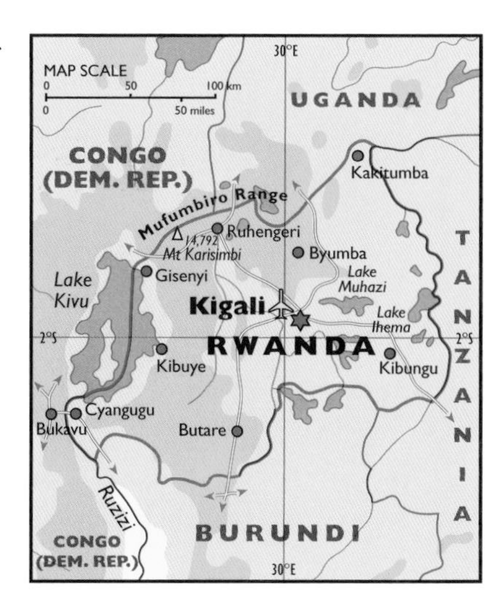

ECONOMY

According to the World Bank, Rwanda is a "low-income" developing country. Most people are poor farmers. Food crops include bananas, beans, cassava, and sorghum.

The country's most valuable crop is coffee, which accounts for more than 70% of Rwanda's exports. Some tea, pyrethrum, which is used to make insecticides, and tin are other products which are exported. Rwanda has few manufacturing industries.

DID YOU KNOW

- that Rwanda is the most densely populated country in mainland Africa
- that the Tutsi people of Rwanda and neighboring Burundi are one of the world's tallest peoples, while the Twa are some of the shortest; the majority of the people, the Hutu, are of medium height
- that Volcanoes National Park, in Rwanda's Mufumbiro range, is one of the last refuges for the endangered mountain gorilla

This stamp, issued in 1970, shows young mountain gorillas, which survive in a national park in Rwanda. But intense fighting in the area in 1994, together with poaching, threatened the existence of these gentle animals.

SAUDI ARABIA

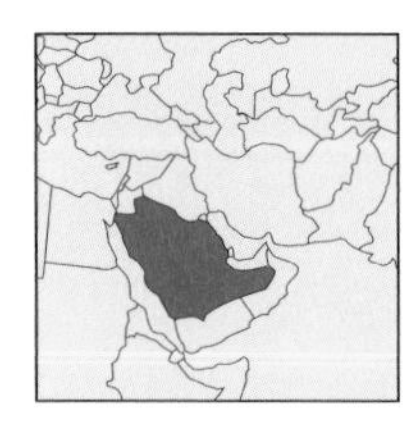

Saudi Arabia's flag was adopted in 1938. It is the only national flag with an inscription as its main feature. The Arabic inscription above the sword means "There is no God but Allah, and Muhammad is the Prophet of Allah."

The Kingdom of Saudi Arabia occupies about three-quarters of the Arabian peninsula in southwest Asia. Deserts cover most of the land. Mountains border the Red Sea plains in the west. The northwest is a region called the Hejaz. The southwest is called the Asir. To the east is the central plateau, or Najd, which descends to the eastern lowlands (Al Hasa), along The Gulf. In the north is the sandy Nafud Desert (An Nafud). In the south is the Rub' al Khali (the "Empty Quarter"), one of the world's bleakest deserts.

AREA 829,995 sq miles [2,149,690 sq km]
POPULATION 22,757,000
CAPITAL (POPULATION) Riyadh (or Ar Riyad, 1,800,000)
GOVERNMENT Absolute monarchy with consultative assembly
ETHNIC GROUPS Arab 90%, Afro-Asian 10%
LANGUAGES Arabic (official)
RELIGIONS Islam 100%
CURRENCY Saudi riyal = 100 halalas

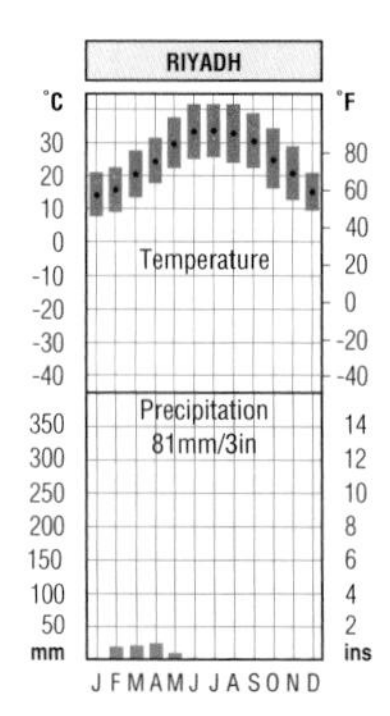

CLIMATE

Saudi Arabia has a hot, dry climate. In the summer months, the temperatures in Riyadh often exceed 104°F [40°C] though the nights are cool. The Asir highlands have an average rainfall of 12 to 20 inches [300 mm to 500 mm]. The rest of the country has less than 4 inches [100 mm].

VEGETATION

Grasses and shrubs provide pasture on the western highlands and parts of the central plateau. The deserts contain few plants except around oases.

HISTORY AND POLITICS

Saudi Arabia contains the two holiest places in Islam – Mecca (or Makka), the birthplace of the Prophet Muhammad in AD 570, and Medina (Al Madinah) where Muhammad went in 622. These places are visited by many pilgrims.

In the early 16th century, the Ottoman Turks won control of parts of Hejaz and Asir, while, in the 19th century, Britain gained areas on the southern and eastern coasts. But the Saud family ruled most of the interior. After the Turks were driven out in World War I (1914–18), the Saud family extended its territory. In 1932, King Ibn Saud proclaimed his country the Kingdom of Saudi Arabia.

Saudi Arabia was poor until the oil industry began to operate on the east-

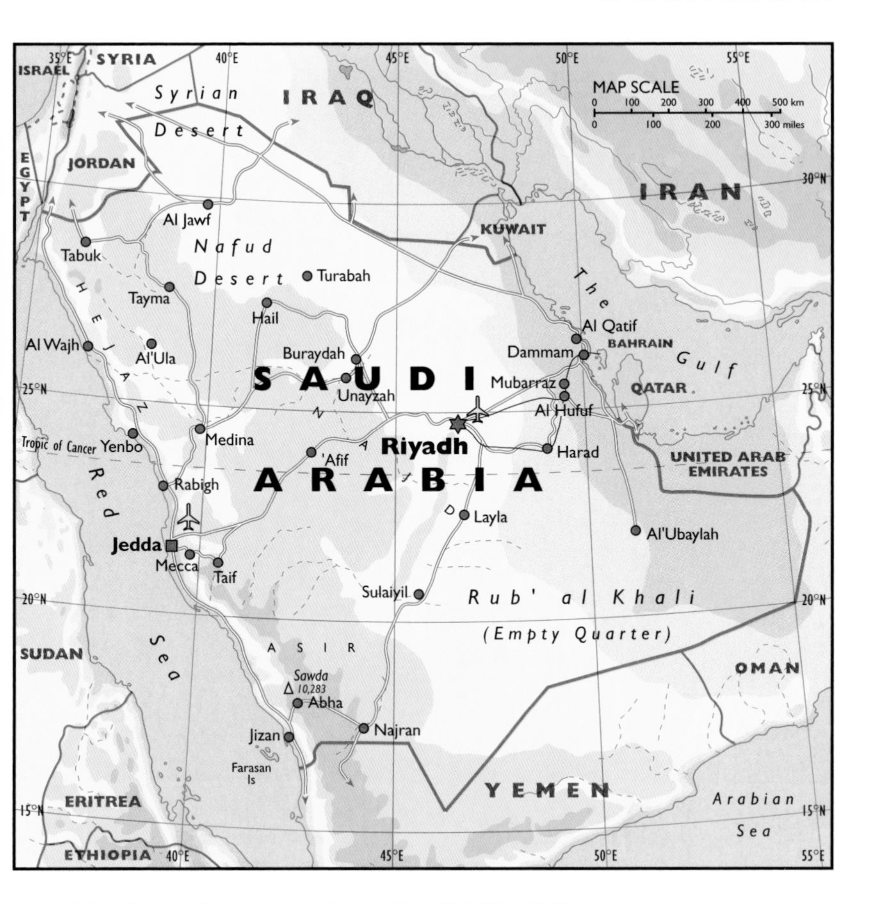

ern plains in 1933. Revenue from oil exports has since been used to develop the country and Saudi Arabia has given aid to poor Arab nations. But the monarch retains absolute authority.

In 1990, when Iraq invaded Kuwait, Saudi Arabia helped to free Kuwait in 1991. The government condemned the attacks on the USA on September 11, 2001, but Saudi-US relations became strained, partly because the alleged terrorist leader Osama bin Laden and many of his people were Saudi-born.

ECONOMY

Saudi Arabia has about 25% of the world's known oil reserves, and oil and oil products make up 85% of its exports. But agriculture remains important, employing 10% of the people. Some are nomadic livestock herders, who rear cattle, goats, sheep, and other animals. Crops grown in Asir and at oases include dates and other fruits, vegetables, and wheat. The government encourages the development of farming and new industries.

SENEGAL

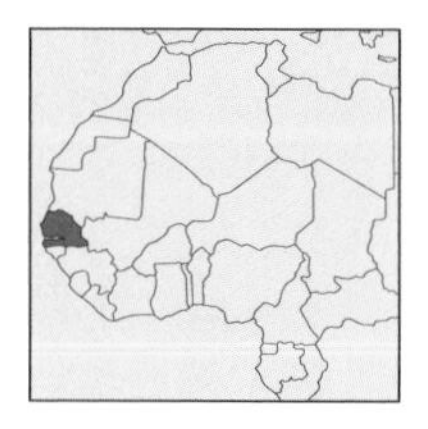

This flag was adopted in 1960 when Senegal became independent from France. It uses the three colors that symbolize African unity. It is identical to the flag of Mali, except for the five-pointed green star. This star symbolizes the Muslim faith of most of the people.

The Republic of Senegal is on the northwest coast of Africa. The volcanic Cape Verde (Cap Vert), on which Dakar stands, is the most westerly point in Africa. Plains cover most of Senegal, though the land rises gently in the southeast. Wind-blown sand dunes are common in the north and on the coast. The main rivers are the Sénégal, which forms the northern border, and the Casamance in the south. The River Gambia flows from Senegal into The Gambia, a small country almost enclosed by Senegal.

AREA 75,954 sq miles [196,720 sq km]
POPULATION 10,285,000
CAPITAL (POPULATION) Dakar (1,905,000)
GOVERNMENT Multiparty republic
ETHNIC GROUPS Wolof 44%, Pular 24%, Serer 15%
LANGUAGES French (official), tribal languages
RELIGIONS Islam 92%, traditional beliefs and others 6%, Christianity (mainly Roman Catholic) 2%
CURRENCY CFA franc = 100 centimes

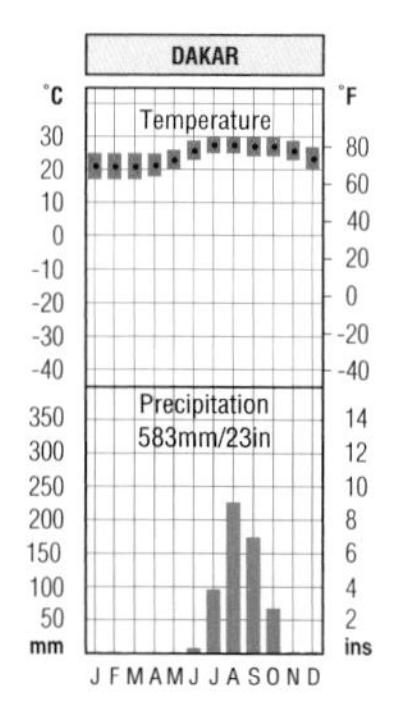

CLIMATE

Dakar has a tropical climate, with a short rainy season between July and October when moist winds blow from the southwest. In the dry season, hot northeasterly winds blow from the Sahara. Temperatures are higher inland, while the rainfall is greatest in the south.

VEGETATION

Desert and semidesert cover northeastern Senegal. In central Senegal, dry grasslands and scrub predominate. The far south is a region of savanna (tropical grassland), with forests along rivers. Although large areas in the south have been cleared for farming, some areas are protected. The largest sanctuary is the Niokolo-Kobo Wildlife Park, which is rich in animal species. Mangrove swamps border parts of the southern coast.

HISTORY AND POLITICS

Between the 4th and late 16th centuries, eastern Senegal formed part of three major African empires – Ancient Ghana, Mali, and Songhai. Islam was introduced in the 14th century, when Zenega Berbers invaded the area from the north. It is from the term Zenega that Senegal got its name.

Portuguese sailors, seeking a route to Asia around Africa, reached Cape Verde in 1444. In the 17th century,

England, France, and the Netherlands competed for control of the slave trade on the Senegal coast. In 1765, England set up a colony called Senegambia, which included parts of present-day Senegal and The Gambia. It handed this colony over to France in 1783.

In 1882, Senegal became a French colony, and from 1895 it was ruled as part of a large region called French West Africa, which also included present-day Benin, Burkina Faso, Guinea, Ivory Coast, Mauritania, Mali, and Niger. The capital of this huge territory was Dakar, which developed as a major city.

In 1959, Senegal joined French Sudan (now Mali) to form the Federation of Mali. But Senegal withdrew in 1960 and became the separate Republic of Senegal. Its first president, Léopold Sédar Senghor, was a noted African poet. Senghor was succeeded by Abdou Diouf in 1981. Diouf was surprisingly beaten by Abdoulaye Wade in elections in 2000.

Senegal and The Gambia have always enjoyed close relations, despite their different colonial traditions, one French and the other British. In 1981, Senegalese troops put down an attempted coup in The Gambia. In 1982, the two countries set up a defense alliance, called the Confederation of Senegambia. But the confederation was dissolved in 1989.

ECONOMY

According to the World Bank, Senegal is a "lower-middle-income" developing country. It was badly hit in the 1960s and 1970s by droughts, which caused starvation. Agriculture still employs 65% of the population though many farmers produce little more than they need to feed their families. Food crops include cassava, millet, and rice, but the most valuable product is groundnuts. Fishing is also important. Phosphates are the country's chief resource, but Senegal also refines oil which it imports from Gabon and Nigeria. Dakar is a busy port and has many industries.

Senegal exports manufactures, fish products, groundnuts, oil products, and phosphates.

This stamp, issued in 1976, shows the cultivation of tomatoes, one of the major crops grown in Senegal for both home consumption and for export.

SERBIA AND MONTENEGRO

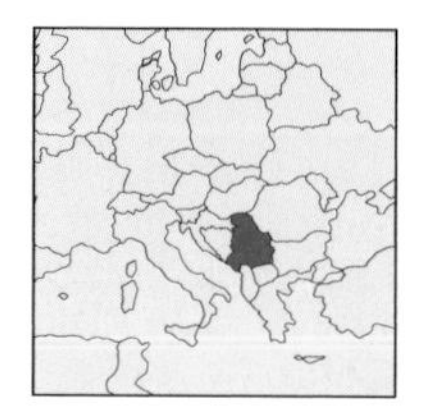

This flag was adopted in 1992. Serbia and Montenegro is a country formerly known as Yugoslavia. Before it broke apart in the early 1990s, it was part of the socialist Federal People's Republic of Yugoslavia, which also included Bosnia-Herzegovina, Croatia, Macedonia, and Slovenia.

AREA 39,449 sq miles [102,170 sq km]

POPULATION 10,677,000

CAPITAL (POPULATION) Belgrade (or Beograd, 1,598,000)

GOVERNMENT Federal republic

ETHNIC GROUPS Serb 62%, Albanian 17%, Montenegrin 5%, Hungarian, Muslim, Croat

LANGUAGES Serbo-Croatian (official), Albanian

RELIGIONS Christianity (mainly Serbian Orthodox, with Roman Catholic and Protestant minorities), Islam

CURRENCY New dinar = 100 paras

Serbia and Montenegro are two of the six republics which made up the former Communist Federal Republic of Yugoslavia, which broke up in 1991 and 1992. Behind Montenegro's short coastline along the Adriatic Sea lies a mountainous region, including the Dinaric Alps and part of the Balkan Mountains. The Pannonian Plains make up the northern part of the country. This fertile region is drained by the River Danube and its tributaries. The river enters neighboring Romania through a scenic gorge called the Iron Gate.

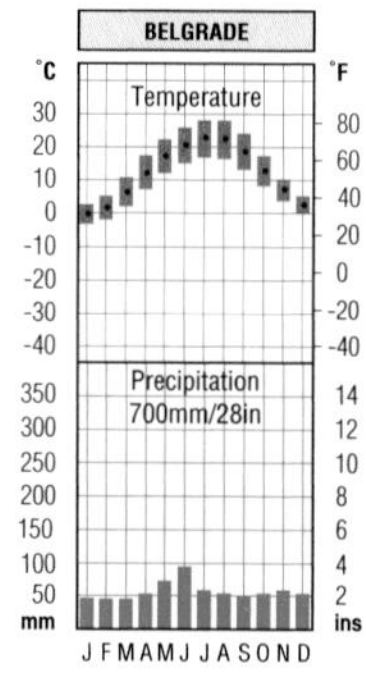

CLIMATE

The coast has a Mediterranean climate. The interior highlands have bitterly cold winters and cool summers. The capital, Belgrade, in the north of the country, has hot summers and cold winters. The wettest season is the summer, but there is also plenty of sunshine.

VEGETATION

Forests of beech, hornbeam, and oak, together with firs and pines at higher levels, cover about a quarter of the country. Farmland and pasture cover more than half of the country.

HISTORY

People who became known as the South Slavs began to move into the region around 1,500 years ago. Each group, including the Serbs and Croats, founded its own state. But, by the 15th century, foreign countries controlled the region. Serbia and Montenegro were under the Turkish Ottoman empire.

In the 19th century, many Slavs worked for independence and Slavic unity. In 1914, Austria-Hungary declared war on Serbia, blaming it for the assassination of Archduke Francis Ferdinand of Austria-Hungary. This led to World War I and the defeat of Austria-Hungary. In 1918, the South Slavs united in the Kingdom of the Serbs, Croats, and Slovenes, which consisted of Bosnia-Herzegovina, Croatia, Dalmatia, Montenegro, Serbia, and Slovenia. The country was renamed Yugoslavia in 1929. Germany occupied Yugoslavia during World War II (1939–45), but resistance groups, including the Communist partisans led by Josip Broz Tito, fought the invaders.

POLITICS

From 1945, the Communists controlled the country, which was called the Federal People's Republic of Yugoslavia. In 1990, non-Communist parties were permitted and non-Communists won majorities in elections in all but Serbia and Montenegro, where Socialists (former Communists) won control. In 1991–2, the six republics which formed Yugoslavia split apart. Bosnia-Herzegovina, Croatia, Macedonia, and Slovenia proclaimed their independence. The fifth country, consisting of Serbia and Montenegro, kept the name Yugoslavia.

Fighting broke out in Croatia and Bosnia-Herzegovina. In 1992, the UN withdrew recognition of Yugoslavia because it failed to halt atrocities committed by Serbs in Croatia and Bosnia. Yugoslavia played a part in the Dayton Peace Accord of 1995 which brought peace to Bosnia-Herzegovina. But when Yugoslavia tried to suppress Kosovar nationalism through "ethnic cleansing" in southern Serbia, NATO launched air strikes against the Serbs in March–June 1999. Serb troops then withdrew. In

2003, with many Montenegrins favoring secession, the two republics officially became a loose union named Serbia and Montenegro. Both states became semi-independent, though they share a seat in the United Nations.

ECONOMY

Manufacturing developed in the region after 1945. But in the 1990s, the World Bank classified Serbia and Montenegro as a "lower-middle-income" economy. Resources include bauxite, coal, copper, and other metals, together with oil and natural gas. Manufactures include aluminum, cars, machinery, steel, and textiles.

Manufactures dominate the exports, but agriculture is important. Crops include fruits, maize, potatoes, tobacco, and wheat, while cattle, pigs, and sheep are raised.

SIERRA LEONE

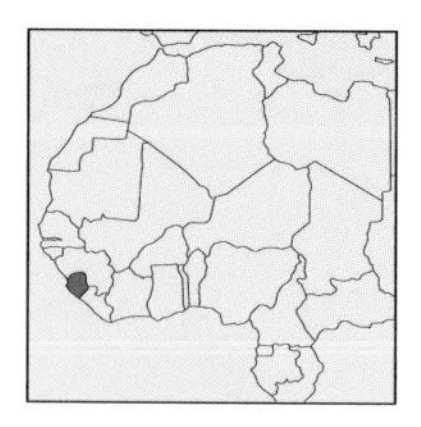

Sierra Leone's flag was adopted when the country became independent from Britain in 1961. The colors come from the country's coat of arms. The green represents agriculture, the white stands for peace, and the blue represents the Atlantic Ocean.

The Republic of Sierra Leone in West Africa is about the same size as the Republic of Ireland. The coast contains several deep estuaries in the north, with lagoons in the south. The most prominent feature is the mountainous Freetown (or Sierra Leone) peninsula. North of the peninsula is the River Rokel estuary, West Africa's best natural harbor. Behind the coastal plain, the land rises to plateaux and mountains. The highest peak, Loma Mansa, reaches 6,393 ft [1,948 m] near the northeastern border.

AREA 27,699 sq miles [71,740 sq km]
POPULATION 5,427,000
CAPITAL (POPULATION) Freetown (505,000)
GOVERNMENT Single-party republic
ETHNIC GROUPS Mende 35%, Temne 30%, Creole
LANGUAGES English (official), Mende, Temne, Krio
RELIGIONS Islam 60%, traditional beliefs 30%, Christianity 10%
CURRENCY Leone = 100 cents

CLIMATE

Sierra Leone has a tropical climate, with heavy rainfall which is brought by southwesterly winds blowing from the sea. The rainfall is heaviest between April and November. The south is dry in January and February. The north's dry season runs from December to March.

VEGETATION

Swamps cover large areas near the coast. Inland, much of the original rain forest has been destroyed and replaced by low bush and coarse grassland. The north contains savanna (grassland with scattered woodland).

HISTORY

Portuguese sailors reached the coast in 1460 and, in the 16th century, the area became a source of slaves. In 1787, Freetown was founded as a home for freed slaves. The British made the Sierra Leone peninsula a colony in 1808, and in 1896 they declared the interior a protectorate. Sierra Leone became independent in 1961.

POLITICS

A former British territory, Sierra Leone became independent in 1961 and a republic in 1971. It became a one-party state in 1978, but, in 1991, the people voted for the restoration of democracy. A military group seized power in 1992 and a civil war in 1994–5 caused much destruction. Elections in 1996 were followed by a military coup.

In 1998, the West African Peace Force restored the deposed President Ahmed Tejan Kabbah. But conflict in 1999 forced Kabbah to agree an accord with the rebels. The peace agreement soon collapsed, but another ceasefire was agreed in 2000. By 2002, the war seemed to be over.

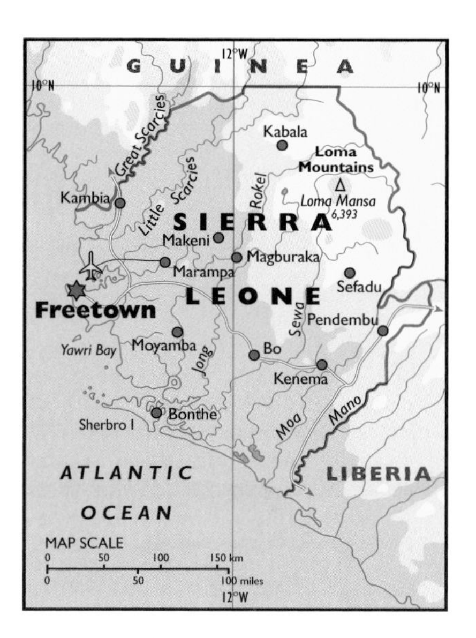

ECONOMY

The World Bank classifies Sierra Leone as a "low-income" economy. Agriculture provides a living for 64% of the people, though many farmers grow little more than they need to feed their families. The chief food crops include cassava, maize, and rice, the staple food. Export crops include cocoa and coffee. But the most valuable exports are minerals, including diamonds, bauxite (aluminum ore), and rutile (titanium ore). The country has few manufacturing industries.

DID YOU KNOW

- that Sierra Leone is the world's ninth leading producer of diamonds; 70% of the stones are of gem quality
- that Sierra Leone means "lion mountain," possibly because early Portuguese sailors confused the sound of heavy rain with a lion's roar, or possibly because they thought that a mountain on the Freetown peninsula resembled a lion
- that Freetown got its name because it was founded in 1787 as a settlement for freed slaves
- that the World Wide Fund for Nature has helped Sierra Leone to set up animal and plant reserves
- that Sierra Leone was once called the "white man's grave" because of its unpleasant climate and the many tropical diseases common to the area

To commemorate the International Year of Peace in 1986, Sierra Leone issued a set of four stamps, including one depicting its fishing industry. Fish are important in the diet of coastal people.

SINGAPORE

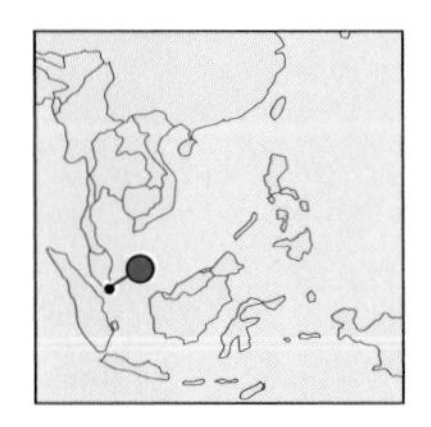

Singapore's flag was adopted in 1959 and it was retained when Singapore became part of the Federation of Malaysia in 1963. The crescent stands for the nation's ascent. The stars stand for Singapore's aims of democracy, peace, progress, justice, and equality.

The Republic of Singapore is an island country at the southern tip of the Malay peninsula. It consists of one large island, called Singapore Island, and 58 small ones, 20 of which are inhabited. Singapore Island is about 26 miles [42 km] wide and 14 miles [28 km] across. It is linked to the Malay peninsula by a causeway, 3,465 ft [1,056 m] long, which carries a road, railroad, and water pipeline. The land is mostly low-lying. The highest point, Bukit Timah, is only 578 ft [176 m] above sea level.

AREA 239 sq miles [618 sq km]
POPULATION 4,300,000
CAPITAL (POPULATION) Singapore City (3,866,000)
GOVERNMENT Multiparty republic
ETHNIC GROUPS Chinese 77%, Malay 14%, Indian 8%
LANGUAGES Chinese, Malay, Tamil and English (all official)
RELIGIONS Buddhism (Chinese), Islam (Malays), Christianity, Hinduism, Sikhism
CURRENCY Singapore dollar = 100 cents

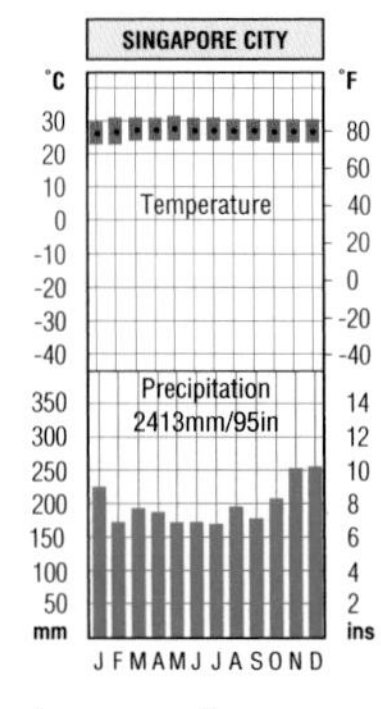

CLIMATE

Singapore has a hot and humid climate, typical of places near the Equator. The temperatures are high and the rainfall is heavy throughout the year. Thunderstorms occur on average on 40 days each year. Rain occurs on an average of more than 180 days per year.

VEGETATION

Rain forest once covered Singapore, but forests now grow on only 5% of the land. Today, about 50% of Singapore is built-up. Most of the rest consists of open spaces, including parks, granite quarries, and inland waters. Farmland covers 4% of the land and plantations of permanent crops make up 7%.

HISTORY

According to legend, Singapore was founded in 1299. It was first called Temasak ("sea town"), but it was named Singapura ("city of the lion") when an Indian prince thought he saw a lion there. Singapore soon became a busy trading center, but Javanese raiders destroyed it in 1377.

Singapore's modern history began in 1819 when Sir Thomas Stamford Raffles (1781–1826), agent of the British East India Company, made a treaty with the Sultan of Johor. This treaty allowed the British to build a settlement on Singapore Island. All of Singapore came under Britain in 1824, and in 1826 it became part of the Straits Settlement, which also included Pinang, Malaka, and some islands.

Singapore soon became the most important British trading center in Southeast Asia. In the 1920s and 1930s, Britain made Singapore a naval base, but Japanese forces seized the island in 1942. British rule was restored in 1945. In 1946, the Straits Settlement was dissolved and Singapore became a separate British colony.

In 1963, Singapore became part of the Federation of Malaysia, which also included Malaya and the territories of Sabah and Sarawak on the island of Borneo. But, in 1965, Singapore broke away from the Federation and became an independent country.

POLITICS

The People's Action Party (PAP) has ruled Singapore since 1959. Its leader, Lee Kuan Yew, served as prime minister from 1959 until 1990, when he resigned and was succeeded by Goh Chok Tong. Under the PAP, the economy has expanded rapidly, though some people consider that the PAP's rule has been rather dictatorial and oversensitive to criticism.

DID YOU KNOW

- that Singapore is one of the world's most densely populated countries, with more than 17,966 people per sq mile [6,935 people per sq km]
- that Changi Airport in Singapore, completed in 1981, stands partly on land reclaimed from the sea
- that the port of Singapore is the world's largest container port
- that Singapore is one of the most prosperous countries in eastern Asia; by 1995 it had the 10th highest per capita GNP in the world

ECONOMY

With its highly skilled and industrious work force, Singapore has one of the world's fastest growing economies. The World Bank classifies it as a "high-income" economy. Trade and finance provide many jobs, and manufacturing is a major activity. Singapore produces a wide range of goods, including chemicals, electronic products, machinery, metal products, paper, scientific instruments, ships, and textiles. Singapore has a large oil refinery, and petroleum products and manufactures are the main exports. Farming is relatively unimportant, but agricultural products include copra, eggs, fruits, pork, poultry, and vegetables. Most farming is intensive, and farmers use the latest technology and scientific methods.

SLOVAK REPUBLIC

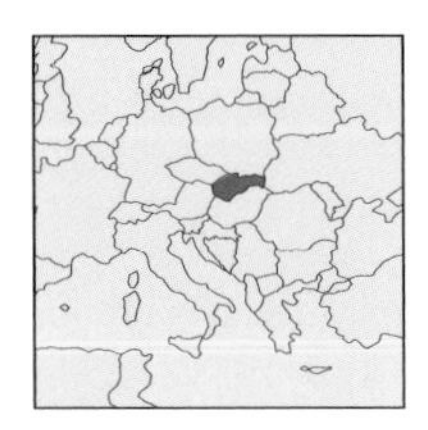

This flag, using the typical red, white, and blue Slavonic colors, dates back to 1848. The Slovak Republic adopted it in September 1992, prior to independence on January 1, 1993. The three blue mounds in the shield represent three mountain ranges.

The Slovak Republic, or Slovakia, is a new nation which emerged on January 1, 1993, when it broke away from Czechoslovakia and became a separate republic. The land is mainly mountainous, consisting of part of the Carpathian range. The highest peak is on the northern boundary in the Tatra Mountains. This peak reaches 8,714 ft [2,655 m].

By contrast, the south is a fertile lowland, drained by the River Danube, which forms part of the southern border. Several rivers flow from the mountainous north into the Danube river basin.

AREA 18,932 sq miles [49,035 sq km]
POPULATION 5,415,000
CAPITAL (POPULATION) Bratislava (451,000)
GOVERNMENT Multiparty republic
ETHNIC GROUPS Slovak 86%, Hungarian 11%, Roma 2%, Czech, Ukrainian, German, Polish
LANGUAGES Slovak (official), Hungarian
RELIGIONS Roman Catholic 60%, Atheist 10%, Protestant 8%, Orthodox 4%
CURRENCY Koruna = 100 halierov

CLIMATE

The Slovak Republic has cold winters and warm summers. Kosice, in the east, has average temperatures ranging from 27°F [–3°C] in January to 68°F [20°C] in July. The highland areas are much colder. Snow or rain falls throughout the year. Kosice has an average annual rainfall of 24 inches [600 mm], the wettest months being July to August. The highland areas have an average annual rainfall of 39 inches [1,000 mm] or more.

VEGETATION

The Slovak Republic has large areas of forest. Evergreen trees, including fir and spruce, are found in mountainous areas. Beech, birch, linden, and oak are more characteristic of lowland areas.

HISTORY

Slavic people, who are also called Slavs, settled in what is now the Slovak Republic in the 5th and 6th centuries AD. Hungarians conquered the area in 907 and ruled it for nearly 1,000 years. The Hungarians tried to suppress Slovakian culture and stifle Slovak development. From the late 18th century, Slovaks began to conserve their culture.

In 1867, Hungary and Austria united to form Austria-Hungary, of which the present-day Slovak Republic was a part. Austria-Hungary collapsed at the end of World War I (1914–18).

The Czech and Slovak people then united to form a new nation,

Czechoslovakia. The Czechs dominated in this union, and many Slovaks became dissatisfied. In 1938, Hungary forced Czechoslovakia to give up some areas with large Hungarian populations. These areas included Kosice in the east.

In 1939, Slovakia declared itself independent, but Germany occupied the entire country. At the end of the war, Slovakia again became part of Czechoslovakia. The Communist party took over the government of Czechoslovakia in 1948. In the 1960s, many Slovaks and Czechs tried to reform the Communist system, but the Russians crushed the reformers. In the late 1980s, changes in the policies of the Soviet Union led the people of Eastern Europe to demand democracy.

POLITICS

Elections in 1992 led to a victory for the Movement for a Democratic Slovakia, led by a former Communist and staunch nationalist, Vladimir Meciar. In September 1992, the Slovak National Council voted to create a separate, independent Slovak Republic on January 1, 1993.

Following the country's independence, the Slovaks kept close relations with the Czech Republic. In 2002, both republics were invited to become members of the European Union in May 2004.

DID YOU KNOW

- that Bratislava, capital of the Slovak Republic, is one of four European capitals on the River Danube – the others are Vienna, Budapest, and Belgrade
- that the Tatra (or Tatry) Mountains, which contain the country's highest peak, form a major protected tourist area
- that the Slovensky Raj, meaning "Slovak Paradise," is a region of great beauty in the eastern part of the Slovak Republic. It contains wild limestone and dolomite landscapes

ECONOMY

Before 1948, Slovakia's economy was based on farming, but Communist governments developed manufacturing industries, producing such things as chemicals, machinery, steel, and weapons. Since the late 1980s, the non-Communist governments have tried to hand over state-run businesses to private ownership.

Farming employs about 9% of the people. Major crops include barley, grapes for wine-making, maize, sugar beet, and wheat. Cattle, pigs, and sheep are raised.

Manufacturing employs about 29% of workers. Bratislava and Kosice are the chief industrial cities. The armaments industry is based in Martin, in the northwest. Products include ceramics, machinery, and steel.

SLOVENIA

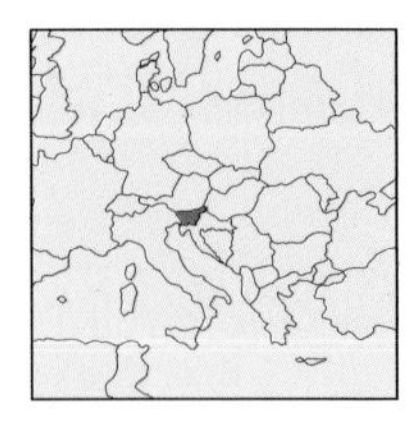

Slovenia's flag, which was based on the flag of Russia, was originally adopted in 1848. Under Communist rule, a red star appeared at the center. This flag, which was adopted in 1991 when Slovenia proclaimed its independence, has a new emblem, the national coat of arms.

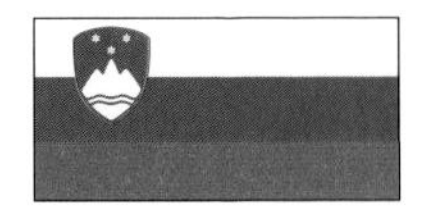

The Republic of Slovenia was one of the six republics which made up the former Yugoslavia. Much of the land is mountainous, rising to 9,396 ft [2,863 m] at Mount Triglav in the Julian Alps (Julijske Alpe) in the northwest. Central and eastern areas contain hills and plains drained by the Drava and Sava rivers. Central Slovenia contains a limestone area, called the Karst region. Here surface water flows downward into deep caves. The Postojna caves near Ljubljana are among the largest in Europe.

AREA 7,817 sq miles [20,251 sq km]
POPULATION 1,930,000
CAPITAL (POPULATION) Ljubljana (280,000)
GOVERNMENT Multiparty republic
ETHNIC GROUPS Slovene 88%, Croat 3%, Serb 2%, Bosnian 1%
LANGUAGES Slovene (official), Serbo-Croat
RELIGIONS Mainly Roman Catholic
CURRENCY Tolar = 100 stotin

CLIMATE

The short coast of Slovenia has a mild Mediterranean climate. Inland, the climate is more continental. The mountains are snow-capped in winter. Eastern Slovenia has cold winters and hot summers. Rain occurs in every month in Ljubljana. Late summer is the rainiest season.

VEGETATION

Forests cover about half of Slovenia. Mountain pines grow on higher slopes, with beech, oak, and hornbeam at lower levels. The Karst region is largely bare of vegetation because of the lack of surface water. Farmland covers about a third of Slovenia.

HISTORY

The ancestors of the Slovenes, the western branch of a group of people called the South Slavs, settled in the area around 1,400 years ago. From the 13th century until 1918, Slovenia was ruled for most of the time by the Austrian Habsburgs. In 1918, Slovenia became part of the Kingdom of the Serbs, Croats, and Slovenes. This country was renamed Yugoslavia in 1929. During World War II (1939–45), Slovenia was invaded and partitioned between Italy, Germany, and Hungary, but after the war, Slovenia again became part of Yugoslavia.

From the late 1960s, some

Slovenes demanded independence, but the central government opposed the breakup of the country. In 1990, when Communist governments had collapsed throughout Eastern Europe, elections were held and a non-Communist coalition government was set up. Slovenia then declared itself independent. This led to fighting between Slovenes and the federal army, but Slovenia did not become a battlefield like other parts of the former Yugoslavia. The European Community recognized Slovenia's independence in 1992. Economic reforms in the 1990s led the European Union in 2002 to invite Slovenia to become a member in May 2004.

DID YOU KNOW

- that there are over 40 distinct dialects in the Slovene language
- that, in the last 2,000 years, the Slovene people have been independent as a nation for less than 50 years; before 1992, the only period of independence was between AD 623 and 658
- that Slovene minorities are found in the southern province of Carinthia (Kärnten) in Austria, and also in the Italian region of Trieste

ECONOMY

The reform of the economy, formerly run by the government, and the fighting in areas to the south have caused problems for Slovenia. In 1992, the World Bank classified Slovenia as an "upper-middle-income" developing country.

Manufacturing is the leading activity and manufactures are the principal exports. Manufactures include chemicals, machinery and transport equipment, metal goods, and textiles. Agriculture employs 10% of the people. Fruits, maize, potatoes, and wheat are major crops, while many farmers raise cattle, pigs, and sheep.

The stamp on the left, showing Slovenia's parliament building, was issued in 1991 to mark the country's independence. The stamp on the right shows the national coat of arms, which appears on Slovenia's flag. It includes an outline of Mount Triglav, the highest peak. The wavy lines beneath the mountain represent the Sava and Drava rivers. Slovenija *is the name of Slovenia in the Slovene language.*

SOMALIA

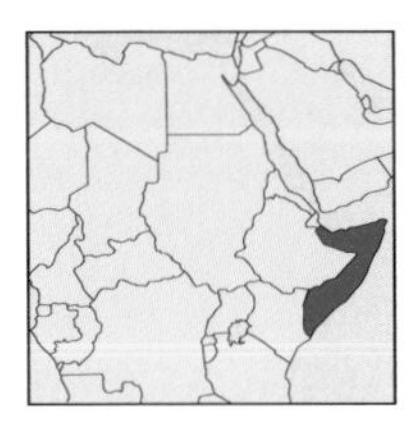

This flag was adopted in 1960, when Italian Somaliland in the south united with British Somaliland in the north to form Somalia. The colors are based on the United Nations flag, and the points of the star represent the five regions of East Africa where Somalis live.

The Somali Democratic Republic, or Somalia, is in a region known as the "Horn of Africa." It is more than twice the size of Italy, the country which once ruled the southern part of Somalia. The most mountainous part of the country is in the north, behind the narrow coastal plains that border the Gulf of Aden.

The south, which consists of low plateaux and plains, is the only area with permanent rivers. These are the Scebeli and the Juba, two rivers which rise in neighboring Ethiopia.

AREA 246,201 sq miles [637,660 sq km]

POPULATION 7,489,000

CAPITAL (POPULATION) Mogadishu (or Muqdisho, 997,000)

GOVERNMENT Single-party republic, military dominated

ETHNIC GROUPS Somali 85%, Bantu and others (including Arabs) 15%

LANGUAGES Somali and Arabic (both official), English, Italian

RELIGIONS Islam 99%

CURRENCY Somali shilling = 100 cents

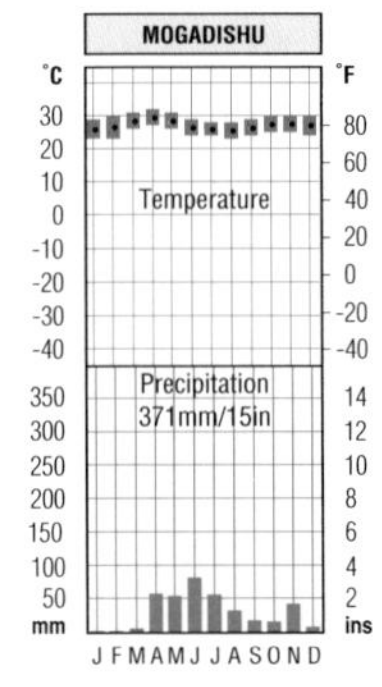

CLIMATE

Rainfall is light throughout Somalia. The wettest regions are the south and the northern mountains, but droughts often occur. Temperatures are high on the low plateaux and plains, but frosts sometimes occur on the highest mountain slopes in the north.

VEGETATION

Much of Somalia is dry grassland or semidesert, but there are also areas of wooded grassland with such trees as acacias and baobabs. Plants are most abundant in the south, especially in the lower Juba valley, which is also rich in wildlife.

HISTORY

European powers became interested in the Horn of Africa in the 19th century. In 1884, Britain made the northern part of what is now Somalia a protectorate, while Italy took the south in 1905. The new boundaries divided the Somalis into five areas: the two Somalilands, Djibouti (which was taken by France in the 1880s), Ethiopia, and Kenya. Since then, many Somalis have longed for the reunification of their people in a Greater Somalia.

Italy entered World War II in 1940 and invaded British Somaliland. But British forces conquered the region in 1941 and ruled both Somalilands until

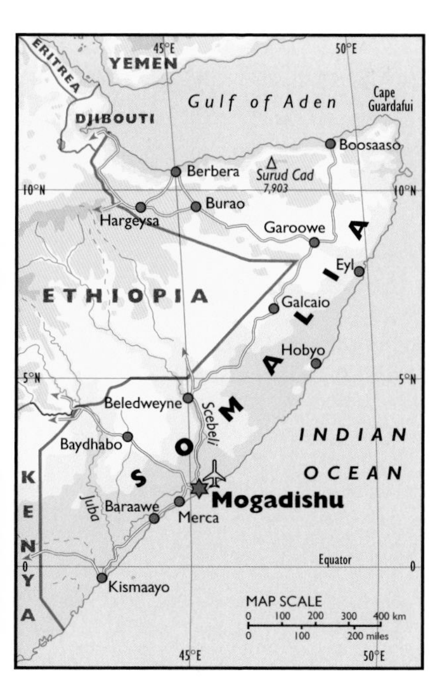

1950, when the United Nations asked Italy to take over the former Italian Somaliland for ten years. In 1960, both Somalilands became independent and united to become Somalia.

POLITICS

Somalia has faced many problems since independence. Economic problems led a military group to seize power in 1969. In the 1970s, Somalia supported an uprising of Somali-speaking people in the Ogaden region of Ethiopia. But Ethiopian forces prevailed and, in 1988, Somalia signed a peace treaty with Ethiopia. The cost of the fighting weakened Somalia's economy.

Further problems occurred when people in the north fought to secede from Somalia. In 1991, they set up the "Somaliland Republic," with its capital at Hargeisa. But the new state was recognized neither internationally nor by Somalia's government. Fighting continued and US troops sent by the UN in 1993 withdrew in 1994. By the early 21st century, Somalia had no effective national government, being divided into three main parts – the north, the northeast, and the south.

ECONOMY

Somalia is a developing country, whose economy has been shattered by drought and war. Many Somalis are nomads who raise livestock. Live animals, meat, and hides and skins are major exports, followed by bananas grown in the wetter south. Other crops include citrus fruits, cotton, maize, and sugarcane. Mining and manufacturing are relatively unimportant in the economy.

This stamp, issued in 1955 and showing the plant Adenium somalense, *was one of a set of nine floral designs. Plant species abound along the country's rivers and also in the far south.*

SOUTH AFRICA

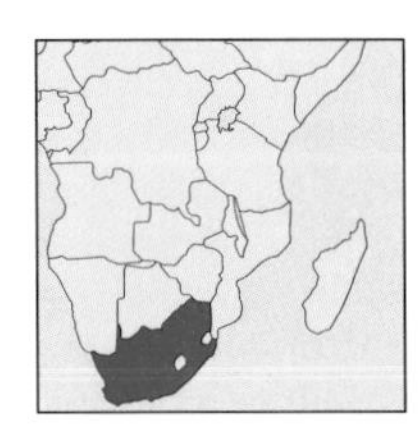

South Africa's flag was first flown in 1994 when the country adopted a new, non-racial constitution. It incorporates the red, white, and blue of former colonial powers, Britain and the Netherlands, together with the green, black, and gold of black organizations.

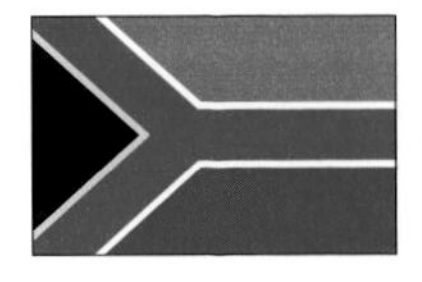

The Republic of South Africa is made up largely of the southern part of the huge plateau which makes up most of southern Africa. The highest peaks are in the Drakensberg range, which is formed by the uptilted rim of the plateau. In the southwest lie the folded Cape Mountain ranges. The coastal plains are mostly narrow. The Namib Desert is in the northwest.

CLIMATE

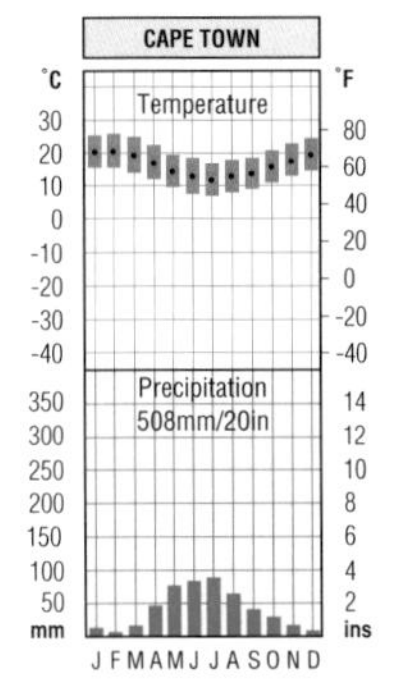

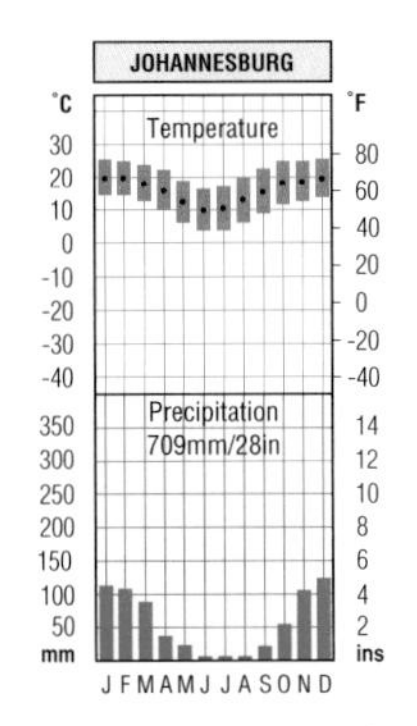

Most of South Africa has a mild, sunny climate. Much of the coastal strip, including the city of Cape Town, has warm, dry summers and mild, rainy winters, just like the Mediterranean lands in northern Africa. Inland, large areas are arid.

VEGETATION

Grassland or savanna covers much of South Africa. Forest and woodland cover only 3% of the land.

AREA 470,566 sq miles [1,219,916 sq km]

POPULATION 43,586,000

CAPITAL (POPULATION) Cape Town (legislative, 2,350,000); Pretoria (administrative, 1,080,000); Bloemfontein (judiciary, 300,000)

GOVERNMENT Multiparty republic

ETHNIC GROUPS Black 76%, White 13%, Colored 9%, Asian 2%

LANGUAGES Afrikaans, English, Ndebele, North Sotho, South Sotho, Swazi, Tsonga, Tswana, Venda, Xhosa and Zulu (all official)

RELIGIONS Christianity 68%, Islam 2%, Hinduism 1%

CURRENCY Rand = 100 cents

HISTORY AND POLITICS

Early inhabitants in South Africa were the Khoisan. In the last 2,000 years, Bantu-speaking people moved into the area. Their descendants include the Zulu, Xhosa, Sotho, and Tswana. The Dutch founded a settlement at the Cape in 1652, but Britain took over in the early 19th century, making the area a colony. The Dutch, called Boers or Afrikaners, resented British rule and moved inland. Rivalry between the groups led to Anglo-Boer Wars in 1880–1 and 1899–1902.

In 1910, the country was united as the Union of South Africa. In 1948, the National Party won power, and introduced a policy known as *apartheid*, under which non-whites had no votes and their human rights were strictly limited. In 1990, Nelson Mandela, leader of the African National Congress (ANC), was released from prison. Multiracial elections were held in 1994, and Mandela became president. After his retirement in 1999, his successor, Thabo Mbeki, led the ANC to victory in the elections.

ECONOMY

South Africa is Africa's most developed country. But most of the black people are poor, with low standards of living. Natural resources include diamonds, gold, and many other metals. Mining and manufacturing are the most valuable activities. Products include chemicals, iron and steel, metal goods, processed food, and vehicles. Major crops include fruits, maize, potatoes, sugarcane, tobacco, and wheat. Livestock products are also important.

SPAIN

The colors on the Spanish flag date back to those used by the old kingdom of Aragon in the 12th century. The present design, in which the central yellow stripe is twice as wide as each of the red stripes, was adopted in 1938, during the Spanish Civil War.

The Kingdom of Spain is the second largest country in Western Europe after France. It shares the Iberian peninsula with Portugal. A large plateau, called the Meseta, covers most of Spain. Much of the Meseta is flat, but it is crossed by several mountain ranges, called *sierras*.

The northern highlands include the Cantabrian Mountains (Cordillera Cantabrica) and the high Pyrenees, which form Spain's border with France. But the highest peak on the Spanish mainland lies in the Sierra Nevada in the southeast.

Spain is bounded by fertile coastal plains. Other important lowlands are the Ebro river basin in the northeast and the Guadalquivir river basin in the southwest.

Spain also includes the Balearic Islands (Islas Baleares) in the Mediterranean Sea, and the **Canary Islands**, which lie off the northwest coast of Africa (*see Atlantic Ocean, page 20*).

AREA 194,896 sq miles [504,780 sq km]
POPULATION 38,432,000
CAPITAL (POPULATION) Madrid (3,030,000)
GOVERNMENT Constitutional monarchy
ETHNIC GROUPS Castilian Spanish 72%, Catalan 16%, Galician 8%, Basque 2%
LANGUAGES Castilian Spanish (official), Catalan, Galician, Basque
RELIGIONS Roman Catholic 99%
CURRENCY Euro = 100 cents

CLIMATE

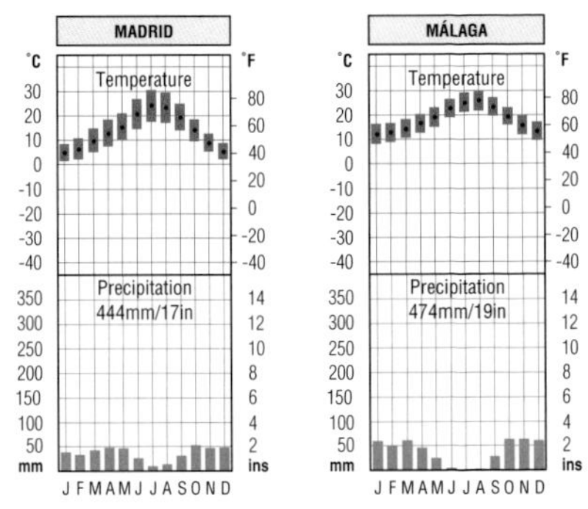

Madrid, in the heart of Spain, has hot summers, but winters are cold and temperatures may fall below freezing point. Snow often covers the higher mountain ranges on the Meseta.

The coastal regions also have hot, dry summers, but the winters are mild. Most of the rain, which averages 19 inches [474 mm] a year in the port of Málaga, is usually confined to the winter months. Málaga has a typical Mediterranean climate and is the center of a major tourist region.

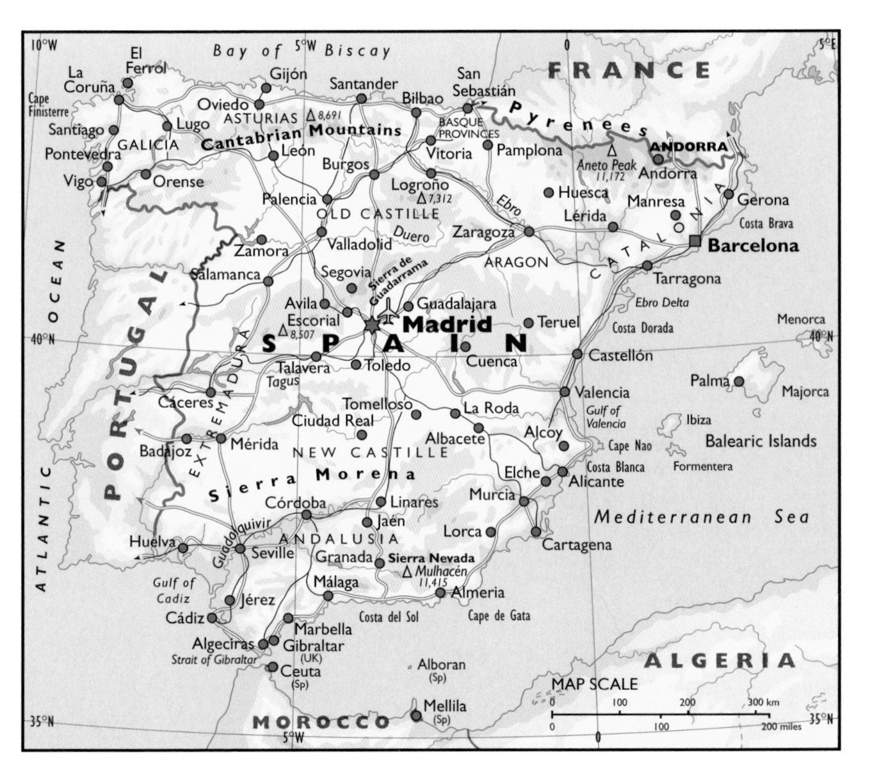

VEGETATION

Much of the original forest in Spain has been cut down. Forests still grow on mountain slopes, with more open woodland in other areas. Grassland and scrub, called *matorral*, cover large areas. The *matorral* resembles the *maquis* in France.

HISTORY

The first inhabitants of Spain, the Iberians, began to trade with the Phoenicians from the eastern Mediterranean Sea around 3,000 years ago. In about 600 BC, Greek merchants set up trading posts in Spain. The Carthaginians conquered Spain in the 5th century BC, but they were later defeated by the Romans. Spain became a Roman province and many Roman ruins can still be seen.

From about AD 400, Germanic tribes swept into Spain. But from 711, Moors from North Africa conquered Spain. They built great mosques and palaces, some of which stand today. From the 11th century, Christian Spaniards began a long struggle against the Moors. The last Moorish fortress, at Granada in southern Spain, fell in 1492. The dominant kingdom in Spain in the 15th century was that of Castile. After Princess Isabella of Castile married Prince

SPAIN

The Tower of San Pedro, in Teruel in eastern Spain, has been designated a World Heritage Site. The stamp was issued in 1990.

Ferdinand of Aragon in 1469, most of Spain was united under their rule.

In the 16th century, Spain became a world power. At its peak, it controlled much of Central and South America, parts of Africa, and the Philippines in Asia. Spain began to decline in the late 16th century. Its sea power was destroyed by a British fleet under Lord Nelson in the Battle of Trafalgar (1805). By the 20th century, it was a poor country.

Spain became a republic in 1931, but the republicans were defeated in the Spanish Civil War (1936–9). General Francisco Franco (1892–1975) became the country's dictator, though, technically, it was a monarchy. When Franco died, the monarchy was restored. Prince Juan Carlos became king.

POLITICS

Spain has several groups with their own languages and cultures. Some of these people want to run their own regional affairs. In the Basque region, in the north, the Basque ETA movement has waged a terrorist campaign for independence. A truce in 1998 was ended in 1999 when talks failed to produce results.

Since the late 1970s, regional parliaments have been set up in the Pais Vasco (Basque Country), in Catalonia in the northeast, and Galicia in the northwest. All these regions have their own languages.

ECONOMY

The revival of Spain's economy, which was shattered by the Civil War, began in the 1950s and 1960s, especially through the growth of tourism and manufacturing. Since the 1950s, Spain has changed from a poor country, dependent on agriculture, to a fairly prosperous industrial nation.

By 2000, agriculture employed 7% of the people, as compared with manufacturing and mining 18%, and services, including tourism, 75%. Farmland, including pasture, makes up about two-thirds of the land, with forest making up most of the rest. Major crops include barley, citrus fruits, grapes for wine-making, olives, potatoes, and wheat. Sheep are the most important farm animals.

Spain has some high-grade iron ore in the north, though otherwise it lacks natural resources. But it has many manufacturing industries. Manufactures include cars, chemicals, clothing, electronics, processed food, metal goods, steel, and textiles.

DID YOU KNOW

- that *gazpacho* is a Spanish soup which is served cold
- that Spain's highest peak at 12,202 ft [3,718 m], called the Teide, is on Tenerife, in the Canary Islands
- that the famous flamenco dance was invented by Andalusian gypsies
- that Spain ranks fifth among the world's producers of cars

ANDORRA

Andorra is a tiny state sandwiched between France and Spain. It lies in the high Pyrenees Mountains. Most Andorrans live in the sheltered valleys.

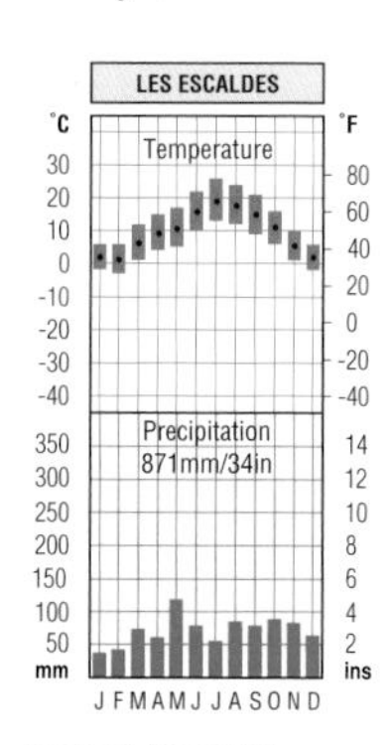

CLIMATE

Andorra has cold and fairly dry winters. The summers are a little rainier, but they are pleasantly cool. The climate graph is for Les Escaldes, a town east of Andorra's capital city.

ECONOMY

Tourism is Andorra's chief activity in both winter, for winter sports, and summer. There is some farming in the valleys and tobacco is the main crop. Cattle and sheep are grazed on the mountain slopes.

AREA 175 sq miles [453 sq km]
POPULATION 68,000
CAPITAL (POPULATION) Andorra La Vella (22,000)
GOVERNMENT Co-principality
ETHNIC GROUPS Spanish 43%, Andorran 33%, Portuguese 11%, French 7%
LANGUAGES Catalan (official)
RELIGIONS Mainly Roman Catholic
CURRENCY Euro = 100 cents

GIBRALTAR

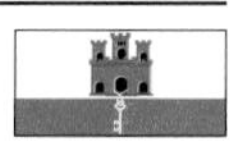

Gibraltar is a tiny British dependency on the south coast of Spain. It occupies a strategic position, overlooking the narrow Strait of Gibraltar, which links the Mediterranean Sea to the Atlantic Ocean. Most people work for the government or in the tourist trade.

Most of the land consists of a huge mass of limestone, known as the Rock of Gibraltar. Between AD 711 and 1309, and again between 1333 and 1462, Gibraltar was held by Moors from North Africa. Spaniards retook the area in 1462, but it became a British territory in 1713. Gibraltar became a vital British military base, but Spain continued to claim the area.

In 1967, the Gibraltarians voted to remain British. Between 1969 and 1985, Spain closed its border with Gibraltar. In 1991, Britain withdrew its military forces. In 2002, proposals that Gibraltar should come under joint Anglo-Spanish sovereignty were rejected by nearly all Gibraltarians.

AREA 2.5 sq miles [6.5 sq km]
POPULATION 28,000
CAPITAL (POPULATION) Gibraltar Town (28,000)
GOVERNMENT British dependency
ETHNIC GROUPS English, Spanish, Maltese, Italian, Portuguese
LANGUAGES English (official), Spanish, Italian, Portuguese
RELIGIONS Mainly Roman Catholic
CURRENCY Gibraltar pound

SRI LANKA

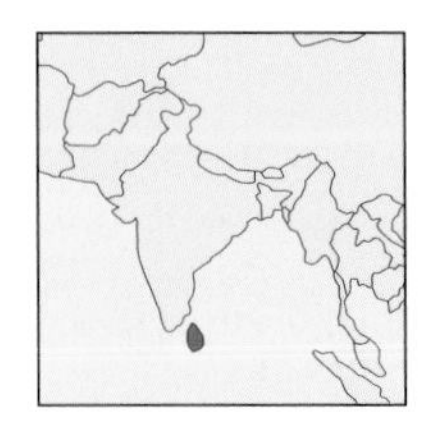

Sri Lanka's unusual flag was adopted in 1951, three years after the country, then called Ceylon, became independent from Britain. The lion banner represents the ancient Buddhist kingdom. The stripes symbolize the minorities – Muslims (green) and Hindus (orange).

The Democratic Socialist Republic of Sri Lanka is an island nation. It is separated from the southeast coast of India by the Palk Strait. The land is mostly low-lying, but a mountain region dominates the south-central part of the country. The highest peak is Pidurutalagala, at 8,284 ft [2,524 m] above sea level. The nearby Adam's Peak, at 7,359 ft [2,243 m], is a center of pilgrimage.

The coastline is varied. Cliffs overlook the sea in the southwest, while lagoons line the coast in many other areas.

AREA 25,332 sq miles [65,610 sq km]
POPULATION 19,409,000
CAPITAL (POPULATION) Colombo (1,863,000)
GOVERNMENT Multiparty republic
ETHNIC GROUPS Sinhalese 74%, Tamil 18%, Sri Lankan Moor 7%
LANGUAGES Sinhala and Tamil (both official)
RELIGIONS Buddhism 69%, Hinduism 16%, Christianity 8%, Islam 7%
CURRENCY Sri Lankan rupee = 100 cents

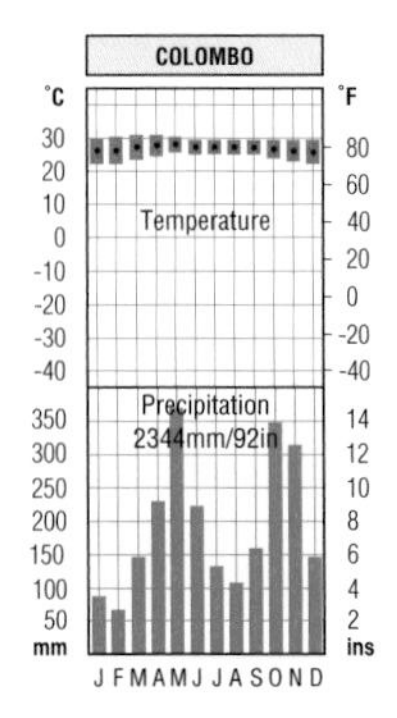

CLIMATE

The western part of Sri Lanka has a wet equatorial climate. Temperatures are high and the rainfall is heavy. The wettest months are May and October, marking the advance and the retreat of the summer monsoon. Eastern Sri Lanka is drier than the west.

VEGETATION

Forests cover nearly two-fifths of the land in Sri Lanka, with open grasslands in the eastern highlands. Farmland, including pasture, covers another two-fifths of the country.

HISTORY

The ancestors of the Sinhalese people probably came from northern India and settled on the island around 2,400 years ago. They pushed the Veddahs, descendants of the earliest inhabitants, into the interior. The Sinhalese founded the city of Anuradhapura, which was their center from the 3rd century BC to the 10th century AD.

Tamils arrived around 2,100 years ago and the early history of Ceylon, as the island was known, was concerned with a struggle between the Sinhalese and the Tamils. Victory for the Tamils led the Sinhalese to move south.

Portugal ruled the island from 1505 to 1655, when the Dutch took

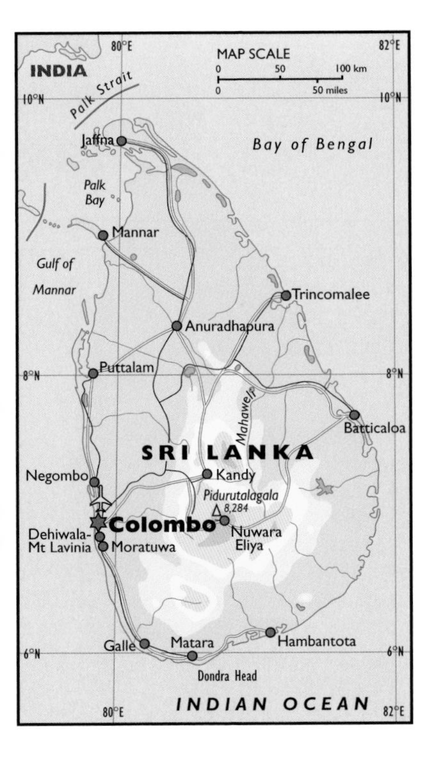

over. Britain captured the Dutch colonies in the 1790s and made Ceylon a colony in the early 19th century. Ceylon became independent in 1948. In 1972, it adopted the name Sri Lanka ("resplendent island").

POLITICS

After independence, rivalries between the Tamils and the Sinhalese marred progress. In the 1950s, the government made Sinhala the official language. Following protests, the prime minister made provisions for Tamil to be used in some areas. In 1959, he was assassinated by a Sinhalese extremist. He was succeeded in 1960 by his wife, Sirimavo Bandanaraike, who was the world's first woman prime minister.

Conflict between Tamils and Sinhalese continued in the 1970s and 1980s. In 1987, India helped to engineer a ceasefire. Indian troops arrived to enforce the agreement, but withdrew in 1990 after failing to subdue the Tamil Tigers, who wanted to set up an independent Tamil homeland in northern Sri Lanka. Conflict continued until 2002, when peace talks led to an agreement that autonomy would be granted to a Tamil homeland, which would remain part of Sri Lanka, within a federal system.

ECONOMY

The World Bank classifies Sri Lanka as a "low-income" developing country. Agriculture employs half of the work force and coconuts, rubber, and tea are exported. Rice is the chief food crop. Textiles and clothing, petroleum products, and jewelry are exported.

Like all thickly populated countries in Asia, Sri Lanka faces problems in conserving its superb plant and animal life. Forest conservation was the theme of two stamps issued in 1987. One of them, shown here, depicted Sri Lanka's national tree, Mesua nagassarium.

SUDAN

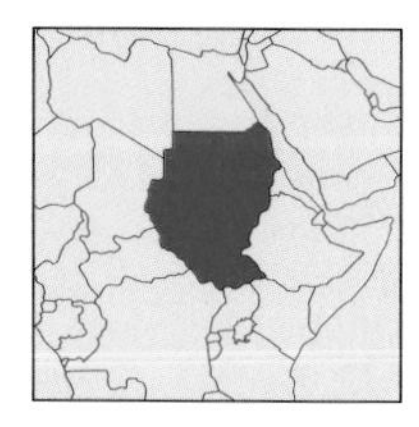

Adopted in 1969, Sudan's flag uses colors associated with the Pan-Arab movement. The Islamic green triangle symbolizes prosperity and spiritual wealth. The flag is based on the one used in the Arab revolt against Turkish rule in World War I (1914–18).

The Republic of Sudan is the largest country in Africa. From north to south, it spans a vast area extending from the arid Sahara in the north to the wet equatorial region in the south. The land is mostly flat, with the highest mountains in the far south. The main physical feature is the River Nile. In the south, where it is called the Bahr el Jebel, the river crosses a swampy region called the Sudd. North of the Sudd, the river is called the White Nile. It meets the Blue Nile, which rises in Ethiopia, at Khartoum.

AREA 967,493 sq miles [2,505,810 sq km]
POPULATION 36,080,000
CAPITAL (POPULATION) Khartoum (925,000)
GOVERNMENT Military regime
ETHNIC GROUPS Sudanese Arab 49%, Dinka 12%, Nuba 8%, Beja 6%, Nuer 5%, Azande 3%
LANGUAGES Arabic (official), Nubian, Dinka
RELIGIONS Islam (mainly Sunni) 70%, traditional beliefs 25%, Christianity 5%
CURRENCY Dinar = 10 Sudanese pounds

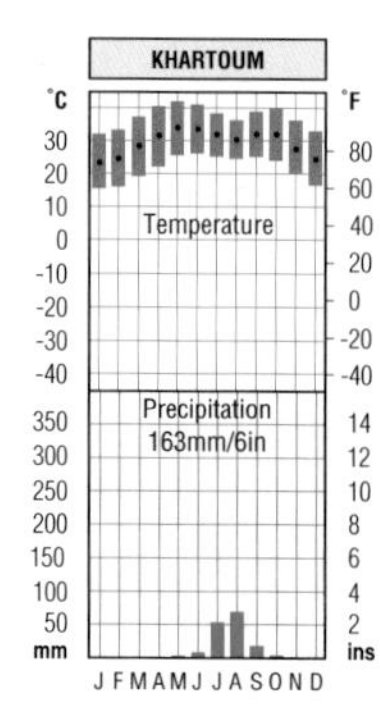

CLIMATE

The climate of Khartoum represents a transition between the virtually rainless northern deserts and the equatorial lands in the south. Some rain falls in Khartoum in summer. Summer in Khartoum is also a time of high temperatures, when dust storms, called *haboobs*, occur.

VEGETATION

From the bare deserts of the north, the land merges into dry grasslands and savanna. In the rainy south, where the average annual rainfall reaches 39 inches [1,000 mm] or more, dense rain forests grow, especially in waterlogged areas.

HISTORY

One of the earliest civilizations in the Nile region of northern Sudan was Nubia, which came under Ancient Egypt around 4,000 years ago. Another Nubian civilization, called Kush, developed from about 1,000 BC, finally collapsing in AD 350. Christianity was introduced to northern Sudan in the 6th century, but Islam later became the dominant religion. In the 19th century, Egypt gradually took over Sudan. In 1881, a Muslim religious teacher, the *Mahdi*

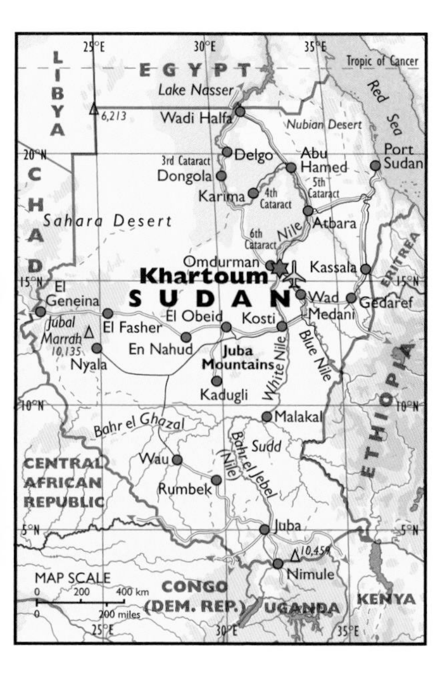

("divinely appointed guide"), led an uprising. Britain and Egypt put the rebellion down in 1898. In 1899, they agreed to rule Sudan jointly as a condominium. Sudan became independent in 1952.

POLITICS

After independence, the black Africans in the south, who were either Christians or followers of traditional beliefs, feared domination by the Muslim northerners. For example, they objected to the government declaring that Arabic was the only official language. In 1964, civil war broke out and continued until 1972, when the south was given regional self-government, though executive power was still vested in the military government in Khartoum.

In 1983, the government established Islamic law throughout the country. This sparked off further conflict when the Sudan People's Liberation Army in the south launched attacks on government installations. Despite attempts to restore order, the fighting continued into the 1990s. In 2002, the government and the rebels reached agreement on the principles of self-determination for the south, though many problems remained.

ECONOMY

The World Bank classifies Sudan as a "low-income" economy. Agriculture employs 62% of the people. The leading crop is cotton. Other crops include groundnuts, gum arabic, millet, sesame, sorghum, and sugarcane, while many people raise livestock.

Minerals include chromium, gold, gypsum, and oil. Manufacturing industries process foods and produce such things as cement, fertilizers, and textiles. The main exports are cotton, gum arabic, and sesame seeds.

This stamp was issued in 1948 to commemorate the opening of Sudan's Legislative Assembly (parliament). Depicted in the center is an Arab postman riding across the northern desert on a camel.

SURINAME

Suriname's flag was adopted in 1975, when the country became independent from the Netherlands. The flag features the colors of the political parties. The yellow star symbolizes unity and a golden future. The red is twice the width of the green and four times that of the white.

The Republic of Suriname, which was formerly known as Dutch Guiana, is sandwiched between French Guiana and Guyana in northeastern South America. The narrow coastal plain was once swampy, but it has been drained and now consists mainly of farmland. Inland lie hills and low mountains, which rise to 4,201 ft [1,280 m] at Julianatop, a peak in the Wilhelmina Mountains in the southwest. Many rivers flow from the interior to the Atlantic Ocean. Their courses are interrupted by rapids and waterfalls.

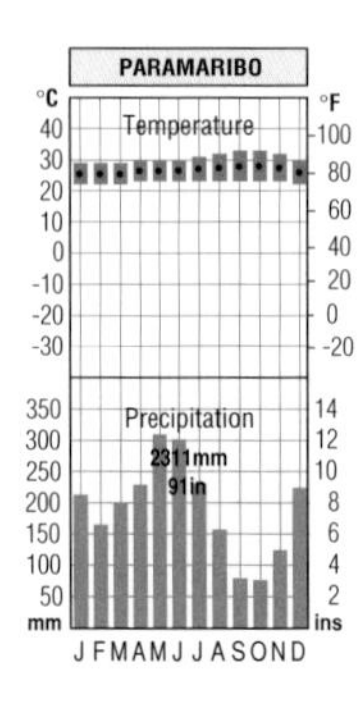

CLIMATE

Suriname has a hot, wet and humid climate. Temperatures are high throughout the year and the rainfall is heavy, although there is a comparatively dry season between the months of August and November. Inland, the height of the land moderates temperatures.

VEGETATION

Mangrove forests line parts of the coast, with rain forest and highland forest in the interior. Forests cover more than 90% of Suriname. The forests contain many valuable species, including balata and greenheart.

AREA 63,039 sq miles [163,270 sq km]

POPULATION 434,000

CAPITAL (POPULATION) Paramaribo (201,000)

GOVERNMENT Multiparty republic

ETHNIC GROUPS Asian Indian 37%, Creole (mixed White and Black), 31%, Indonesian 14%, Black 9%, Amerindian 3%, Chinese 3%, Dutch 1%

LANGUAGES Dutch (official), Sranantonga

RELIGIONS Hinduism 27%, Roman Catholic 23%, Islam 20%, Protestant 19%

CURRENCY Suriname guilder = 100 cents

HISTORY

Christopher Columbus sighted the coast of Suriname in 1498, but the British established the first settlement in 1651. The British brought in black African slaves to work their plantations, producing cocoa, coffee, cotton, and sugar. In 1667, the British handed the territory to the Dutch in return for New Amsterdam, an area that is now the state of New York.

In the 18th century, slave revolts and Dutch neglect slowed down the area's development. In the early 19th century, Britain and the Netherlands disputed the ownership of the area. The British gave up their claim in 1815. Slavery was abolished in 1863. Soon afterward, laborers were brought to the area from India and Indonesia to work on the plantations.

Suriname became fully independent on November 25, 1975. But the country's economy was weakened when thousands of skilled people emigrated from Suriname to the Netherlands.

POLITICS

In 1980, a military group seized power and abolished the parliament. A National Military Council was set up to rule Suriname. Elections were held in 1987 and the country returned to democratic rule in 1988. Another military coup occurred in 1990, but further elections were held in 1991. Ronald Venetiaan, leader of a coalition group called the New Front, became president.

> **DID YOU KNOW**
> - that Suriname's official name in Dutch is Republiek Suriname
> - that the most commonly used language in Suriname is called Sranantongo, or Taki-Taki; it combines words from Dutch, English and several African languages
> - that Suriname is the world's seventh largest producer of bauxite (aluminum ore)
> - that the so-called "Bush Negroes," or *boschneger,* in Suriname are descendants of black Africans who escaped from slavery; most of them live in the forests and follow African customs

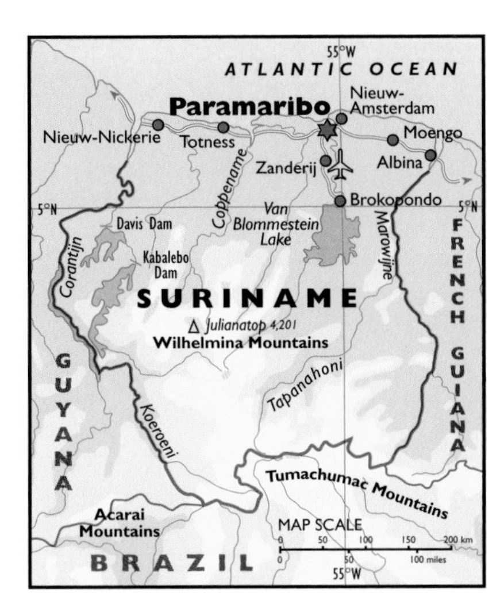

In 1992, the government negotiated a peace agreement with the *boschneger*, descendants of African slaves, who had launched a struggle against the government in 1986. This rebellion had disrupted the area where bauxite, the main export, was mined. But instability continued, especially among the military. But stability was restored and elections were held in 1996. In 2000, Venetiaan was re-elected president, and the New Front won a majority in parliament.

ECONOMY

The World Bank classifies Suriname as an "upper-middle-income" developing country. Its economy is based on mining and metal processing. Suriname is a leading producer of bauxite, from which the metal aluminum is made. Bauxite and aluminum make up more than 80% of the exports. Other exports include shrimps, rice, bananas, and timber, including logs and plywood.

SWAZILAND

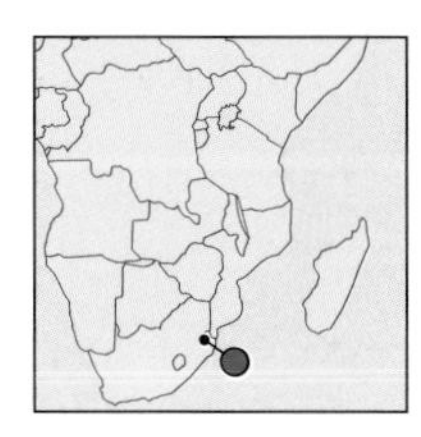

Swaziland has used this distinctive flag since it became independent from Britain in 1968. It is based on the flag of the Swazi Pioneer Corps of World War II. The emblem shows a warrior's weapons – an ox-hide shield, two assegais (spears), and a fighting stick.

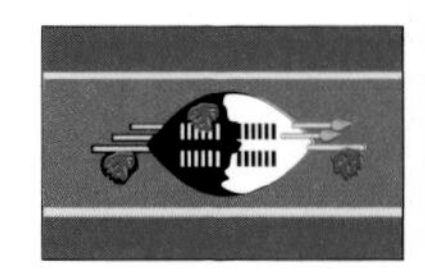

The Kingdom of Swaziland is a small, landlocked country in southern Africa. The country has four regions which run north–south. In the west, the Highveld, with an average height of 3,950 ft [1,200 m], makes up 30% of Swaziland. The Middleveld, between 1,150 ft and 3,280 ft [350 m to 1,000 m], covers 28% of the country. The Lowveld, with an average height of 886 ft [270 m], covers another 33%. The Lebombo Mountains, the fourth region, reach 2,600 ft [800 m] along the eastern border.

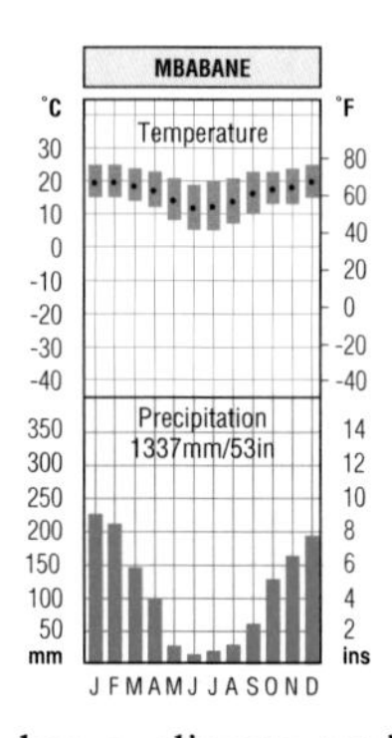

CLIMATE

The Lowveld is almost tropical, with an average temperature of 72°F [22°C] and a low rainfall of about 20 inches [500 mm] a year. The altitude moderates the climate in the west of the country. Mbabane has a climate typical of the Highveld with warm summers and cool winters.

VEGETATION

Meadows and pasture cover about two-thirds of Swaziland. Arable land covers 8% of the land and forests only 6%.

AREA 6,703 sq miles [17,360 sq km]

POPULATION 1,104,000

CAPITAL (POPULATION) Mbabane (42,000)

GOVERNMENT Monarchy

ETHNIC GROUPS Swazi 84%, Zulu 10%, Tsonga 2%

LANGUAGES Siswati and English (both official)

RELIGIONS Zionist (a blend of Christianity and traditional beliefs) 40%, Roman Catholic 20%, Islam 10%

CURRENCY Lilangeni = 100 cents

HISTORY

According to tradition, in the 18th century, a group of Bantu-speaking people, under the Swazi chief Ngwane II, crossed the Lebombo range and united with local African groups to form the Swazi nation. Under attack from Zulu armies, the Swazi people were forced to seek British protection in the 1840s. Gold was discovered in the 1880s and many Europeans sought land concessions from the king, who did not realize that in doing this he was losing control of the land.

In 1894, Britain and the Boers of South Africa agreed to put Swaziland under the control of the South African Republic (the Transvaal). But

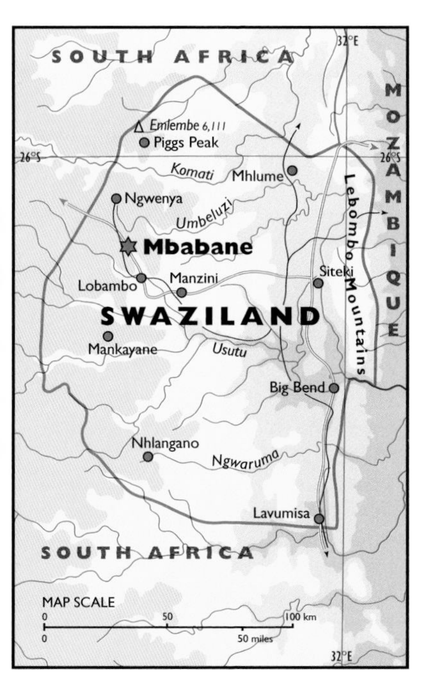

at the end of the Anglo-Boer War (1899–1902), Britain took control of the country. In 1968, when Swaziland became fully independent as a constitutional monarchy, the head of state was King Sobhuza II.

POLITICS

In 1973, Sobhuza suspended the constitution and took over supreme power in 1976. He banned political parties in 1978, though the people elected the 80 members of an electoral college, which chose representatives to serve in parliament.

Sobhuza died in 1982 after a reign of 82 years. In 1983, one of his sons, Prince Makhosetive (born 1968), was chosen as his heir. In 1986, he was installed as king, taking the name Mswati III. Elections were held in 1993 and 1998, but political parties were illegal.

ECONOMY

The World Bank classifies Swaziland as a "lower-middle-income" developing country, with 74% of the population living in rural areas in 1999. Many farmers live at subsistence level, producing little more than they need to feed their families. The leading exports are sugar, and wood and wood products.

Mining has declined in importance in recent years. Swaziland's high-grade iron ore reserves were used up in 1978, while the world demand for its asbestos has fallen. Swaziland is heavily dependent on South Africa and the two countries are linked through a customs union.

Swaziland is rich in flowering plants and ferns. A set of four stamps entitled "Flowers" was issued in 1971. It included the Bauhinia galpinii.

SWEDEN

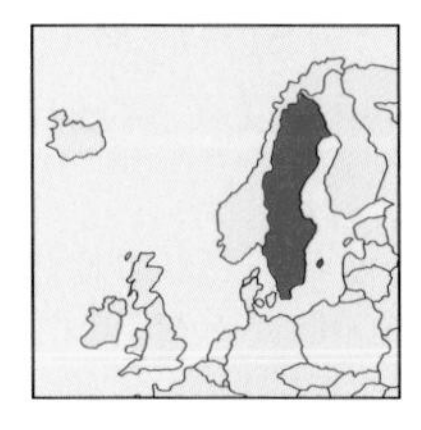

Sweden's flag was adopted in 1906, though it had been in use since the time of King Gustavus Vasa (reigned 1523–60). This king won many victories for Sweden and laid the foundations of the modern nation. The colors on the flag come from a coat of arms dating from 1364.

AREA 173,730 sq miles [449,960 sq km]

POPULATION 8,875,000

CAPITAL (POPULATION) Stockholm (727,000)

GOVERNMENT Constitutional monarchy

ETHNIC GROUPS Swedish 91%, Finnish 3%

LANGUAGES Swedish (official), Finnish

RELIGIONS Lutheran 89%, Roman Catholic 2%

CURRENCY Swedish krona = 100 öre

The Kingdom of Sweden is the largest of the countries of Scandinavia in both area and population. It shares the Scandinavian peninsula with Norway. The western part of the country, along the border with Norway, is mountainous. The highest point is Kebnekaise, which reaches 6,948 ft [2,117 m] in the northwest.

But Sweden is less mountainous than Norway and much of its land is hilly or flat. The lowlands of southern Sweden contain two of Europe's largest lakes, Vänern and Vättern.

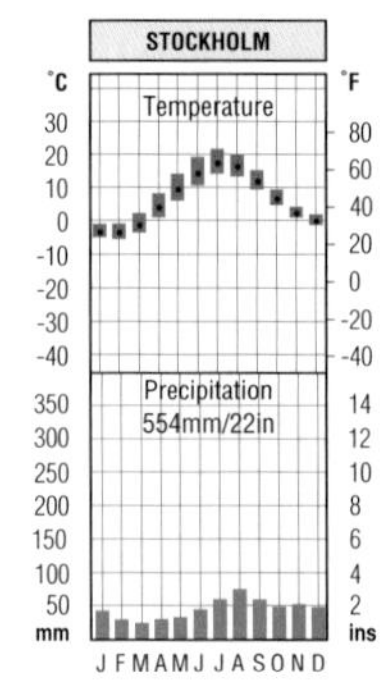

CLIMATE

The climate of Sweden becomes more severe from south to north. Stockholm has cold winters and cool summers. The average number of days with temperatures below freezing is 120, while, at the Arctic Circle, it is 180 or more. The far south of the country, warmed by the Gulf Stream, is much milder.

VEGETATION

Forest and woodland cover about 64% of the country. The south has mixed woodland with oak and beech. In the north, the trees are conifers, including birch, pine, and spruce.

HISTORY AND POLITICS

Swedish Vikings plundered areas to the south and east between the 9th and 11th centuries. Sweden, Denmark, and Norway were united in 1397, but Sweden regained its independence in 1523. In 1809, Sweden lost Finland to Russia, but, in 1814, it gained Norway from Denmark. The union between Sweden and Norway was dissolved in 1905. Sweden was neutral in World Wars I and II. Since 1945, Sweden has become a prosperous country. It was a founder member of the European Free Trade Association, but in 1994 the people voted to join the European Union on January 1, 1995.

Sweden has wide-ranging welfare services. But many people are concerned about the high cost of these services and the high taxes they must pay. In 1991, the Social Democrats, who had built up the welfare state,

DID YOU KNOW

- that Sweden has not fought in a war since 1814
- that the Swedish language resembles Danish and Norwegian and people from the three countries can usually understand each other
- that Alfred Nobel, the Swedish chemist who founded the Nobel prizes, was the inventor of dynamite
- that Swedes spend more on vacations than any other nationality in Europe
- that Sweden was the first country to appoint an *ombudsman*, an official who hears complaints from individuals against government officials and departments
- that Stockholm, which is built on 14 islands, is called "the Venice of the North"

were defeated. They were re-elected in 1994, 1998, and 2002, when they sought to control public spending and expand the economy.

ECONOMY

Sweden is a highly developed industrial country. Major products include steel and steel goods. Steel is used in the engineering industry to manufacture aircraft, cars, machinery, and ships. Sweden has some of the world's richest iron ore deposits. They are located near Kiruna in the far north. But most of this ore is exported, and Sweden imports most of the materials needed by its industries. Forestry is important, as also is fishing. Farmland covers 10% of Sweden. Livestock and dairy farming are important, and crops include barley, oats, potatoes, sugar beet, and wheat.

SWITZERLAND

Switzerland has used this square flag since 1848, though the white cross on the red shield has been Switzerland's emblem since the 14th century. The flag of the International Red Cross, which is based in Geneva, was derived from this flag.

The Swiss Confederation is a landlocked country in Western Europe. Much of the land is mountainous. The Jura Mountains lie along Switzerland's western border with France, while the Swiss Alps make up about 60% of the country in the south and east. Between these scenic regions lies the Swiss plateau, a hilly zone between Lake Constance (Bodensee) and Lake Geneva (Lac Léman). Four-fifths of the people of Switzerland live on this fertile plateau, which contains most of Switzerland's large cities.

AREA 15,942 sq miles [41,290 sq km]
POPULATION 7,283,000
CAPITAL (POPULATION) Bern (942,000)
GOVERNMENT Federal republic
ETHNIC GROUPS German 65%, French 18%, Italian 10%, Romansch 1%
LANGUAGES French, German, Italian and Romansch (all official)
RELIGIONS Roman Catholic 46%, Protestant 40%
CURRENCY Swiss franc = 100 centimes

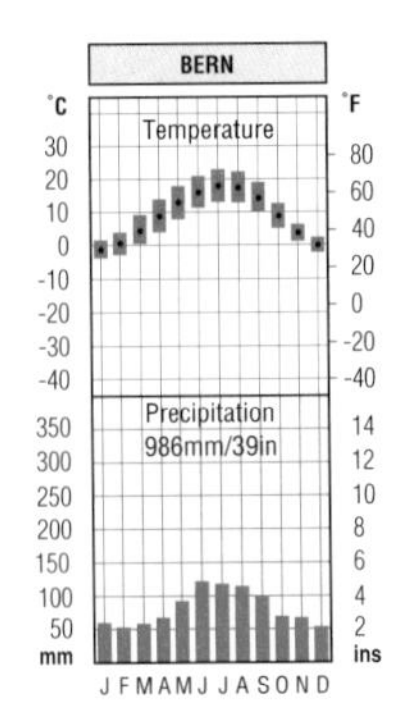

CLIMATE

The climate of Switzerland varies greatly according to the height of the land. The plateau region has a central European climate with warm summers, but cold and snowy winters. Rain occurs all through the year. The rainiest months are in summer.

VEGETATION

Grassy meadows cover about two-fifths of the country and farmland about 11%. Forests cover about a quarter of Switzerland. Forests help to reduce the risk of avalanches.

HISTORY AND POLITICS

In 1291, three small cantons (states) united to defend their freedom against the Habsburg rulers of the Holy Roman Empire. They were Schwyz, Uri, and Unterwalden, and they called the confederation they formed "Switzerland." Switzerland expanded and, in the 14th century, defeated Austria in three wars of independence. After a defeat by the French in 1515, the Swiss adopted a policy of neutrality, which they still follow.

In 1815, the Congress of Vienna expanded Switzerland to 22 cantons and guaranteed its neutrality. Switzerland's 23rd canton, Jura, was created in 1979 from part of Bern. Neutrality combined with the vigor and indepen-

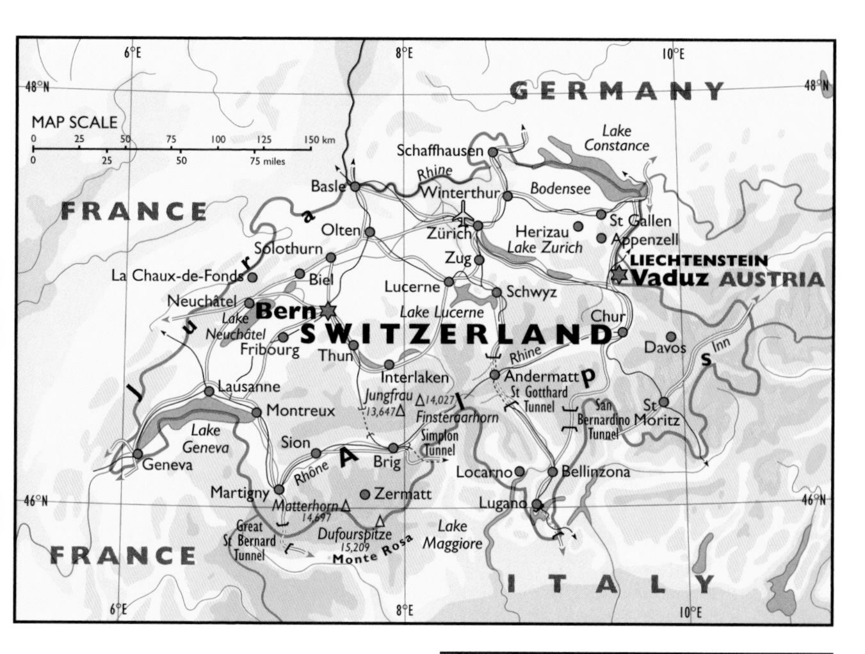

dence of its people have made Switzerland prosperous. But, in 2002, it surrendered some of its neutrality by joining the United Nations.

ECONOMY

Although lacking in natural resources, Switzerland is a wealthy, industrialized country. Many workers are highly skilled. Major products include chemicals, electrical equipment, machinery and machine tools, precision instruments, processed food, watches, and textiles.

Farmers produce about three-fifths of the country's food – the rest is imported. Livestock raising, especially dairy farming, is the chief agricultural activity. Crops include fruits, potatoes, and wheat. Tourism and banking are also important. Swiss banks attract investors from all over the world.

LIECHTENSTEIN

Liechtenstein is a tiny monarchy ruled by a prince. Independent since 1806 and neutral since 1866, it has many ties with its neighbor Switzerland. The people speak a German dialect. The country is highly industrialized and prosperous.

AREA 61 sq miles [157 sq km]
POPULATION 33,000
CAPITAL (POPULATION) Vaduz (4,900)
GOVERNMENT Principality
ETHNIC GROUPS Alemannic 86%
CURRENCY Swiss franc = 100 centimes

SYRIA

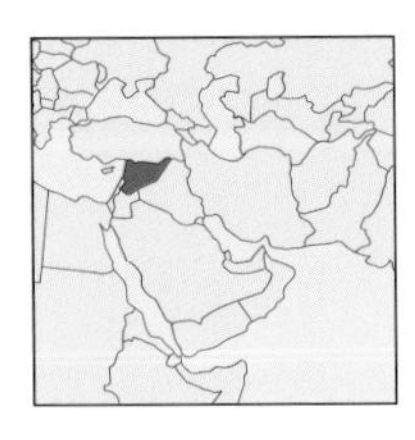

Syria has used this flag since 1980. The colors are those used by the Pan-Arab movement. This flag is the one that was used by the United Arab Republic between 1958 and 1961, when Syria was linked with Egypt and North Yemen.

The Syrian Arab Republic is a country in southwestern Asia. The narrow coastal plain is overlooked by a low mountain range which runs north–south. Another range, the Jabal ash Sharqi, runs along the border with Lebanon. South of this range is a region called the Golan Heights. Israel has occupied this region since 1967. East of the mountains, the bulk of Syria consists of fertile valleys, grassy plains, and large sandy deserts. This region contains the valley of the River Euphrates (Nahr al Furat).

AREA 71,498 sq miles [185,180 sq km]
POPULATION 16,729,000
CAPITAL (POPULATION) Damascus (or Dimashq, 1,394,000)
GOVERNMENT Multiparty republic
ETHNIC GROUPS Arab 90%, Kurdish, Armenian
LANGUAGES Arabic (official)
RELIGIONS Islam 90%, Christianity 9%
CURRENCY Syrian pound = 100 piastres

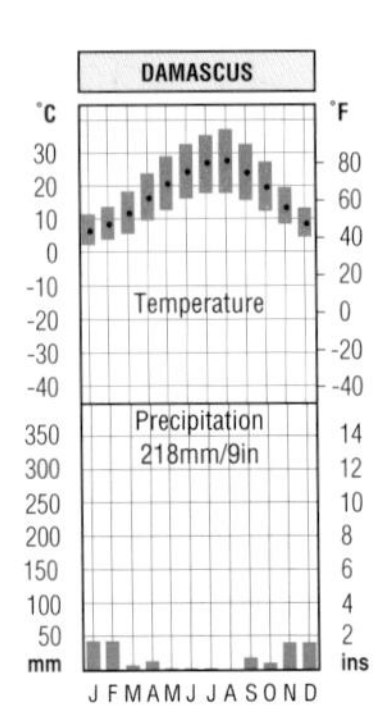

CLIMATE

The coast has a Mediterranean climate, with dry, warm summers and wet, mild winters. The low mountains cut off Damascus from the sea. It has less rainfall than the coastal areas. To the east, the land becomes drier. Winters are colder and the summers are hotter.

VEGETATION

Forests cover only 3% of Syria. Fir, lime, and yew trees grow on the mountains, and palm trees are common in the Euphrates valley. Farmland covers about three-tenths of Syria, with grasslands making up nearly half of the country.

HISTORY

In early times, Syria lay at the crossroads of Asia, Africa and Europe. As a result, it is rich in fascinating historic sites. Conquerors of Syria included the Akkadians, Arameans, Persians, Greeks, and Romans. Muslim Arabs invaded Syria in AD 637. They made Damascus the capital of their empire and their Muslim culture survives to this day. From the late 11th century, Christian crusaders invaded Syria, but, by the end of the 12th century, the Muslim leader Saladin ruled most of the land.

In 1516 the Ottoman Turks conquered the area and made it part of their empire. During World War I (1914–18), Syrian and other Arabs revolted and helped Britain to defeat the Turks. After the war, France ruled the area, but Syria became fully independent in 1946.

POLITICS

Since independence, Syria has supported the Arab cause in southwestern Asia. In 1947 and 1948, it was involved in the first of the Arab-Israeli wars, and in 1967 it lost a strategic border area, the Golan Heights, to Israel. Internationally, Syria sought to create Arab unity through joining the United Arab Republic with Egypt and North Yemen in 1958. This union fell apart in 1961, as did other short-lived attempts to unite with other Arab nations.

At home, Syria suffered from instability after independence and the country suffered many military revolts and changes of government. In 1970, Lieutenant-General Hafez al-Assad took power. His stable but repressive regime provoked much international criticism. In 1991, Syria's reputation was enhanced by its support for the alliance which defeated Iraq and freed the state of Kuwait. In 1999, Syria held talks with Israel over the future of the Golan Heights, but without success. President Hafez al-Assad died in 2000 and was succeeded by his son, Bashar al-Assad.

DID YOU KNOW

- that Damascus (Dimashq) is the world's oldest continuously inhabited city
- that Syria contains the ruins of several castles built by Christian crusaders; the best known is the Castle of the Knights (Crac des Chevaliers), west of Homs (Hims)
- that the oldest known alphabet was developed in Syria around 1500 BC
- that Palmyra (or Tudmur) in central Syria was once a great trading center linking the ancient Silk Road from China to Europe

ECONOMY

The World Bank classifies Syria as a "lower-middle-income" developing country. But it has great potential for development. Its main resources are oil, hydroelectricity from the dam at Lake Assad (Bahret Assad), and fertile land. Oil is the leading export, but Syria also exports farm products, textiles, and phosphates. Agriculture employs up to 23% of the people. The chief crops are cotton and wheat, and the main manufactures are textiles. Other manufactures include cement, fertilizers, processed food, and sugar.

TAJIKISTAN

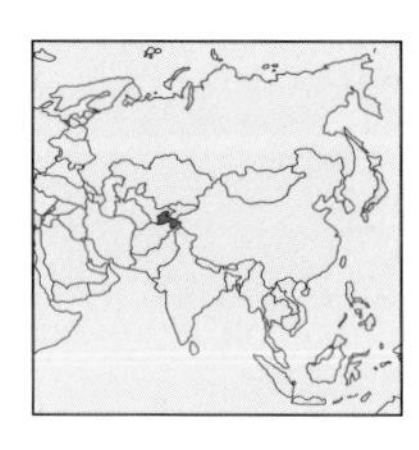

Tajikistan's flag was adopted in 1993. It replaced the flag used during the Communist period which showed a hammer and sickle. The new flag shows an unusual gold crown under an arc of seven stars on the central white band.

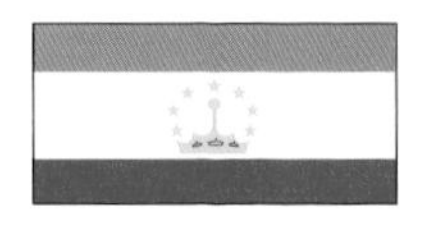

The Republic of Tajikistan is one of the five central Asian republics that formed part of the former Soviet Union. Only 7% of the land is below 3,280 ft [1,000 m], while almost all of eastern Tajikistan is above 9,840 ft [3,000 m]. The highest point is Communism Peak (Pik Kommunizma), which reaches 24,599 ft [7,495 m]. The main ranges are the westward extension of the Tian Shan range in the north, and the snow-capped Pamirs in the southeast. Earthquakes are common throughout the country.

AREA 55,520 sq miles [143,100 sq km]
POPULATION 6,579,000
CAPITAL (POPULATION) Dushanbe (524,000)
GOVERNMENT Transitional democracy
ETHNIC GROUPS Tajik 65%, Uzbek 25%, Russian 3%, Tatar, Kyrgyz, Ukrainian, German
LANGUAGES Tajik (official), Uzbek, Russian
RELIGIONS Islam (mainly Sunni) 80%
CURRENCY Somoni = 100 dirams

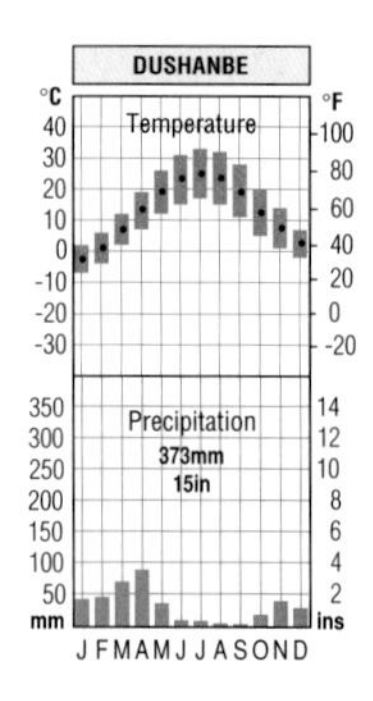

CLIMATE

Tajikistan is a landlocked country with a continental climate. Summers are hot and dry in the lower valleys, but winters are long and bitterly cold in the mountain areas. Much of the country is arid with sparse rainfall, but the southeast has heavy snowfalls.

VEGETATION

Much of Tajikistan consists of desert or rocky mountain landscapes, capped by snow and ice. The plant life, which consists mainly of bushes, grasses, shrubs, and many kinds of flowering plants, varies according to the altitude.

HISTORY AND POLITICS

The ancestors of the people of Tajikistan were Persians who had settled in the area by about 2,500 years ago. The area was conquered many times. One important group of invaders were the Arabs who introduced Islam in the 7th century. From the 16th to the 19th centuries, Uzbeks ruled the area. But in the late 19th century, Russia conquered part of Tajikistan and it extended its rule after the Russian Revolution of 1917. By 1920, all of Tajikistan was under Russian control.

In 1924, the area was ruled as part of the Uzbek Soviet Socialist Republic. But in 1929, it was expanded, taking in areas containing Uzbeks, and it became the Tajik Soviet Socialist

Republic within the Soviet Union. Under Communism, all features of Tajik life were strictly controlled and religious worship was discouraged. But many aspects of life were improved.

While the Soviet Union began to introduce reforms in the 1980s, many Tajiks demanded freedom. In 1989, the Tajik government made Tajik the official language instead of Russian and, in 1990, it stated that its local laws overruled Soviet laws. Tajikistan became fully independent in 1991, following the breakup of the Soviet Union. As the poorest of the former Soviet republics, Tajikistan faced numerous problems in trying to reorganize its economy along free-enterprise lines.

In 1992, civil war broke out between the government, which was run by former Communists, and an alliance of democrats and Islamic forces. A ceasefire was agreed in 1996, and in 1997 representatives of the opposition were brought into the government. But some groups threatened the peace process by continuing to commit acts of violence.

ECONOMY

The World Bank classifies Tajikistan as a "low-income" developing country. Agriculture, mainly on irrigated land, is the main activity and cotton is the chief product. Other crops include fruits, grains, and vegetables. The country has large hydroelectric power resources and it produces aluminum. Otherwise, manufacturing is mostly on a small scale. The main exports are aluminum, uranium, cotton, fruits, vegetable oils, and textiles.

DID YOU KNOW

- that the official name of Tajikistan in the Tajik language is *Jumhurii Tojikistan*; the Russians called it Tadzhikistan
- that the salty Lake Kara-Kul in northeastern Tajikistan is devoid of life
- that Dushanbe was called Stalinabad, after the Soviet dictator Joseph Stalin, between 1929 and 1962

The new flag of Tajikistan (also spelled Tadzikistan) is the subject of this stamp issued in 1993. In the background is an outline map of the new country.

TANZANIA

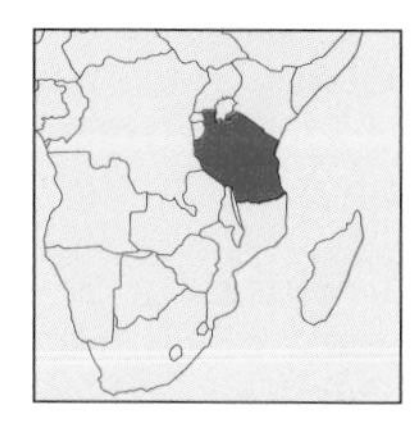

Tanzania's flag was adopted in 1964 when mainland Tanganyika joined with the island nation of Zanzibar to form the United Republic of Tanzania. The green represents agriculture and the yellow minerals. The black represents the people, while the blue symbolizes Zanzibar.

The United Republic of Tanzania consists of the former mainland country of Tanganyika and the island nation of Zanzibar, which also includes the island of Pemba. Behind a narrow coastal plain, most of Tanzania is a plateau between 2,950 ft and 4,920 ft [900 m to 1,500 m]. The plateau is broken by arms of the Great African Rift Valley. In the west, this valley contains lakes Nyasa and Tanganyika. The highest peak is Kilimanjaro, Africa's tallest mountain. Zanzibar and Pemba are coral islands.

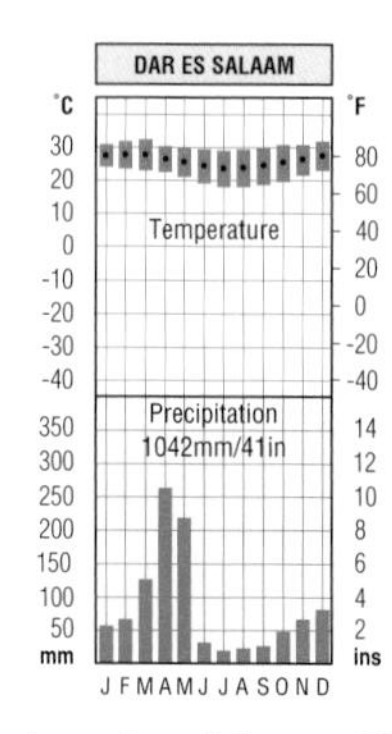

CLIMATE

Dar es Salaam, on the coast of Tanzania, has a hot and humid climate. The rainfall is heaviest in April and May. Inland on the plateaux and mountains, it is cooler and less humid. The Rift Valley is hot, but Mount Kilimanjaro is permanently covered by snow and ice.

VEGETATION

Mangrove swamps and palm groves grow along the coasts. Inland, open savanna grassland or savanna woodland, called *miombo,* cover vast areas. Tanzania's rich wildlife is now protected in national parks and reserves which cover over 12% of the land.

AREA 364,899 sq miles [945,090 sq km]

POPULATION 36,232,000

CAPITAL (POPULATION) Dodoma (204,000)

GOVERNMENT Multiparty republic

ETHNIC GROUPS Nyamwezi and Sukuma 21%, Swahili 9%, Hehet and Bena 7%, Makonde 6%, Haya 6%

LANGUAGES Swahili and English (both official), Arabic

RELIGIONS Christianity (mostly Roman Catholic) 45%, Islam 35% (99% in Zanzibar), traditional beliefs and others 20%

CURRENCY Tanzanian shilling = 100 cents

HISTORY AND POLITICS

Around 2,000 years ago, Arabs, Persians and even Chinese probably traded on the Tanzanian coast. The Portuguese took control of the coastal trade in the early 16th century, but Arabs regained control in the 17th century. In the 19th century, explorers and missionaries were active, mapping the country and working to stop the slave trade.

Mainland Tanganyika became a German territory in the 1880s, while

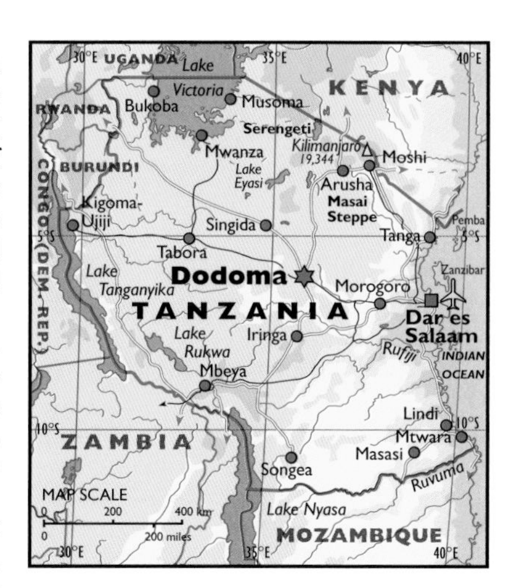

Britain made Zanzibar and Pemba a British protectorate in 1890. After Germany's defeat in World War I (1914–18), Britain took control of Tanganyika, which remained under British rule until its independence in 1961. In 1964, Tanganyika and Zanzibar united to form the United Republic of Tanzania.

Tanzania's leader, Julius Nyerere, pursued socialist policies of self-help (*ujamaa*) and egalitarianism. Tanzania achieved success in social reforms, including education, but it failed to make economic progress.

Nyerere stepped down as president in 1985, although he retained much influence until his death in 1999. His successors, Ali Hassan Mwinyi and, from 1995, Benjamin Mkapa, introduced more liberal economic policies.

ECONOMY

Tanzania is one of the world's poorest countries. Although crops are grown on only 5% of the land, agriculture employs 80% of the people. Most farmers grow only enough to feed their families. Food crops include bananas, cassava, maize, millet, rice, and vegetables. Export crops include coffee, cotton, tea, and tobacco. Some diamonds are mined, but manufacturing is mostly small scale.

DID YOU KNOW

- that Lake Tanganyika is the world's longest freshwater lake; it is over 400 miles [645 km] long
- that Zanzibar was a thriving center of the slave and ivory trade in the 18th and 19th centuries
- that Tanzania's people are divided into about 120 ethnic and language groups
- that Swahili, the official language of Tanzania, is a Bantu language that originated on the coasts of Kenya and Tanzania

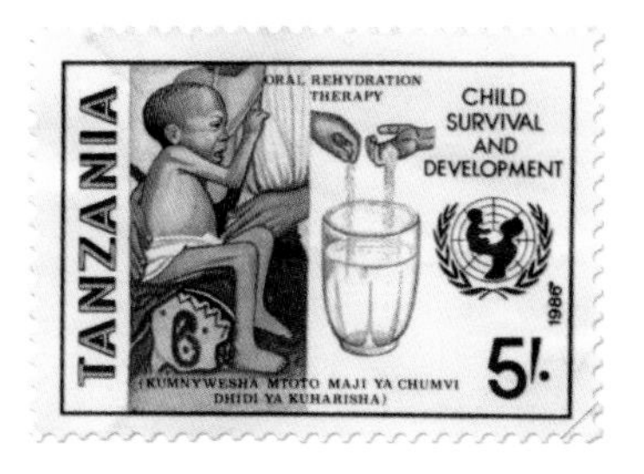

Many children in Africa die of dehydration caused by infectious diarrhoea. This Tanzanian stamp, issued in 1986, illustrates the use of a simple and cheap form of treatment, oral rehydration therapy, where doctors use a solution containing glucose to restore lost water and salt. The use of this therapy is part of a campaign sponsored by UNICEF (United Nations Children's Fund).

THAILAND

Thailand's flag was adopted in 1917. In the late 19th century, it featured a white elephant on a plain red flag. In 1916, white stripes were introduced above and below the elephant, but in 1917 the elephant was dropped and a central blue band was added.

The Kingdom of Thailand is one of the ten countries in Southeast Asia. The highest land is in the north, where Doi Inthanon, the highest peak, reaches 8,517 ft [2,595 m], southwest of Chiang Mai. The Khorat Plateau, in the northeast, makes up about 30% of the country. It is the most heavily populated part of Thailand. The central plains, drained by the Chao Phraya and other rivers, have fertile soils. In the south, Thailand shares the finger-like Malay Peninsula with Burma and Malaysia.

AREA 198,116 sq miles [513,120 sq km]
POPULATION 61,798,000
CAPITAL (POPULATION) Bangkok (7,507,000)
GOVERNMENT Constitutional monarchy
ETHNIC GROUPS Thai 75%, Chinese 14%, Malay 4%, Khmer 3%
LANGUAGES Thai (official), Chinese, Malay, English
RELIGIONS Buddhism 94%, Islam 4%, Christianity 1%
CURRENCY Thai baht = 100 satang

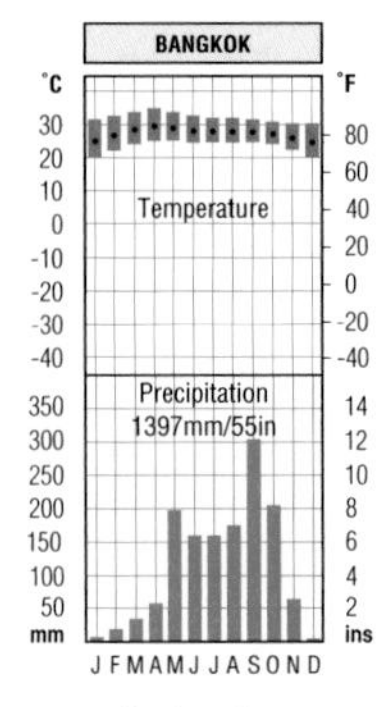

CLIMATE

Thailand has a tropical climate. Monsoon winds from the southwest bring heavy rains between the months of May and October. The rainfall in Bangkok is lower than in many other parts of Southeast Asia, because mountains shelter the central plains from the rain-bearing winds.

VEGETATION

Forests containing such valuable trees as teak cover nearly half of Thailand. Grass, shrub, and swamp make up a fifth of the land area, while farms cover just over a third.

HISTORY AND POLITICS

The first Thai state was set up in the 13th century. By 1350, it included most of what is now Thailand. European contact began in the early 16th century. But, in the late 17th century, the Thais, fearing interference in their affairs, forced all Europeans to leave. This policy continued for 150 years. In 1782, a Thai General, Chao Phraya Chakkri, became king, founding a dynasty which continues today. The country became known as Siam, and Bangkok became its capital. From the mid-19th century, contacts with the West were restored.

In World War I (1914–18), Siam

supported the Allies against Germany and Austria-Hungary. But, in 1941, Japanese forces invaded Thailand, which became Japan's ally. After the end of World War II, Thailand became an ally of the United States.

Between 1946 and 1957, Thailand was a dictatorship. The military continued to play a role in government in the 1960s and 1970s. From 1979, Thailand had elected governments, but the military again took over in 1991. However, elections were held in 1992, 1995, and 2001.

ECONOMY

Since 1967, when Thailand became a member of ASEAN (the Association of Southeast Asian Nations), its economy has grown, especially its manufacturing and service industries. However, along with the other fast-developing countries in eastern Asia, Thailand experienced a recession in 1997–8 and its economic policies had to be modified. The economy still depends on agriculture, which employs over two-fifths of the people. Rice is the chief crop. Cassava, cotton, maize, rubber, sugarcane, and tobacco are also grown. Thailand also mines tin and other minerals, but the chief exports are manufactures, including food products, machinery, timber products, and textiles. Tourism is another major source of income.

> **DID YOU KNOW**
> - that Thailand used to be called Siam; it has been called Thailand since 1949
> - that Bangkok is often called the "Venice of the East"
> - that Thailand's official name, *Muang Thai*, means "Land of the Free"
> - that in Thai-style boxing people can fight with their feet as well as their hands
> - that Thailand is the only Southeast Asian country never ruled by a European power
> - that Thailand is the world's leading exporter of rice

TOGO

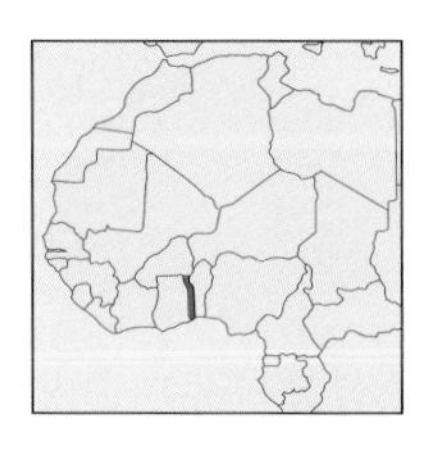

Togo's flag uses the colors that symbolize African unity. The green stands for agriculture and the future, the yellow for mineral wealth and the value of hard work, and the red for the blood shed during the struggle for independence. The white star stands for national purity.

The Republic of Togo is a long, narrow country in West Africa. From north to south, it extends about 310 miles [500 km]. Its coastline on the Gulf of Guinea is only 40 miles [64 km] long and it is only 90 miles [145 km] at its widest point. The Togo-Atacora Mountains in southern Togo, which rise to 3,236 ft [986 m], overlook a plateau. The land descends to a densely populated coastal plain in the south. In the north lies another plateau, which descends northward to the border with Burkina Faso.

AREA 21,927 sq miles [56,790 sq km]
POPULATION 5,153,000
CAPITAL (POPULATION) Lomé (590,000)
GOVERNMENT Multiparty republic
ETHNIC GROUPS Ewe-Adja 43%, Tem-Kabre 26%, Gurma 16%
LANGUAGES French (official), Ewe, Kabiye
RELIGIONS Traditional beliefs 50%, Christianity 35%, Islam 15%
CURRENCY CFA franc = 100 centimes

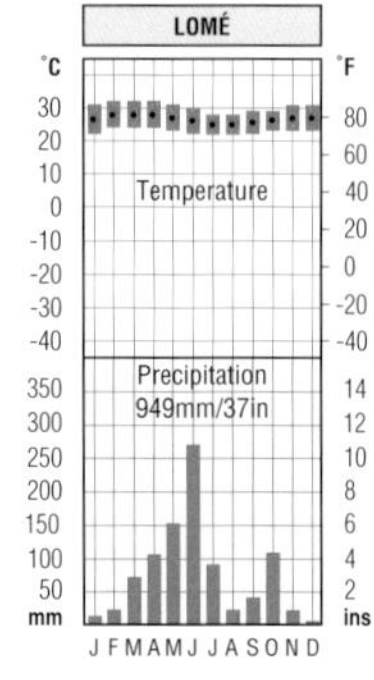

CLIMATE

Togo has high temperatures all through the year. The main wet season is from March to July, with a minor wet season in October and November. The central mountains are cooler and rainier than Lomé, on the coast. The far north is hot, with a single rainy season.

VEGETATION

Behind the sandy coastal plain, with its many palm trees, are areas of grassland with clumps of hardwood trees. Some rain forest grows in central Togo, but the north is savanna country with many animal species, including antelopes, buffaloes, elephants, leopards, and lions.

HISTORY AND POLITICS

Portuguese explorers reached the Togo coast in the late 15th century. Between the 17th and 19th centuries, the area became known as "the coast of slaves." German missionaries and traders began to operate in Togo in the mid-19th century. In 1884, Germany set up a protectorate over the area.

French and British troops occupied Togoland in World War I (1914–18). In 1919, Britain took over the western one-third of Togo, while France took the eastern two-thirds. In 1956, the people of British Togoland voted to join Ghana. French Togoland became

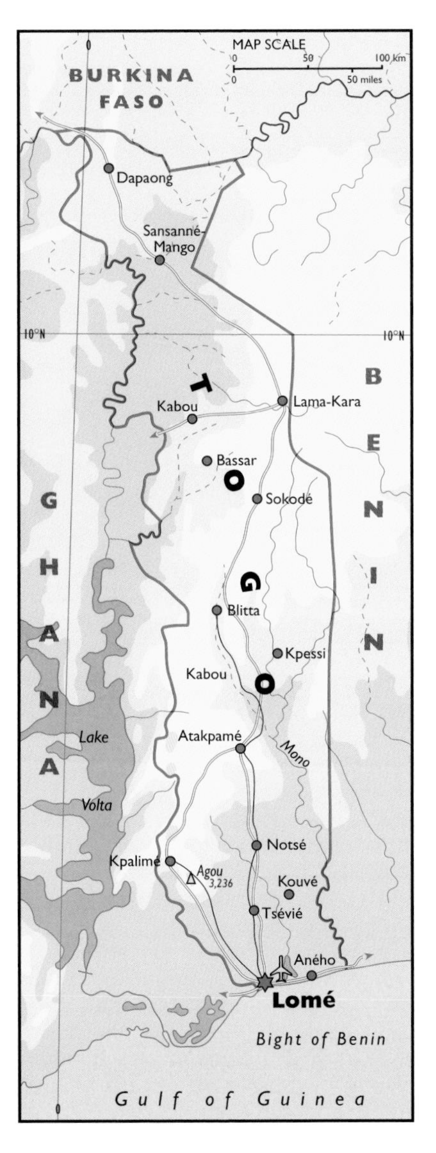

an independent republic in 1960.

A military regime took power in 1963. In 1967, General Gnassingbe Eyadema, became head of state and he suspended the constitution. Under a new constitution adopted in 1992, multiparty elections were held in 1994. However, in 1998, paramilitary policies stopped the count in the elections when it became clear that Eyadema was defeated. As a result, leading opposition parties boycotted the general elections in 1999.

ECONOMY

Togo is a poor developing country. Farming employs 65% of the people, but most farmers grow little more than they need to feed their families. Major food crops include cassava, maize, millet, and yams.

The chief export crops are cocoa, coffee, and cotton. But the leading export is phosphate rock, which is used to make fertilizers. Togo's small-scale manufacturing and mining industries employ about 6% of the people.

This Togolese stamp shows a chimpanzee, which is found in the rain forests of West and Central Africa. Forest clearance for farming, mining, and the building of roads is reducing the animals' habitats.

TUNISIA

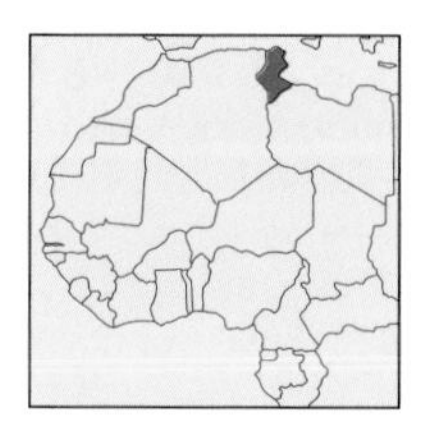

Tunisia's flag originated in about 1835 when the country was officially under Turkish rule. It became the national flag in 1956, when Tunisia became independent from France. The flag contains two traditional symbols of Islam, the crescent and the star.

The Republic of Tunisia is the smallest country in North Africa. The mountains in the north are an eastward and comparatively low extension of the Atlas Mountains, which run through Morocco and Algeria.

To the north and east of the mountains lie fertile plains, especially between Sfax, Tunis, and Bizerte. South of the mountains lie broad plateaux which descend toward the south. This low-lying region contains a large salt pan, called the Chott Djerid, and part of the Sahara.

AREA 63,170 sq miles [163,610 sq km]
POPULATION 9,705,000
CAPITAL (POPULATION) Tunis (1,827,000)
GOVERNMENT Multiparty republic
ETHNIC GROUPS Arab 98%, Berber 1%, French and other
LANGUAGES Arabic (official), French
RELIGIONS Islam 99%
CURRENCY Dinar = 1,000 millimes

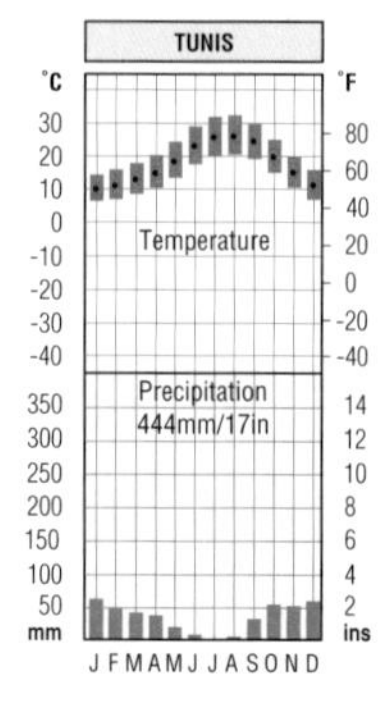

CLIMATE

Northern Tunisia has a Mediterranean climate, with dry, sunny summers, and mild winters with a moderate rainfall. The average yearly rainfall decreases toward the south. For example, Gafsa, in south-central Tunisia, has an average annual rainfall of only 6 inches [150 mm].

VEGETATION

Some cork oak forests grow in the northern mountains. To the south lie steppes with coarse grasses. In the south, the land is barren, except around oases where palm trees grow.

HISTORY AND POLITICS

According to legend, Carthage was founded near Tunis in 814 BC, but the Romans destroyed it in 146 BC. The Arabs invaded the land in AD 647, introducing Islam and the Arabic language to the local Berber people. In 1547, Tunisia became part of the Turkish Ottoman empire.

In 1881, France established a protectorate over Tunisia and ruled the country until 1956. The new parliament abolished the monarchy and declared Tunisia to be a republic in 1957, with the nationalist leader, Habib Bourguiba, as president.

In 1975, Bourguiba was elected president for life. His government introduced many reforms, including votes for women. But problems arose,

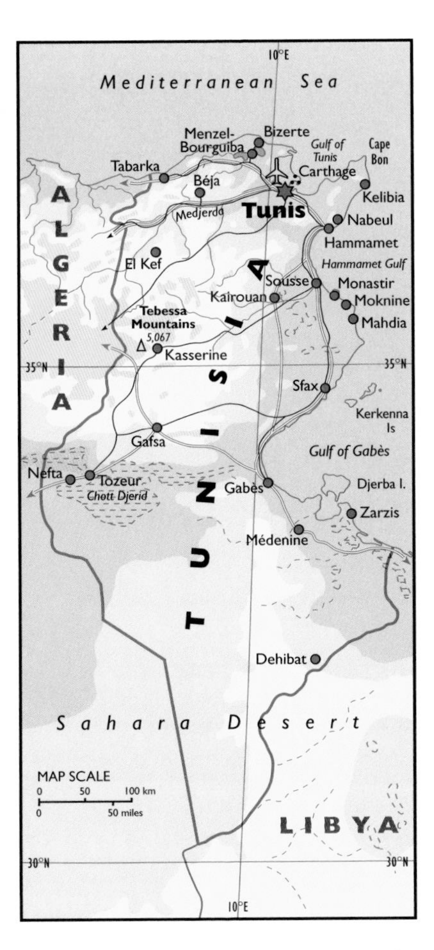

such as unemployment among the educated groups and fears that Western influences might damage traditional values, although Islamic fundamentalism never proved to be as effective in Tunisia as it has done in neighboring Algeria. Finally, however, the prime minister, Zine el Abidine Ben Ali, removed the aging Bourguiba from office in 1987 and became president. He was elected in 1989, and re-elected in 1994 and 1999.

DID YOU KNOW

- that the ruins of the ancient city of Carthage are close to Tunis
- that more than 3 million tourists visited Tunisia in 1991
- that Kairouan, founded in Tunisia in AD 671, is the fourth holiest Muslim city, after Mecca (Makkah), Medina (Al Madinah), and Jerusalem
- that Tunisia has some of North Africa's finest Roman ruins at such places as Dougga, southwest of Tunis, and El Djem, south of Sousse

ECONOMY

The World Bank classifies Tunisia as a "middle-income" developing country. The main resources and chief exports are phosphates and oil. Most industries are concerned with food processing. Agriculture employs 22% of the people. Major crops include barley, dates, grapes for wine-making, olives, and wheat. Fishing is important, as also is tourism.

*This stamp, issued in 1992, shows a bee-eater (*Merops apiaster*).*

TURKEY

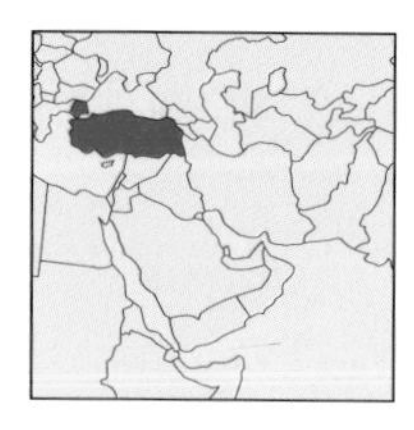

Turkey's flag was adopted when the Republic of Turkey was established in 1923. The crescent moon and the five-pointed star are traditional symbols of Islam. They were used on earlier Turkish flags used by the Turkish Ottoman empire.

The Republic of Turkey lies in two continents. The European section lies west of a waterway between the Black and Mediterranean seas. This waterway consists of the Bosporus, on which the city of Istanbul stands, the Sea of Marmara, and a narrow strait called the Dardanelles. European Turkey, also called Thrace, is a fertile, hilly region. Most of the Asian part of Turkey consists of plateaux and mountains, which rise to 16,952 ft [5,165 m] at Mount Ararat (Agri Dagi) near the border with Armenia.

AREA 300,946 sq miles [779,450 sq km]
POPULATION 66,494,000
CAPITAL (POPULATION) Ankara (3,294,000)
GOVERNMENT Multiparty republic
ETHNIC GROUPS Turkish 80%, Kurdish 20%
LANGUAGES Turkish (official), Kurdish
RELIGIONS Islam 99%
CURRENCY Turkish lira = 100 kurus

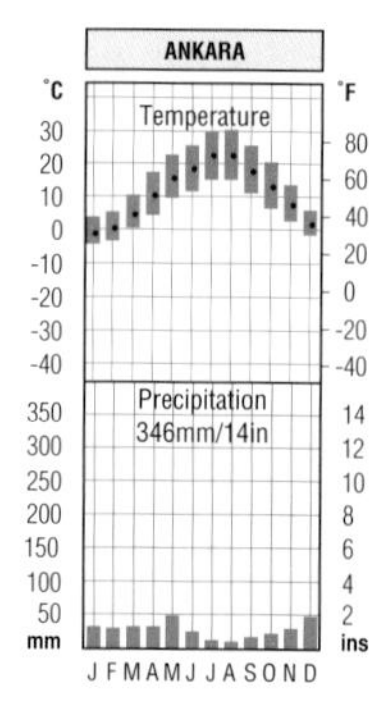

CLIMATE

Central Turkey has a dry climate, with hot, sunny summers and cold winters. The driest part of the central plateau lies south of the city of Ankara, around Lake Tuz. Western Turkey has a Mediterranean climate, while the Black Sea coast has cooler summers.

VEGETATION

Mediterranean scrub vegetation is common in the west, south, and northwest. Deciduous forests grow inland, with conifers on the mountains. Steppes cover much of the plateaux.

HISTORY AND POLITICS

In AD 330, the Roman empire moved its capital to Byzantium, which it renamed Constantinople. Constantinople became capital of the East Roman (or Byzantine) empire in 395. Muslim Seljuk Turks from central Asia invaded Anatolia in the 11th century. In the 14th century, another group of Turks, the Ottomans, conquered the area. In 1453, the Ottoman Turks took Constantinople, which they called Istanbul.

The Ottoman Turks built up a large empire which finally collapsed during World War I (1914–18). In 1923, Turkey became a republic. Its leader Mustafa Kemal, who was called Atatürk ("father of the Turks"), launched policies to modernize the country.

Since the 1940s, Turkey has sought

to strengthen its ties with Western powers. It joined NATO (North Atlantic Treaty Organization) in 1951 and it applied to join the EEC (European Economic Community) in 1987. But conflict with Greece and Turkey's invasion of northern Cyprus in 1974 have led many Europeans to treat Turkey's aspirations with caution. Political instability, military coups and conflict with Kurdish nationalists in eastern Turkey are other problems. Turkey has enjoyed democracy since 1983, but many people are concerned about its record on human rights. Although Turkey no longer occupies a key strategic position as it did during the Cold War, it remains an important bridge between Europe and Asia.

DID YOU KNOW

- that Turkey's highest peak, Mount Ararat, was supposed to be the resting place of Noah's Ark after the Biblical Flood
- that about 97% of Turkey is in Asia and 3% is in Europe
- that Istanbul was once known as Byzantium and later as Constantinople
- that the Asian part of Turkey is called Asia Minor or Anatolia

ECONOMY

The World Bank classifies Turkey as a "lower-middle-income" developing country. Agriculture employs 40% of the people, and barley, cotton, fruits, maize, tobacco, and wheat are major crops. Livestock farming is important and wool is a leading product.

Turkey is a major producer of chromium, but manufacturing is the most valuable activity. The chief manufactures are processed farm products and textiles. Also important are cars, fertilizers, iron and steel, machinery, metal products, and paper products.

TURKMENISTAN

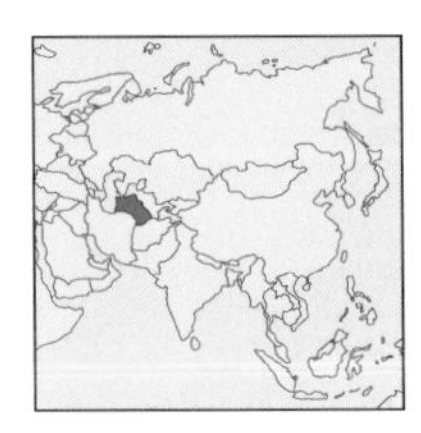

Turkmenistan's flag was adopted in 1992. It incorporates a typical Turkmen carpet design. The crescent is a symbol of Islam, while the five stars and the five elements in the carpet represent the traditional tribal groups of Turkmenistan.

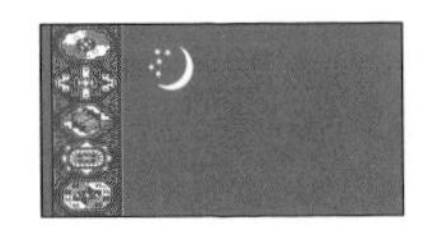

The Republic of Turkmenistan is one of the five central Asian republics which once formed part of the former Soviet Union. Most of the land is low-lying, with mountains lying on the southern and south-western borders.

In the west lies the salty Caspian Sea. A depression called the Kara Bogaz Gol Bay contains the country's lowest point, at 102 ft [31 m] below sea level. Most of Turkmenistan is arid and the Garagum, Asia's largest sand desert, covers about 80% of the country.

AREA 188,450 sq miles [488,100 sq km]
POPULATION 4,603,000
CAPITAL (POPULATION) Ashgabat (536,000)
GOVERNMENT Single-party republic
ETHNIC GROUPS Turkmen 77%, Russian 17%, Uzbek, Kazakh, Tatar
LANGUAGES Turkmen (official), Russian, Uzbek, Kazakh
RELIGIONS Islam
CURRENCY Manat = 100 tenesi

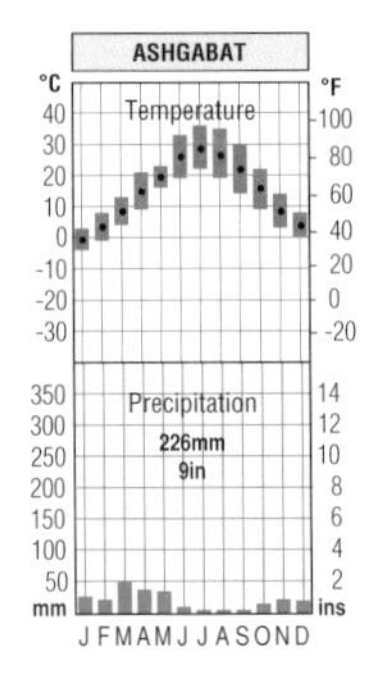

CLIMATE

Turkmenistan has a continental climate. The average annual rainfall varies from 3 inches [80 mm] in the desert to 12 inches [300 mm] in the southern mountains. The summer months are hot. In the winter, temperatures sometimes drop well below freezing point.

VEGETATION

The vegetation in Turkmenistan is sparse and practically non-existent in sandy areas. Plants grow around desert oases and in the mountains, where such trees as almond, fig, and walnut are found.

HISTORY

Just over 1,000 years ago, Turkic people settled in the lands east of the Caspian Sea and the name "Turkmen" comes from this time. Mongol armies conquered the area in the 13th century and Islam was introduced in the 14th century.

Russia took over the area in the 1870s and 1880s. After the Russian Revolution of 1917, the area came under Communist rule and, in 1924, the area became the Turkmen Soviet Socialist Republic. The Communists strictly controlled all aspects of life and, in particular, they discouraged religious worship. But they also improved such services as education, health, housing, and transport.

Turkmenistan remained a one-party state. In 1992, Saparmurad Niyazov, the former Communist and now Democratic party leader, was the only candidate. In 1999, Niyazov was made president for life and, in 2002, he survived an attempt on his life.

Faced with many economic problems, Turkmenistan began to look south for support. To this end, it joined the Economic Cooperation Organization, which had been set up in 1985 by Iran, Pakistan, and Turkey.

POLITICS

In the 1980s, when the Soviet Union began to introduce reforms, the Turkmen began to demand more freedom. In 1990, the Turkmen government stated that its laws overruled Soviet laws. In 1991, Turkmenistan became independent after the breakup of the Soviet Union, but it kept ties with Russia through the Commonwealth of Independent States.

In 1992, Turkmenistan adopted a new constitution, allowing for the setting up of political parties, providing that they were not ethnic or religious in character. But, effectively,

ECONOMY

The World Bank classifies Turkmenistan as a "lower-middle-income" developing country. The country's chief natural resources are oil and natural gas, but the main activity is agriculture. The chief crop, which is grown on irrigated land, is cotton. Grains and vegetables are also important. Various animals are raised and wool is another major product. Its manufactures include cement, glass, petrochemicals, and textiles. The country's leading exports include natural gas, oil, chemicals, cotton, textiles, and carpets.

Saparmurad Niyazov, an ex-Communist leader, who was named Turkmenistan's president for life in 1999, appears on this stamp.

The Kulan, a rare form of the Asiatic wild ass, is now found only in a reserve in Turkmenistan. It is the subject of this stamp which was issued in 1992.

UGANDA

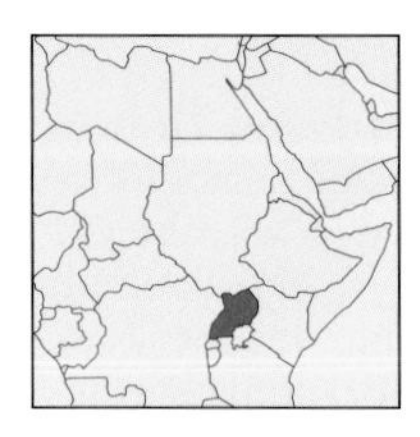

The flag used by the party that won the first national election was adopted as the national flag when Uganda became independent from Britain in 1962. The black represents the people, the yellow the sun, and the red brotherhood. The crested crane is the country's emblem.

AREA 91,073 sq miles [235,880 sq km]
POPULATION 23,986,000
CAPITAL (POPULATION) Kampala (954,000)
GOVERNMENT Republic in transition
ETHNIC GROUPS Baganda 17%, Karamojong 12%, Basogo 8%, Iteso 8%, Langi 6%, Rwanda 6%, Bagisu 5%, Acholi 4%, Lugbara 4%
LANGUAGES English and Swahili (both official), Ganda
RELIGIONS Roman Catholic 33%, Protestant 33%, traditional beliefs 18%, Islam 16%
CURRENCY Uganda shilling = 100 cents

The Republic of Uganda is a landlocked country in East Africa. It contains part of Lake Victoria, Africa's largest lake and a source of the River Nile. Lake Victoria occupies a depression in the African plateau. The plateau varies in height from about 4,920 ft [1,500 m] in the south to 2,950 ft [900 m] in the north. Mountains in the west border an arm of the Great African Rift Valley. This valley contains lakes Edward and Albert (Mobutu Sese Seko). Other mountains border Uganda in parts of the east.

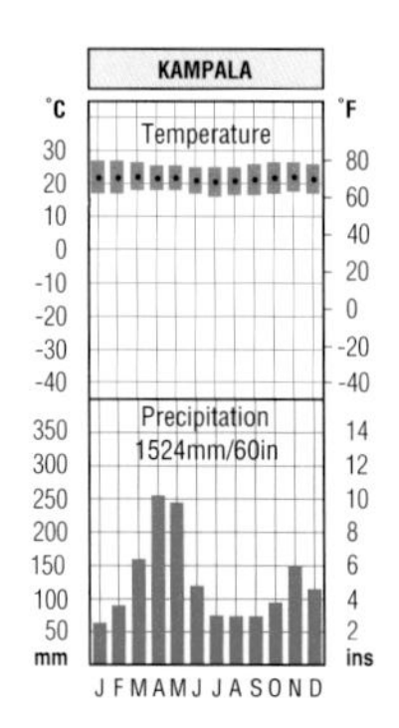

CLIMATE

The Equator runs through Uganda and the country is warm throughout the year, though the high altitude moderates the temperature. The wettest regions are the lands to the north of Lake Victoria, where Kampala is situated, and the western mountains, especially the high Ruwenzori range.

VEGETATION

Some rain forest is found in the south of the country, though most of the trees have been cleared for farming. Central and northern Uganda are covered largely by wooded savanna. The high slopes of the western mountains contain some strange plants, such as giant lobelias, senecios, and tree heathers.

HISTORY

Little is known of the early history of Uganda. When Europeans first reached the area in the 19th century, many of the people were organized in kingdoms, the most powerful of which was Buganda, the home of the Baganda people. Britain took over the country between 1894 and 1914, and ruled it until independence in 1962.

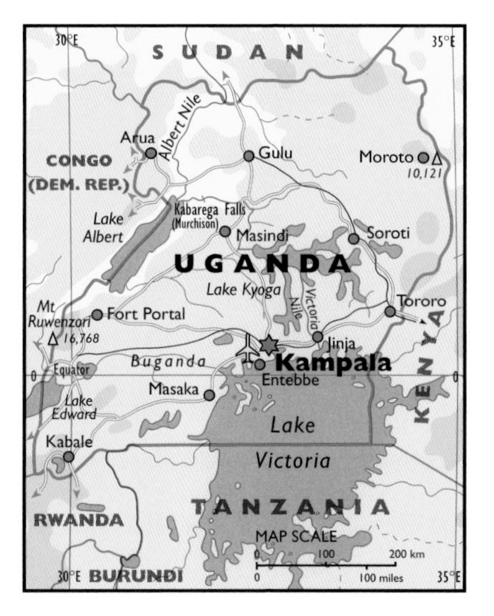

POLITICS

In 1967, Uganda became a republic and Buganda's Kabaka (king), Sir Edward Mutesa II, was made president. Tensions between the Kabaka and the prime minister, Apollo Milton Obote, led to the dismissal of the Kabaka in 1966. Obote also abolished the traditional kingdoms, including Buganda. Obote was overthrown in 1971 by an army group led by General Idi Amin Dada. Amin ruled as a dictator. He forced most of the Asians who lived in Uganda to leave the country and had many of his opponents killed.

In 1978, a border dispute between Uganda and Tanzania led Tanzanian troops to enter Uganda. With help from Ugandan opponents of Amin, they overthrew Amin's government. In 1980, Obote led his party to victory in national elections. But after charges of fraud, Obote's opponents began guerrilla warfare. A military group overthrew Obote in 1985, but strife continued until 1986, when Yoweri Museveni's National Resistance Movement seized power. In 1993, Museveni restored the traditional kingdoms. Museveni held national elections in 1994, but political parties were not allowed. Museveni was elected president in 1996 and 2001. From the late 1990s, Uganda faced problems in containing rebels in the north and west, and also for its support of rebels in neighboring Congo (Dem. Rep.).

ECONOMY

The strife in Uganda since the 1960s has damaged the economy, but it grew during a period of stability in the 1990s.

Agriculture employs 80% of the people. Food crops include bananas, cassava, maize, and millet. Some copper is mined and there is some manufacturing. Coffee is, by far, the leading export, followed by cotton, fish, and tea.

*This 50-shilling stamp shows a secretary bird (*Sagittarius serpentarius*). These birds of the African savanna fly well, but they are most often seen running or walking. They kill snakes by stamping their feet on them or by dropping them from a great height. They also eat small mammals, insects, and birds.*

UKRAINE

Ukraine's flag was first used between 1918 and 1922. It was readopted in September 1991. The colors were first used in 1848. They are heraldic in origin and were first used on the coat of arms of one of the Ukrainian kingdoms in the Middle Ages.

Ukraine is the second largest country in Europe after Russia. It was formerly part of the Soviet Union, which split apart in 1991. This mostly flat country faces the Black Sea in the south. The Crimean peninsula includes a highland region overlooking Yalta. The highest point, 6,764 ft [2,061 m], is in the eastern Carpathian Mountains, which extend into the Slovak Republic and Romania. The largest land region, the central plateau, descends to the north to the Dnipro (Dnieper)-Pripet lowlands. A low plateau occupies the northeast.

AREA 233,100 sq miles [603,700 sq km]
POPULATION 48,760,000
CAPITAL (POPULATION) Kyyiv (or Kiev, 2,621,000)
GOVERNMENT Multiparty republic
ETHNIC GROUPS Ukrainian 73%, Russian 22%, Jewish 1%, Belarussian 1%, Moldovan, Bulgarian, Polish
LANGUAGES Ukrainian (official), Russian
RELIGIONS Mostly Ukrainian Orthodox
CURRENCY Hryvnia = 100 kopiykas

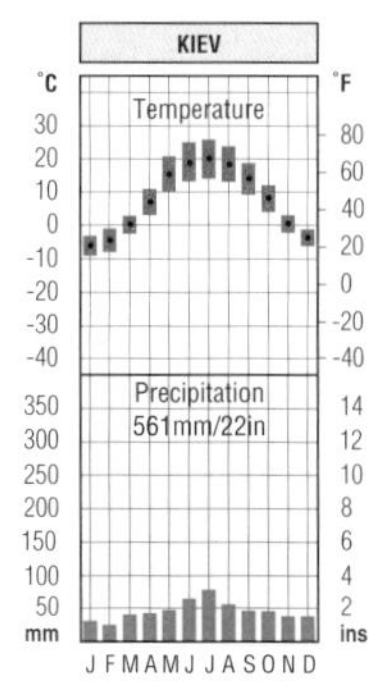

CLIMATE

Ukraine has warm summers, but the winters are cold, becoming more severe from west to east. In the summer, the east of the country is often warmer than the west. The rainfall is heaviest in the summer months. The highlands have a more severe climate than the Black Sea coastal plains.

VEGETATION

Woodland with such trees as ash, oak, and pine grows in the north, while pine forests swathe the slopes of the Carpathians and Crimean mountains. Grassy steppe once covered central Ukraine, but much of it is now farmed.

HISTORY AND POLITICS

In the 9th century AD, a group of people, called the East Slavs, founded a civilization called Kievan Rus, with its capital at Kiev (Kyyiv). Russians took over the area in 980 and the region became prosperous. In the 13th century, Mongol armies ravaged the area. Later, the region was split into small kingdoms and large areas fell under foreign rule. In the 17th and 18th centuries, parts of Ukraine came under Polish and Russian rule. But Russia gained most of Ukraine in the late 18th century. In 1918, Ukraine became independent, but in 1922 it became part of the Soviet Union.

Millions of people died in the 1930s as the result of Soviet policies. Millions more died during the Nazi occupation between 1941 and 1944. In 1945, the Soviet Union took control of parts of Ukraine that had been in Czechoslovakia, Poland, and Romania.

In the 1980s, Ukrainian people demanded more say over their affairs. The country finally became independent when the Soviet Union broke up in 1991. Ukraine continued to work with Russia through an organization named the Commonwealth of Independent States. But Ukraine differed with Russia on several issues, including control over Crimea and the Soviet fleet.

ECONOMY

The World Bank classifies Ukraine as a "lower-middle-income" economy. Agriculture is important. Crops include wheat and sugar beet, which are the major exports, together with barley, maize, potatoes, sunflowers, and tobacco. Livestock rearing and fishing are also important industries.

Manufacturing is the chief economic activity. Major manufactures include iron and steel, machinery, and vehicles. The country has large coalfields. The country imports oil and natural gas, but it has hydroelectric and nuclear power stations. In 1986, an accident at the Chernobyl (Chornobyl) nuclear power plant caused widespread nuclear radiation.

UNITED KINGDOM

The flag of the United Kingdom was officially adopted in 1801. The first Union flag, combining the cross of St George (England) and the cross of St Andrew (Scotland), dates from 1603. In 1801, the cross of St Patrick, Ireland's emblem, was added to form the present flag.

The United Kingdom (or UK) is a union of four countries. Three of them – England, Scotland, and Wales – make up Great Britain. The fourth country is Northern Ireland. The Isle of Man and the Channel Islands, including Jersey and Guernsey, are not part of the UK. They are self-governing British dependencies.

The land is highly varied. Much of Scotland and Wales is mountainous, and the highest peak is Scotland's Ben Nevis at 4,408 ft [1,343 m]. England has some highland areas, including the Cumbrian Mountains (or Lake District) and the Pennine range in the north. But England also has large areas of fertile lowland. Northern Ireland is also a mixture of lowlands and uplands. It contains the UK's largest lake, Lough Neagh.

AREA 94,202 sq miles [243,368 sq km]
POPULATION 59,648,000
CAPITAL (POPULATION) London (8,089,000)
GOVERNMENT Constitutional monarchy
ETHNIC GROUPS White 94%, Asian Indian 1%, Pakistani 1%, West Indian 1%
LANGUAGES English (official), Welsh, Gaelic
RELIGIONS Anglican 57%, Roman Catholic 13%, Presbyterian 7%, Methodist 4%, Baptist 1%, Islam 1%, Judaism, Hinduism, Sikhism
CURRENCY Pound sterling = 100 pence

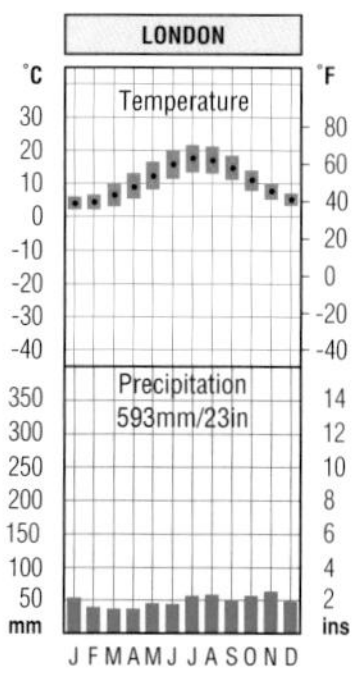

CLIMATE

The UK has a mild climate, influenced by the warm Gulf Stream which flows across the Atlantic from the Gulf of Mexico, then past the British Isles. Moist winds from the southwest bring rain, but the rainfall decreases from west to east. Winds from the east and north bring cold and sometimes snowy weather in winter.

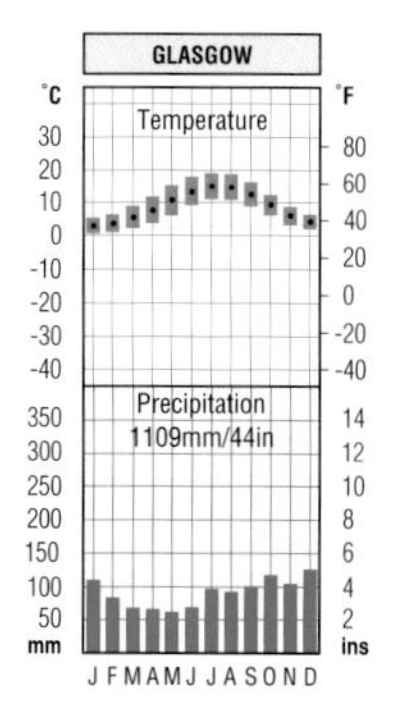

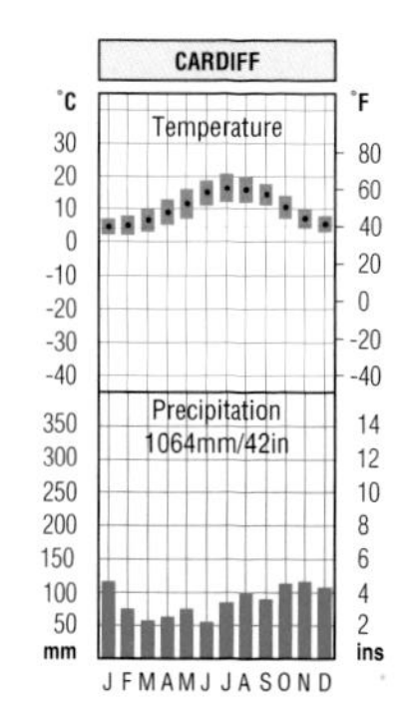

DID YOU KNOW

- that the UK once ruled the largest empire in history
- that the UK is the world's fifth most important trading nation, after the USA, Germany, Japan, and France
- that the UK was joined to mainland Europe until about 5500 BC
- that the UK's most visited historic buildings are Westminster Abbey, the Tower of London, and York Minster
- that the highest shade temperature in the UK, 98.8°F [37.1°C], was recorded at Cheltenham, Gloucestershire, in 1990
- that the lowest air temperature in the UK, –17°F [–27°C], was recorded at Braemar, Scotland, in 1895 and again in 1982

VEGETATION

The original vegetation in much of the United Kingdom was deciduous forest, with oak being the dominant tree. But human activity has greatly modified the landscape and only small patches of original woodland survive. Most of the land is farmed or used as pasture for livestock. Moorlands and heathlands occur in many upland areas. Together, they cover about a third of the country.

HISTORY

In ancient times, Britain was invaded by many peoples, including Iberians, Celts, Romans, Angles, Saxons, Jutes, Norsemen, Danes, and Normans, who arrived in 1066. The evolution of the United Kingdom spanned hundreds of years. The Normans finally overcame Welsh resistance in 1282, when King Edward I annexed Wales and united it with England. Union with Scotland was achieved by the Act of Union of 1707. This created a country known as the United Kingdom of Great Britain.

Ireland came under Norman rule in the 11th century, and much of its later history was concerned with a struggle against English domination. In 1801, Ireland became part of the United Kingdom of Great Britain and Ireland. But in 1921, southern Ireland broke away to become the Irish Free State. Most of the people in the Irish Free State were Roman Catholics. In Northern Ireland, where the majority of the people were Protestants, most people wanted to remain citizens of the United Kingdom. As a result, the country's official name changed to the United Kingdom of Great Britain and Northern Ireland.

From the late 18th century, the United Kingdom led the world in industrializing its economy. In the second half of the 20th century, the country faced the problem of renewing the decaying hearts of many industrial and other ancient cities, many of which had become slums. In 1984, a set of four stamps entitled "Urban Renewal" recorded the changes in British cities, including the Scottish city of Perth (above).

UNITED KINGDOM

The United Kingdom's official mapping organization, the Ordnance Survey, was founded in 1791 as a branch of the army, but it is now a civilian organization. The stamp, showing part of a village in Kent, was published in 1991. It is one of a set of four issued to commemorate the bicentenary of the Ordnance Survey.

The modern history of the UK began in the 18th century when the British empire began to develop, despite the loss in 1783 of its 13 North American colonies which became the core of the modern United States. The other major event occurred in the late 18th century, when the UK became the first country to industrialize its economy.

POLITICS

The British empire broke up after World War II (1939–45), though the UK still administers many small, mainly island, territories around the world. The empire was transformed into the Commonwealth of Nations, a free association of independent countries, which numbered 54 in 2003.

But while the UK retained a world role through the Commonwealth and the United Nations, it recognized that its economic future lay within Europe. As a result, it became a member of the European Economic Community (now the European Union) in 1973. However, it retained its traditional ties with the USA. In the early 21st century, many people feared a loss of British identity should the European Union evolve into a political federation. Anxiety about the future role contributed to Britain's decision not to adopt the euro, the single currency of the European Union.

ECONOMY

The UK is a major industrial and trading nation. It lacks natural resources apart from coal, iron ore, oil, and natural gas. It imports most of the materials it needs for industry and about a third of the food it needs.

In the first half of the 20th century, the UK became known for exporting such products as cars, ships, steel, and textiles. However, many traditional industries have suffered from increased competition from other countries, whose lower labor costs enable them to produce goods more cheaply. Today, many industries use sophisticated high-technology in order to compete on the world market.

The UK is one of the world's most urbanized countries, and agriculture employs only 1.2% of the people. Yet production is high because farms use scientific methods and modern machinery. Major crops include barley, potatoes, sugar beet, and wheat. Sheep are the leading livestock, but beef cattle, dairy cattle, pigs, and poultry are also important. Fishing is another major activity.

Service industries play a major part in the UK's economy. Financial and insurance services bring in much-needed foreign exchange, while tourism has become a major earner.

Orkney Islands
Mainland
Kirkwall
C. Wrath
Thurso
John o' Groats
Wick
Stornoway
Lewis
Outer Hebrides
St Kilda
Ullapool
North West Highlands
Moray Firth
Inner Hebrides
Loch Ness
Inverness
Spey
Skye
Grampians
Aberdeen
SCOTLAND
Dee
Fort William
4,408
Ben Nevis
ATLANTIC OCEAN
Mull
Oban
Dundee
Perth
Dunfermline
Stirling
Firth of Forth
Jura
Islay
Glasgow
Edinburgh
Berwick-upon-Tweed
Arran
Firth of Clyde
Ayr
Southern Uplands
Malin Head
North Channel
Hawick
Londonderry
Coleraine
Dumfries
Newcastle-upon-Tyne
Larne
Ballymena
NORTHERN IRELAND
L. Neag
Stranraer
Carlisle
Sunderland
Bangor
Belfast
Lurgan
Solway Firth
Middlesbrough
LAKE DISTRICT
Sca Fell 3,209
Enniskillen
Portadown
Armagh
Kendal
Scarborough
Pennines
Newry
Douglas
Barrow-in-Furness
Lancaster
Isle of Man
Bradford
York
UNITED KINGDOM
Blackpool
Preston
Leeds
Hull
Irish Sea
Huddersfield
Humber
Grimsby
IRELAND
Anglesey
Liverpool
Manchester
Doncaster
Holyhead
Chester
Sheffield
Lincoln
Trent
Conwy
The Wash
Snowdon 3,561
Stoke-on-Trent
Derby
Nottingham
King's Lynn
Great Yarmouth
Stafford
Leicester
Ouse
EAST ANGLIA
Norwich
Cardigan Bay
Telford
Peterborough
Lowestoft
Birmingham
Coventry
St George's Channel
Aberystwyth
WALES
Severn
ENGLAND
Cambridge
Worcester
Northampton
Bedford
Fishguard
Cheltenham
Milton Keynes
Ipswich
Carmarthen
Hereford
Cotswolds
Luton
Colchester
Gloucester
Oxford
Chelmsford
Thames
Newport
Slough
Southend
Swansea
Swindon
London
Margate
Cardiff
Reading
Bristol
Bath
Canterbury
Bristol Channel
Guildford
Dover
Salisbury
Crawley
Folkestone
Taunton
EXMOOR
Winchester
Southampton
Hastings
Exeter
Bournemouth
Brighton
Eastbourne
Portsmouth
Poole
FRANCE
DARTMOOR
Weymouth
Isle of Wight
Torquay
Plymouth
English Channel
Land's End
Penzance
Isles of Scilly
Channel Is see page 118
North Sea
MAP SCALE
0
50
100 km
0
50 miles
Shetland Islands
Mainland
Lerwick
Orkney Islands
Kirkwall
John o' Groats
8°W
6°W
4°W
2°W
0
2°E
50°N
52°N
54°N
56°N
58°N
60°N

UNITED STATES OF AMERICA

This flag, known as the "Stars and Stripes," has had the same basic design since 1777, during the War of Independence. The 13 stripes represent the 13 original colonies in the eastern United States. The 50 stars represent the 50 states of the Union.

The United States of America is the world's fourth largest country in area and the third largest in population. It contains 50 states, 48 of which lie between Canada and Mexico, plus Alaska in northwestern North America (*see map, page 66*), and Hawaii, a group of volcanic islands in the North Pacific Ocean (*see map, page 249*).

The densely populated eastern coastal plains lie east and south of the Appalachian Mountains. The central lowlands drained by the Mississippi-Missouri rivers stretch from the Appalachians to the Rocky Mountains, the chief range in the Western Highlands. The Pacific region includes fertile valleys and coastal ranges.

AREA 3,618,765 sq miles [9,372,610 sq km]

POPULATION 278,059,000

CAPITAL (POPULATION) Washington, D.C. (4,466,000)

GOVERNMENT Federal republic

ETHNIC GROUPS White 70%, Hispanic 13%, African American 12.7%, others 4.3%

LANGUAGES English (official), Spanish, more than 30 others

RELIGIONS Protestant 56%, Roman Catholic 28%, Islam 2%, Judaism 2%

CURRENCY US dollar = 100 cents

CLIMATE

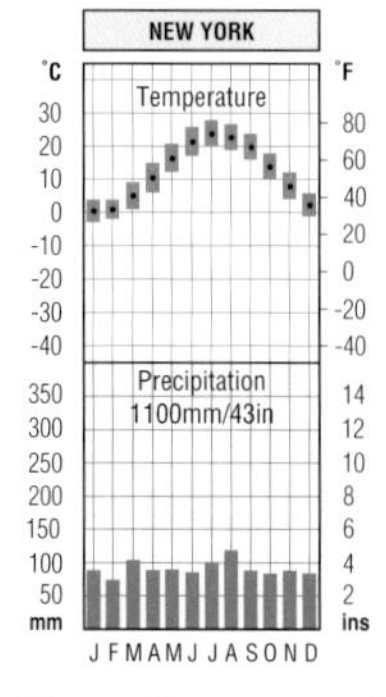

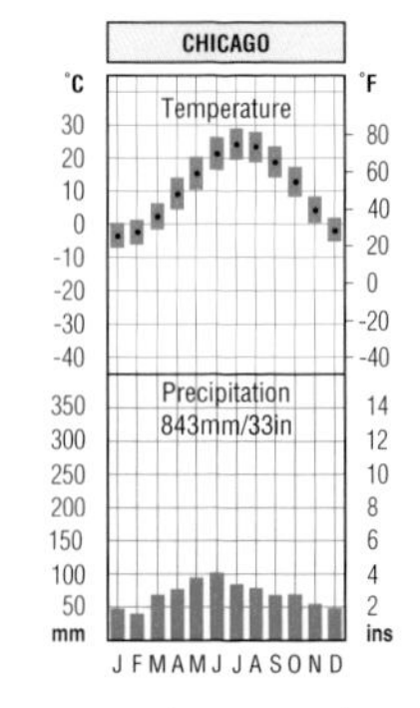

The climate varies greatly, ranging from the Arctic cold of Alaska to the intense heat of Death Valley, a bleak desert in California. Of the 48 states between Canada and Mexico, winters are cold and snowy in the north, but mild in the south, a region which is often called the "Sun Belt."

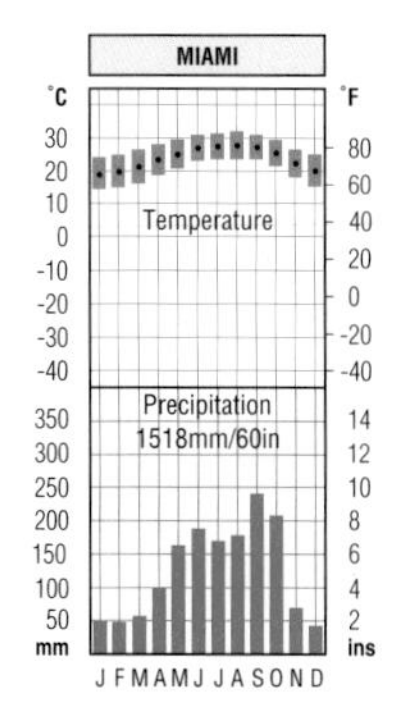

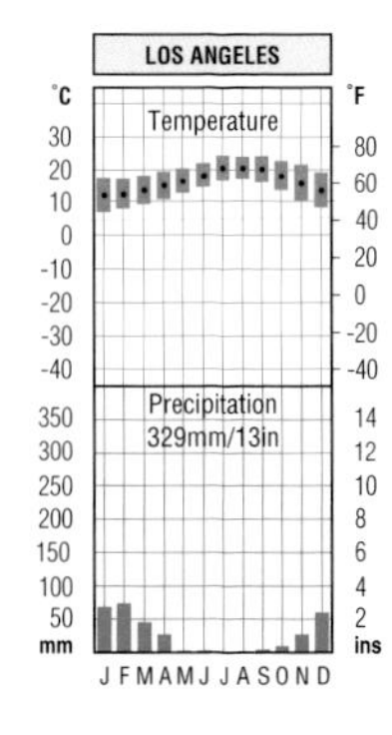

VEGETATION

The vegetation varies as much as the climate. Alaska contains forests of conifers, such as fir, pine, and spruce. In the northernmost of the 48 states, there are forests of hardwoods, such as maple and beech, while huge redwoods grow along the Pacific coast. The dry central plains are a region of grassy prairies, while large areas in the southwest are deserts, where prickly cacti and tumbleweed are common plants.

HISTORY AND POLITICS

The first people in North America, the ancestors of the Native Americans (or American Indians) arrived perhaps 40,000 years ago from Asia. Although Vikings probably reached North America 1,000 years ago, the European exploration of the continent did not begin until the late 15th century.

The first Europeans to settle in large numbers were the British, who founded settlements on the eastern coast in the early 17th century. British rule ended in the War of Independence (1775–83). The country expanded in 1803 when a vast territory in the south and west was acquired through the Louisiana Purchase, while the border with Mexico was fixed in the mid-19th century.

The Civil War (1861–5) ended the threat that the nation might split in two parts. It also ended slavery for the country's many African Americans. In the late 19th century, the West was opened up, while immigrants flooded in from Europe and elsewhere.

During the late 19th and early 20th centuries, industrialization led to the United States becoming the world's leading economic superpower and a pioneer in science and technology. By the 21st century, it was the world's only superpower and it had become champion of the Western world and of democratic government.

This block of four stamps commemorates the cooperation between the United States and the former Soviet Union in the exploration of space.

ECONOMY

The United States has the world's largest economy in terms of the total value of its production. Although agriculture employs only 2% of the people, farming is highly mechanized and scientific, and the United States leads the world in farm production.

UNITED STATES OF AMERICA

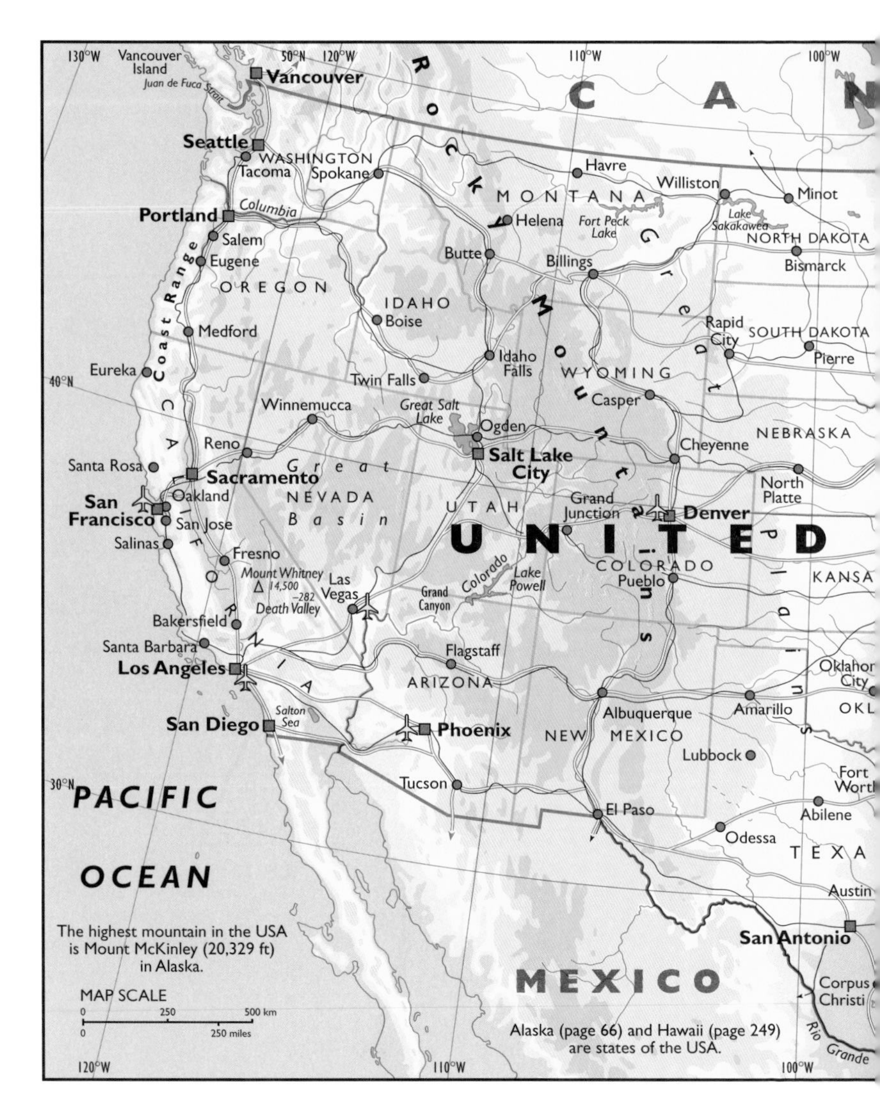

Major products include beef and dairy cattle, together with a wide range of crops, including cotton, fruits, groundnuts, maize, potatoes, soya beans, tobacco, and wheat.

The country's natural resources include oil, natural gas, and coal. There are also a wide range of metal ores which are used in manufacturing industries, together with timber, espe-

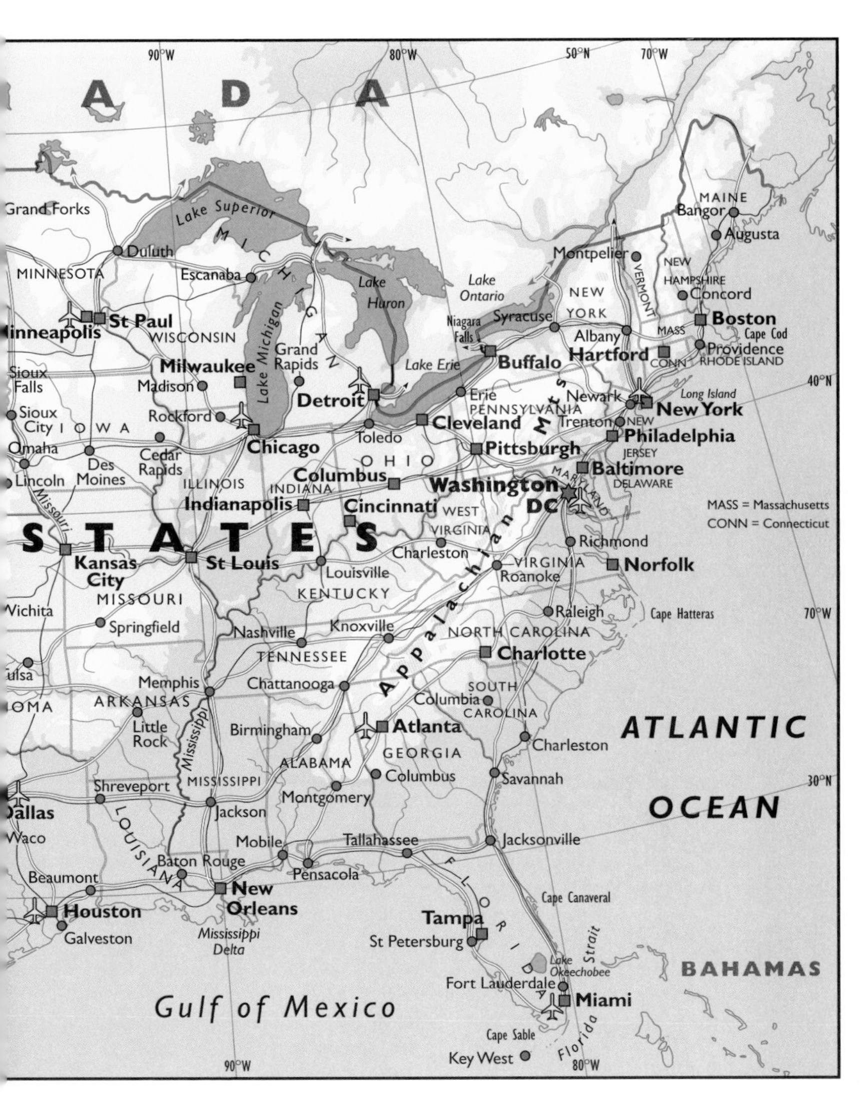

cially from the forests of the Pacific northwest. Manufacturing is the single most important activity, employing about 14% of the population. Major products include vehicles, food products, chemicals, machinery, printed goods, metal products, and scientific instruments. In recent years, California has emerged as the leading manufacturing state.

URUGUAY

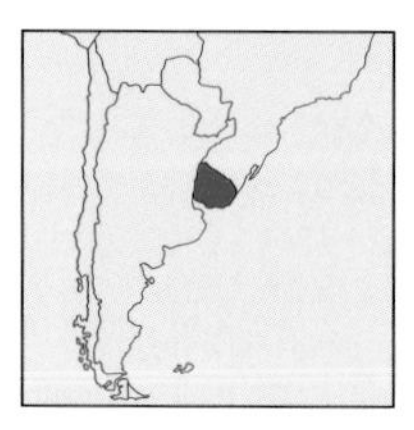

Uruguay has used this flag since 1830. The nine stripes represent the nine provinces which Uruguay contained when it became an independent republic in 1828. The colors and the May Sun had originally been used by Argentina during its struggle against Spanish rule.

The Eastern Republic of Uruguay, as Uruguay is officially known, is South America's second smallest independent country after Suriname. The River Uruguay, which forms the country's western border, flows into the Río de la Plata, a large estuary which leads into the South Atlantic Ocean.

The land consists of low-lying plains and hills. The highest point, Mirador Nacional, which lies south of Minas, is only 1,644 ft [501 m] above sea level. The main river in the interior is the Río Negro.

AREA 68,498 sq miles [177,410 sq km]
POPULATION 3,360,000
CAPITAL (POPULATION) Montevideo (1,379,000)
GOVERNMENT Multiparty republic
ETHNIC GROUPS White 88%, Mestizo 8%, Mulatto or Black 4%
LANGUAGES Spanish (official)
RELIGIONS Roman Catholic 66%, Protestant 2%, Judaism 1%, non-professing or other 31%
CURRENCY Uruguay peso = 100 centésimos

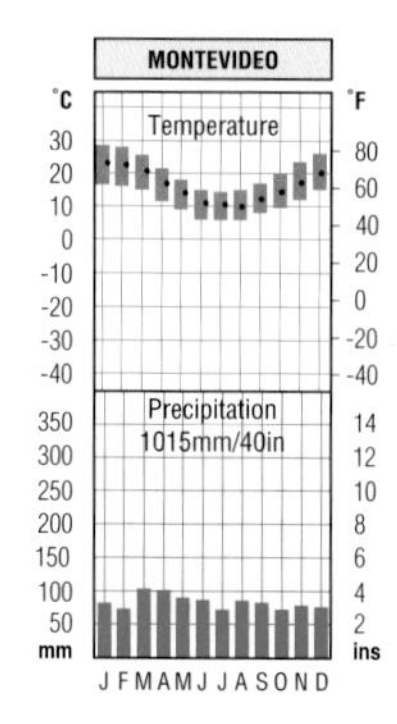

CLIMATE

Uruguay has a mild climate. Rain falls throughout the year, though droughts sometimes occur. The summer months are pleasantly warm, especially near the coast. Winters are less cold than places in North America in similar latitudes.

VEGETATION

Grasslands cover 78% of Uruguay and arable land covers about 8%. Such trees as acacia, aloe, eucalyptus, and willow grow along the river valleys. Uruguay also has commercial tree plantations, including such trees as quebracho, whose wood and bark contain tannin which is used in tanning and dyeing.

HISTORY

The first people of Uruguay were Amerindians. But the Amerindian population has largely disappeared. Many were killed by Europeans, some died of European diseases, while others fled into the interior. The majority of Uruguayans today are of European origin, though there are some mestizos (of mixed European and Amerindian descent).

The first European to arrive in Uruguay was a Spanish navigator in 1516. But few Europeans settled there until the late 17th century. In 1726, Spanish settlers founded Montevideo

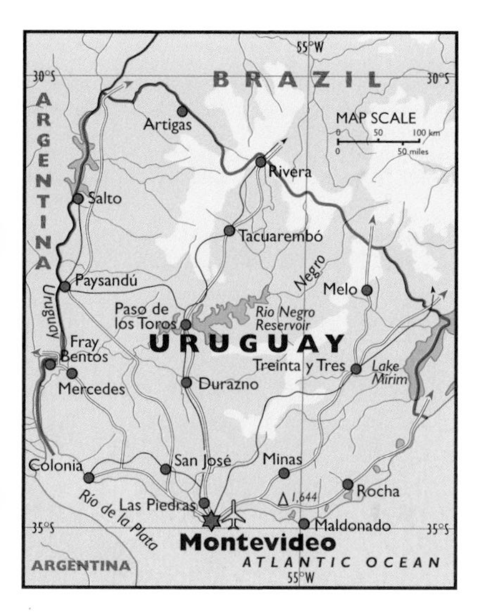

in order to halt the Portuguese gaining influence in the area. By the late 18th century, Spaniards had settled in most of the country. Uruguay became part of a colony called the Viceroyalty of La Plata, which also included Argentina, Paraguay, and parts of Bolivia, Brazil, and Chile.

In 1820 Brazil annexed Uruguay, ending Spanish rule. In 1825, Uruguayans, supported by Argentina, began a struggle for independence. Finally, in 1828, Brazil and Argentina recognized Uruguay as an independent republic. Social and economic developments were slow in the 19th century, but, from 1903, governments made Uruguay a democratic and stable country.

POLITICS

From the 1950s, economic problems caused unrest. Terrorist groups, notably the Tupumaros, carried out murders and kidnappings. The army crushed the Tupumaros in 1972, but the army took over the government in 1973. Military rule continued until 1984 when elections were held. Julio Maria Sanguinetti, who led Uruguay back to civilian rule, was re-elected president in 1994. He was succeeded in 1999 by Jorge Battle.

ECONOMY

The World Bank classifies Uruguay as an "upper-middle-income" developing country. Agriculture employs only 4% of the people, but farm products, notably hides and leather goods, beef, and wool, are the leading exports, while the leading manufacturing industries process farm products. The main crops include maize, potatoes, sugar beet, and wheat. Other manufacturing industries, mainly in and around Montevideo, produce such items as cement, textiles, and tires. Tourism is also an important activity.

Uruguay's main natural resources are its grasslands and rivers. Many rivers are dammed to produce hydroelectric power. This stamp showing a dam and a child, whose hopes for the future, like those of all Uruguayans, depend on economic growth, was one of a set of five issued in 1959, entitled "National Recovery."

UZBEKISTAN

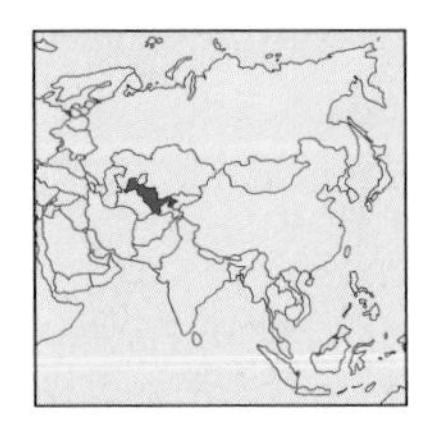

Uzbekistan's flag was adopted in 1991. The blue recalls the flag of Tamerlane, or Timur (a former ruler). The white represents peace, the green nature, and the red vitality. The crescent moon symbolizes Islam. The 12 stars represent the months in the Islamic calendar.

The Republic of Uzbekistan is one of the five republics in Central Asia which were once part of the Soviet Union. Plains cover most of western Uzbekistan, with highlands in the east. The main rivers, the Amu (or Amu Darya) and Syr (or Syr Darya), drain into the Aral Sea. So much water has been taken from these rivers to irrigate the land that the Aral Sea shrank from 25,830 sq miles [66,900 sq km] in 1960 to 12,989 sq miles [33,642 sq km] in 1993. The dried-up lake area has become desert, like much of the rest of the country.

AREA 172,740 sq miles [447,400 sq km]
POPULATION 25,155,000
CAPITAL (POPULATION) Toshkent (or Tashkent, 2,118,000)
GOVERNMENT Socialist republic
ETHNIC GROUPS Uzbek 80%, Russian 5%, Tajik 5%, Kazakh 3%, Tatar 2%, Kara-Kalpak 2%, Crimean Tatar, Kyrgyz, Ukrainian, Turkmen
LANGUAGES Uzbek (official), Russian
RELIGIONS Islam 88%, Eastern Orthodox 9%
CURRENCY Som = 100 tyiyn

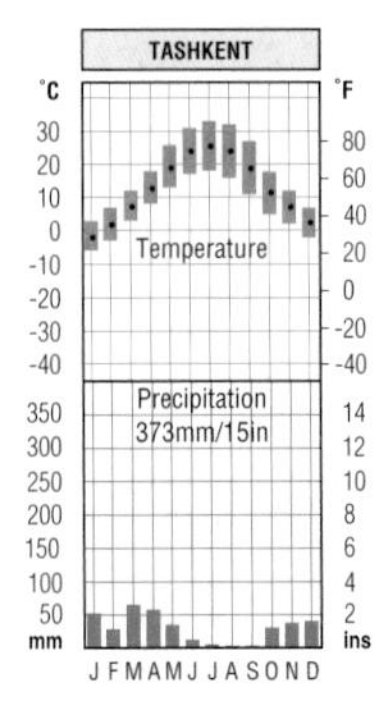

CLIMATE

Uzbekistan has a continental climate. The winters are cold, but the temperatures soar in the summer months. The west is extremely arid, with an average annual rainfall of about 8 inches [200 mm]. But parts of the highlands in the east have three times as much rain.

VEGETATION

Some patches of tough grasses grow in desert regions. Grassy steppe occurs in wetter areas, with forests on the mountain slopes.

HISTORY

Turkic people first settled in the area that is now Uzbekistan about 1,500 years ago and Islam was introduced in the 7th century AD. Mongols invaded the land in the 13th century and, in the late 14th century, the Mongol Turk Tamerlane ruled a great empire from Samarkand (Samarqand). Turkic Uzbek people invaded in the 16th century and gradually the area was divided into states called *khanates*.

Russia took the area in the 19th century. After the Russian Revolution of 1917, the Communists took over and, in 1924, they set up the Uzbek Soviet Socialist Republic. Under Communism, all aspects of Uzbek life were controlled and religious worship was discouraged. But education, health, housing, and transport were improved. The Communists also increased cotton production, but they caused great environmental damage in doing so.

In the 1980s, when reforms were being introduced in the Soviet Union, the people demanded more freedom. In 1990, the government stated that its laws overruled those of the Soviet Union. In 1991, Uzbekistan became independent when the Soviet Union broke up. But it kept links with Russia through the Commonwealth of Independent States.

On December 29, 1991, Islam Karimov, the leader of the People's Democratic Party (formerly the Communist Party), was elected president. By the time that Karimov was re-elected president in 2001, Uzbekistan had become one of the world's most repressive nations, with many political prisoners. However, in 2001, Karimov declared his support for the USA's campaign against the terrorist al Qaida bases in Afghanistan.

DID YOU KNOW

- that 300,000 people were made homeless when an earthquake hit Tashkent (Toshkent) in 1966
- that Uzbekistan was named after Khan Uzbek, one of the Mongol leaders whose armies swept across Asia into central Europe in the 14th century
- that Uzbek marble is famous for its beauty; it was used to face walls in Moscow's underground railroad system
- that Samarkand (Samarqand) was the capital of the Mongol emperor Tamerlane in the 14th century

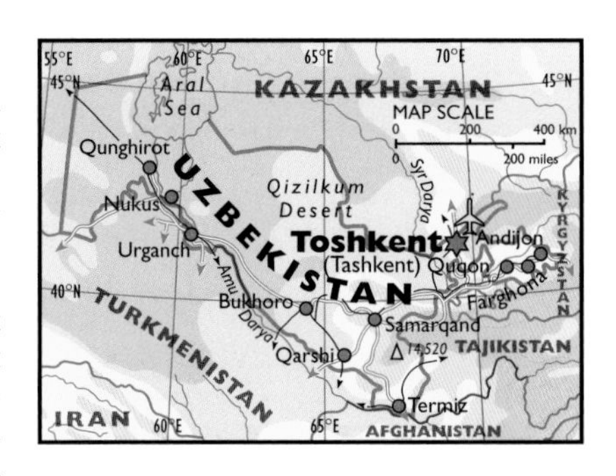

ECONOMY

The World Bank classifies Uzbekistan as a "lower-middle-income" developing country and the government still controls most of the economic activity. The country produces coal, copper, gold, oil, and natural gas. Manufactures include agricultural machinery, chemicals, processed food, and textiles.

Agriculture is important and cotton is the main crop. Other crops include fruits, rice, and vegetables. Cattle, sheep, and goats are also raised. Uzbekistan's exports include cotton, gold, textiles, chemicals, and fertilizers.

The national flag flies over one of the country's many magnificent Muslim buildings on this stamp, which was issued by Uzbekistan in 1992.

VENEZUELA

Venezuela's flag, adopted in 1954, has the same basic tricolor as the flags of Colombia and Ecuador. The colors were used by the Venezuelan patriot Francisco de Miranda. The seven stars represent the provinces in the Venezuelan Federation in 1811.

The Republic of Venezuela, in northern South America, contains the Maracaibo lowlands in the west. The lowlands surround the oil-rich Lake Maracaibo. Arms of the Andes Mountains enclose the lowlands and extend across most of northern Venezuela. Between the northern mountains and the scenic Guiana Highlands in the southeast, where the Angel Falls are found, lie the *llanos* (tropical grasslands), a low-lying region drained by the River Orinoco and its tributaries. The Orinoco is Venezuela's longest river.

AREA 352,143 sq miles [912,050 sq km]

POPULATION 23,917,000

CAPITAL (POPULATION) Caracas (1,975,000)

GOVERNMENT Federal republic

ETHNIC GROUPS Mestizo 67%, White 21%, Black 10%, Amerindian 2%

LANGUAGES Spanish (official), Goajiro

RELIGIONS Roman Catholic 96%

CURRENCY Bolívar = 100 céntimos

CLIMATE

Venezuela has a tropical climate. Temperatures are high throughout the year on the lowlands, though the mountains are much cooler. The rainfall is heaviest in the mountains. But much of the country has a marked dry season between December and April.

VEGETATION

Forests cover about two-fifths of Venezuela. They include dense rain forest in the lower Orinoco basin and in the southern Guiana Highlands. Savanna (tropical grassland) covers much of the lowlands, while mountain grassland occurs in the highlands. Only about 4% of the land area is cultivated.

HISTORY

The Arawak and Carib Amerindians were the main inhabitants of Venezuela before the arrival of Europeans. The first European to arrive was Christopher Columbus, who sighted the area in 1498. Spaniards began to settle in the early 16th century, but economic development was slow.

In the early 19th century, Venezuelans, such as Simón Bolívar and Francisco de Miranda, began a struggle against Spanish rule. Venezuela declared its independence in 1811. But the country did not become truly independent until 1821, when the Spanish were defeated in a battle near Valencia. In 1819, Venezuela became part of Gran Colombia, a republic that also included Colombia, Ecuador, and Panama. Venezuela broke away from Gran Colombia in 1829.

POLITICS

The development of Venezuela in the 19th and the first half of the 20th centuries was marred by instability, violence, and periods of harsh dictatorial rule. But the country has had elected governments since 1958. The country has greatly benefited from its oil resources which were first exploited in 1917. In 1960, Venezuela helped to form OPEC (the Organization of Petroleum Exporting Countries) and, in 1976, the government of Venezuela took control of the entire oil industry. Money from oil exports has helped Venezuela to raise living standards.

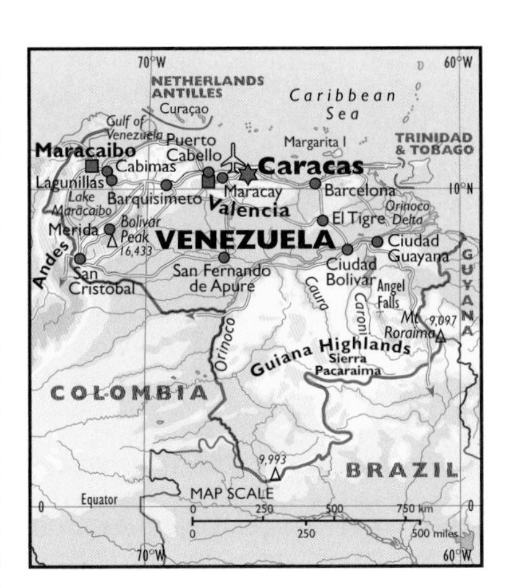

In 1998, a radical politician, Hugo Chávez, was elected president. He introduced a new constitution in 1999. In 2002, he survived a coup attempt.

ECONOMY

The World Bank classifies Venezuela as an "upper-middle-income" developing country. Oil accounts for 80% of the exports. Other exports include bauxite and aluminum, iron ore, and farm products. Agriculture employs 8% of the people. Cattle ranching is important and the major crops include bananas and other fruits, coffee, maize, rice, and sugarcane. Manufacturing has increased greatly in the last 30 years and industry now employs 25% of the population. The chief industry is petroleum refining. Other manufactures include aluminum, cement, processed food, steel, and textiles. The main manufacturing centers include Caracas, Ciudad Guayana, a center of aluminum and steel production, and Maracaibo, with its oil refineries.

DID YOU KNOW

- that the Angel Falls in Venezuela are the world's highest waterfalls; the total drop is 3,212 ft [979 m] – the longest single drop is 2,648 ft [807 m]
- that early Spanish explorers who saw houses on stilts along the coast named the country Venezuela, meaning "Little Venice"
- that Venezuela is one of the world's leading oil exporters
- that Lake Maracaibo is the largest lake in South America
- that the Guiana Highlands, in southeastern Venezuela, were made famous by Sir Arthur Conan Doyle in his novel *Lost World*

VIETNAM

Vietnam's flag was first used by forces led by the Communist Ho Chi Minh during the liberation struggle against Japan in World War II (1939–45). It became the flag of North Vietnam in 1945. It was retained when North and South Vietnam were reunited in 1975.

The Socialist Republic of Vietnam occupies an S-shaped strip of land facing the South China Sea in Southeast Asia. The coastal plains include two densely populated, fertile river delta regions. The Red (Hong) delta faces the Gulf of Tonkin in the north, while the Mekong delta is in the south. Inland are thinly populated highland regions, including the Annam Cordillera, which forms much of the boundary with Cambodia. The highlands in the northwest extend into Laos and China.

AREA 128,065 sq miles [331,689 sq km]

POPULATION 79,939,000

CAPITAL (POPULATION) Hanoi (3,056,000)

GOVERNMENT Socialist republic

ETHNIC GROUPS Vietnamese 87%, Tho (Tay), Chinese (Hoa), Tai, Khmer, Muong, Nung

LANGUAGES Vietnamese (official)

RELIGIONS Buddhism 55%, Roman Catholic 7%

CURRENCY Dong = 10 hao = 100 xu

CLIMATE

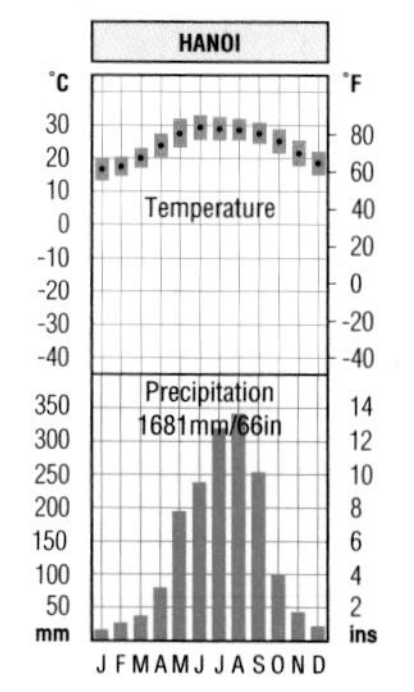

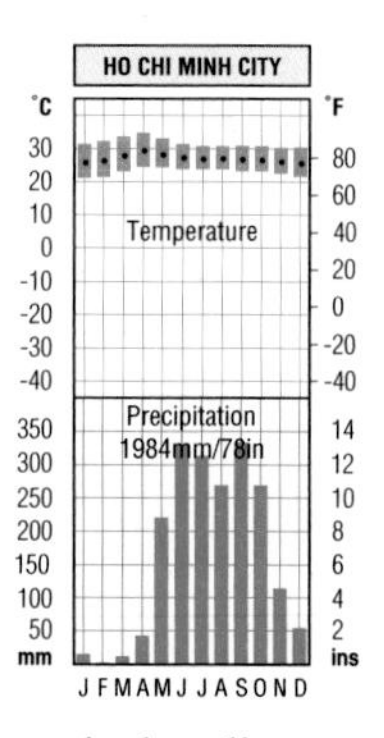

Vietnam has a tropical climate, though the driest months of January to March are a little cooler than the wet, hot summer months, when monsoon winds blow from the southwest. Typhoons (cyclones) sometimes hit the coast, causing much damage.

VEGETATION

Forests, including such valuable trees as ebony and teak, cover about two-fifths of Vietnam. There are also pine forests and mangrove swamps along the coast. Vietnam also has areas of savanna (tropical grassland with scattered trees) and about a quarter of the land is farmed.

HISTORY AND POLITICS

China dominated Vietnam for a thousand years before AD 939, when a Vietnamese state was founded. The French took over the area between the 1850s and 1880s. They ruled Vietnam as part of French Indochina, which also included Cambodia and Laos.

Japan conquered Vietnam during World War II (1939–45). In 1946, war broke out between a nationalist

group, called the Vietminh, and the French colonial government. France withdrew in 1954 and Vietnam was divided into a Communist North Vietnam, led by the Vietminh leader, Ho Chi Minh, and a non-Communist South.

A force called the Viet Cong rebelled against South Vietnam's government in 1957 and a war began, which gradually increased in intensity. The United States aided the South, but after it withdrew in 1975, South Vietnam surrendered. In 1976, the united Vietnam became a Socialist Republic.

Vietnamese troops intervened in Cambodia in 1978 to defeat the Communist Khmer Rouge government, but it withdrew its troops in 1989. Vietnam began to introduce reforms in the 1990s and, in 2000, the USA and Vietnam signed a historic trade pact.

ECONOMY

The World Bank classifies Vietnam as a "low-income" developing country and agriculture employs 67% of the population. The main food crop is rice. Other products include maize and sweet potatoes, while commercial crops include bananas, coffee, groundnuts, rubber, soya beans, and tea. Fishing is also important. Northern Vietnam has most of the country's natural resources, including coal. The country also produces chromium, oil (which was discovered off the south coast in 1986), phosphates, and tin. Manufactures include cement, fertilizers, processed food, machinery, steel, and textiles. The main exports are farm products and handicrafts, coal, minerals, oil, and seafood.

A set of seven stamps entitled "Fishes" was issued by Vietnam in 1987. It included this species, Puntis conchonius.

YEMEN

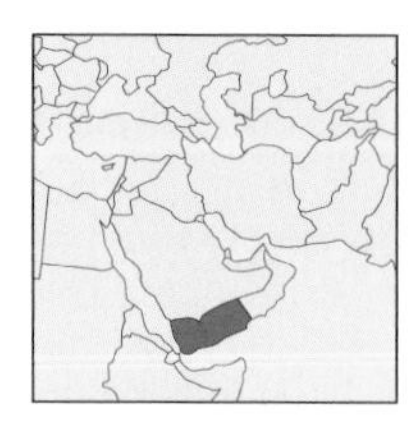

Yemen's flag was adopted in 1990 when the Yemen Arab Republic (or North Yemen) united with the People's Democratic Republic of Yemen (or South Yemen). This simple flag is a tricolor of red, white, and black, colors associated with the Pan-Arab movement.

The Republic of Yemen faces the Red Sea and the Gulf of Aden in the southwestern corner of the Arabian peninsula. Behind the narrow coastal plain along the Red Sea, the land rises to a mountain region called High Yemen. Beyond the mountains, the land slopes down toward the Rub' al Khali desert. Other mountains rise behind the coastal plain along the Gulf of Aden. To the east lies a fertile valley called the Hadramaut.

AREA 203,849 sq miles [527,970 sq km]
POPULATION 18,078,000
CAPITAL (POPULATION) San'a (972,000)
GOVERNMENT Multiparty republic
ETHNIC GROUPS Arab 96%, Somali 1%
LANGUAGES Arabic (official)
RELIGIONS Islam
CURRENCY Rial = 100 fils

CLIMATE

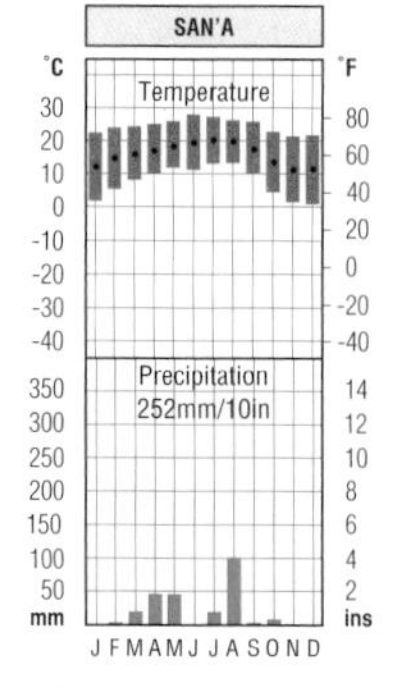

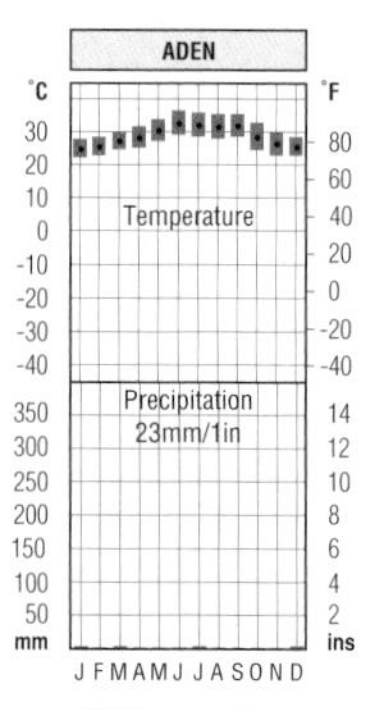

The climate in San'a is moderated by the altitude of more than 6,000 ft [2,000 m]. Temperatures are much lower than in Aden, which is at sea level. In August, southwest monsoon winds bring thunderstorms. But most of Yemen is arid. The south coasts are particularly hot and humid, especially from June to September.

VEGETATION

Palm trees grow on the coast, while such plants as euphorbias, acacias, and eucalyptus flourish in the interior. Thorn shrubs and mountain pasture are found in mountain areas. Inland lies barren desert.

HISTORY AND POLITICS

From about 1400 BC, Yemen lay on important trade routes. But it lost its prosperity in the 4th century AD, when it became divided between warring groups. In 1517, the area came under the Turkish Ottoman empire. After the collapse of the Ottoman empire in World War I (1914–18), the north, which Turkey had ruled, began to evolve into a separate state from the south, where Britain was in control.

Civil war marred South Yemen's progress toward independence. Britain finally withdrew in 1967, and a left-wing government took over. In North Yemen, military officers overthrew the monarchy in 1962 and made the country a republic. War between royalists and republicans broke out, but military forces took control in 1974. Clashes occurred between the traditionalist Yemen Arab Republic and the Marxist People's Democratic Republic of Yemen.

In 1990, the two Yemens merged to form a single republic. But, further conflict occurred during 1994, when southern secessionist forces were defeated. Islamic militants tried to destabilize the country in the 1990s and, in 2000, suicide bombers attacked a US destroyer in Aden harbor, killing 17 sailors.

ECONOMY

The World Bank classifies Yemen as a "low-income" developing country. Agriculture employs nearly half of the people. Herders raise sheep, while farmers grow barley, fruits, wheat, and vegetables in highland valleys and around oases. Cash crops include coffee and cotton.

> **DID YOU KNOW**
> - that San'a is said to have been founded by Shem, the eldest son of Noah
> - that the island of Socotra in the Indian Ocean, formerly a British protectorate, belongs to Yemen
> - that Aden (Al'Adan) was once a major British port, supplying ships on the journey to India
> - that Mocha coffee got its name from the port of Al Mukha

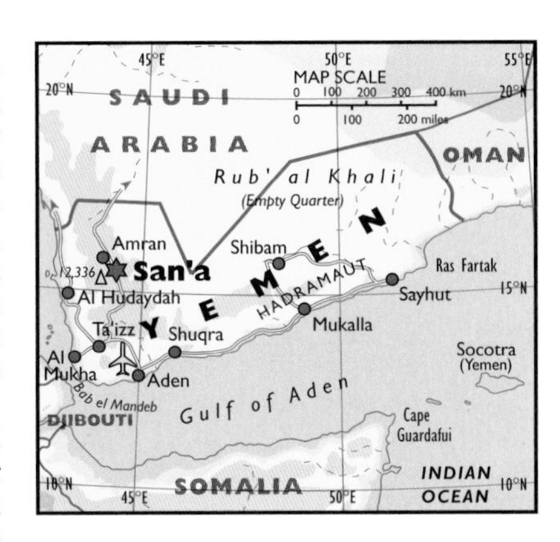

Imported oil is refined at Aden and petroleum extraction began in the northwest in the 1980s. Handicrafts, leather goods, and textiles are manufactured. Remittances from Yemenis abroad are a major source of revenue.

A painting of an Arab scribe is the subject of this stamp issued in 1967 by the royalist side in the 1962–70 civil war in North Yemen. It is one of a set of five stamps using Arab paintings. During the civil war, the republican side also issued its own stamps.

ZAMBIA

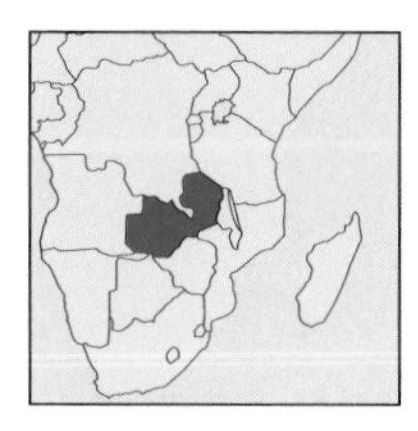

Zambia's flag was adopted when the country became independent from Britain in 1964. The colors are those of the United Nationalist Independence Party, which led the struggle against Britain and ruled until 1991. The flying eagle represents freedom.

The Republic of Zambia is a landlocked country in southern Africa. Zambia lies on the plateau that makes up most of southern Africa. Much of the land is between 2,950 ft and 4,920 ft [900 m to 1,500 m] above sea level. The Muchinga Mountains in the northeast rise above this flat land.

Lakes include Bangweulu, which is entirely within Zambia, together with parts of lakes Mweru and Tanganyika in the north. In the south, it shares the artificial Lake Kariba, on the River Zambezi, with Zimbabwe.

AREA 290,586 sq miles [752,614 sq km]
POPULATION 9,770,000
CAPITAL (POPULATION) Lusaka (982,000)
GOVERNMENT Multiparty republic
ETHNIC GROUPS Bemba 36%, Maravi (Nyanja) 18%, Tonga 15%
LANGUAGES English (official), Bemba, Nyanja, and about 70 others
RELIGIONS Christianity 68%, Islam, Hinduism
CURRENCY Kwacha = 100 ngwee

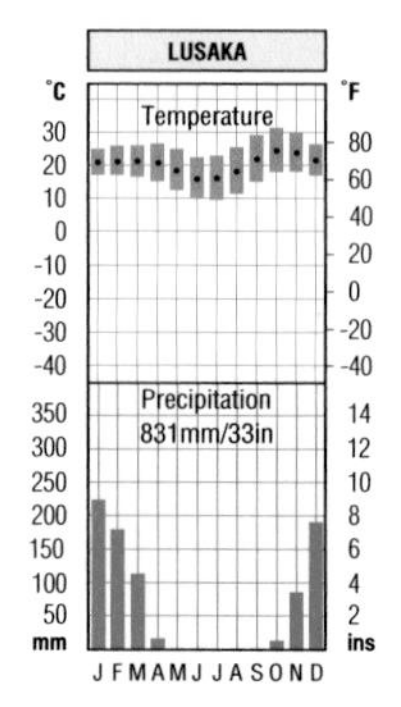

CLIMATE

Zambia lies in the tropics, but temperatures are moderated by the height of the land. The rainy season in Zambia begins in the month of November and carries on until March, when the rivers sometimes flood. Northern Zambia is the wettest region. The average yearly rainfall decreases toward the south.

VEGETATION

Grassland with clumps of trees, or wooded savanna, covers much of the country. There are some areas of swamp, while evergreen forests grow in the drier, sandy southwest.

HISTORY

European contact with Zambia began in the 19th century, when the explorer David Livingstone crossed the River Zambezi. In the 1890s, the British South Africa Company, set up by Cecil Rhodes (1853–1902), the British financier and statesman, made treaties with local chiefs and gradually took over the area. In 1911, the Company named the area Northern Rhodesia. In 1924, Britain took over the government of the country.

In 1953, Britain formed a federation of Northern Rhodesia, Southern Rhodesia (now Zimbabwe), and Nyasaland (now Malawi). But local Africans opposed this step, because they believed that it concentrated power in the hands of the European minority. In 1963 the federation was dissolved. Northern Rhodesia became independent as Zambia in 1964.

POLITICS

The leading opponent of British rule, Kenneth Kaunda, became president of Zambia in 1964. His party, the United Nationalist Independence Party (UNIP), was the only party from 1972 until 1990, when a new constitution was adopted. In elections in 1991, the Movement for Multiparty Democracy (MMD) defeated UNIP. Its leader, Frederick Chiluba, became president. He was followed in 2001 by the new MMD leader, Levy Mwanawasa.

ECONOMY

Copper is the leading export, accounting for 90% of Zambia's total exports

in 1990. Zambia also produces cobalt, lead, zinc, and various gemstones. But dependence on minerals has created problems for Zambia.

Agriculture accounts for 38% of the workers, as compared with 8% in industry, including mining. Maize is the chief crop. Other crops include cassava, coffee, millet, sugarcane, and tobacco.

DID YOU KNOW

- that Zambia got its name from the Zambezi, Africa's fourth longest river after the Nile, Congo, and Niger
- that a favorite dish in Zambia, called *nshima*, is a porridge made from maize meal
- that Zambia is the world's second largest producer of cobalt ore and the fifth largest of copper ore
- that Kafue National Park, west of Lusaka, covers a larger area than Wales
- that "Operation Noah" was a program to evacuate wild animals from the Zambezi valley after the building of the Kariba Dam on the Zambia–Zimbabwe border in 1959

*This stamp was one of a set of three entitled "Fish of Zambia," issued in 1971. The tiger fish (*Hydrocynus vittatus*) is a large predator found in inland waters in southern Africa.*

ZIMBABWE

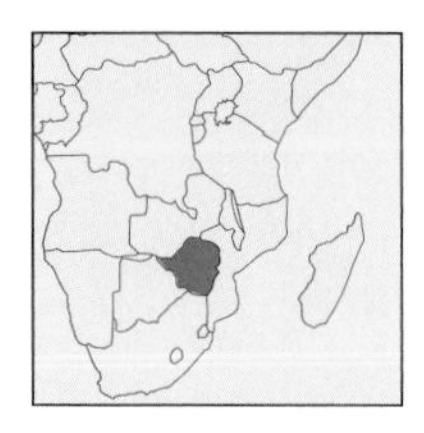

Zimbabwe's flag, adopted in 1980, is based on the colors used by the ruling Zimbabwe African National Union Patriotic Front. Within the white triangle is the Great Zimbabwe soapstone bird, the national emblem. The red star symbolizes the party's socialist policies.

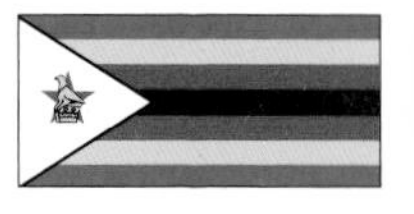

The Republic of Zimbabwe is a landlocked country in southern Africa. Most of the country lies on a high plateau between the Zambezi and Limpopo rivers between 2,950 ft and 4,920 ft [900 m to 1,500 m] above sea level. The main land feature is the High Veld, a ridge crossing Zimbabwe from northeast to southwest. Bordering the High Veld is the Middle Veld, the country's largest region. Below 2,950 ft [900 m] is the Low Veld. The highest land is in the east near the Mozambique border.

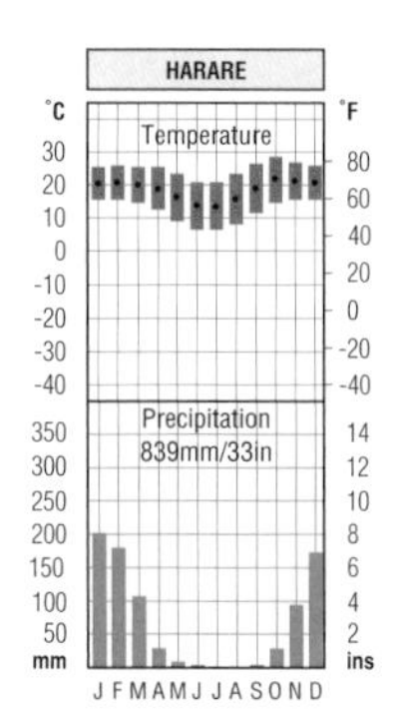

CLIMATE

In the summer, between October and March, the weather is hot and wet. But in the winter, daily temperatures can vary greatly. Frosts have been recorded between June and August. The climate varies according to the altitude. The Low Veld is much warmer and drier than the High Veld.

VEGETATION

Woodland savanna (tropical grassland with scattered trees) covers much of Zimbabwe. Forests grow on the Eastern Highlands and along the rivers.

AREA 150,873 sq miles [390,579 sq km]

POPULATION 11,365,000

CAPITAL (POPULATION) Harare (1,189,000)

GOVERNMENT Multiparty republic

ETHNIC GROUPS Shona 71%, Ndebele 16%, other Bantu-speaking Africans 11%, White 2%, Asian 1%

LANGUAGES English (official), Shona, Ndebele

RELIGIONS Christianity 45%, traditional beliefs 40%

CURRENCY Zimbabwe dollar = 100 cents

HISTORY AND POLITICS

The Shona people became dominant in the region about 1,000 years ago. They built the impressive Great Zimbabwe, a city of stone buildings.

The British South Africa Company, under financier and statesman Cecil Rhodes (1853–1902), occupied the area in the 1890s, after having obtained mineral rights from African chiefs. In 1895, the area was named Rhodesia and later Southern Rhodesia. Southern Rhodesia became a self-governing British colony in 1923. Between 1953 and 1963, Southern and Northern Rhodesia (now Zambia) were joined to Nyasaland (Malawi) in a federation.

In 1965, the European government

of Southern Rhodesia (then called Rhodesia) declared their country independent. But Britain refused to accept Rhodesia's independence. Finally, after a civil war, the country became legally independent in 1980. After independence, rivalries between the Shona and Ndebele people threatened its stability. But order was restored when the Shona prime minister, Robert Mugabe, brought his Ndebele rivals into his government. In 1987, the post of prime minister was abolished, and Mugabe became the country's executive president. The government renounced its Marxist ideology in 1991 and Mugabe was re-elected president in 1990, 1996, and 2002.

During the late 1990s, Mugabe threatened to seize white-owned farms without paying compensation to owners. His announcement caused much disquiet among the white farmers. The plan was eventually modified when the International Monetary Fund (IMF) threatened to withdraw a loan to Zimbabwe. But, the situation became increasingly dangerous during early 2000 when the landless "war veterans" began to occupy white-owned farms, causing deaths and destruction. Mugabe's government was also criticized for its brutal suppression of all opposition.

ECONOMY

The World Bank classifies Zimbabwe as a "low-income" developing country. The country has valuable mineral resources and mining accounts for a fifth of the country's exports. Gold, asbestos, chromium, and nickel are all mined. Zimbabwe also has some coal and iron ore. The country has some metal and food-processing industries.

However, agriculture employs 64% of working people. Maize is the main food crop, and export crops include tobacco, cotton, and sugar.

This stamp, issued in 1990, shows the greater kudu, a large antelope which lives on the savanna lands of eastern and southern Africa. The stamp also contains, top right, the national emblem, a carving of a bird in soapstone, a soft rock which is composed mostly of the mineral talc. Several of these carvings were found during the excavations of the country's major archaeological site, Greater Zimbabwe.

INDEX